Handbook of European Sphagna

Institute of Terrestrial Ecology

Handbook of European Sphagna

London: HMSO

© Copyright Controller of HMSO 1990

First published in 1985 by the
Institute of Terrestrial Ecology
Second impression with minor corrections 1990

ISBN 0 11 701431 1

COVER ILLUSTRATIONS
S. angermanicum (K Dierssen)
S. Fuscum (H J B Birks)
S. russowii (H J B Birks)
S. subsecundum (R E Daniels)

The INSTITUTE OF TERRESTRIAL ECOLOGY (ITE) is one of 15 component and grant-aided research organizations within the NATURAL ENVIRONMENT RESEARCH COUNCIL. The Institute is part of the Terrestrial and Freshwater Sciences Directorate, and was established in 1973 by the merger of the research stations of the Nature Conservancy with the Institute of Tree Biology. It has been at the forefront of ecological research ever since. The six research stations of the Institute provide a ready access to sites and to environmental and ecological problems in any part of Britain. In addition to the broad environmental knowledge and experience expected of the modern ecologist, each station has a range of special expertise and facilities. Thus, the Institute is able to provide unparallelled opportunities for long-term, multidisciplinary studies of complex environmental and ecological problems.

ITE undertakes specialist ecological research on subjects ranging from micro-organisms to trees and mammals, from coastal habitats to uplands, from derelict land to air pollution. Understanding the ecology of different species of natural and man-made communities plays an increasingly important role in areas such as improving productivity in forestry, rehabilitating disturbed sites, monitoring the effects of pollution, managing and conserving wildlife, and controlling pests.

The Institute's research is financed by the UK Government through the science budget, and by private and public sector customers who commission or sponsor specific research programmes. ITE's expertise is also widely used by international organizations in overseas collaborative projects.

The results of ITE research are available to those responsible for the protection, management and wise use of our natural resources, being published in a wide range of scientific journals, and in an ITE series of publications. The Annual Report contains more general information.

Dr R E Daniels
Institute of Terrestrial Ecology
Furzebrook Research Station
Wareham
Dorset
BH20 5AS
0929 551518

Mr A Eddy
British Museum (Natural History)
Cromwell Road
London
SW7 5BD

01 589 6323

CONTENTS

	Page
PREFACE	5
ACKNOWLEDGEMENTS	6
INTRODUCTION	7
Taxonomic relationships	8
Morphology and life history	12
Distribution	20
Ecology	22
Collection, preservation and examination	30
KEY TO EUROPEAN SPECIES OF *SPHAGNUM*	32
SPECIES DESCRIPTIONS	43
Section Sphagnum	44
1. *S. palustre*	46
1a. *S. palustre* var. *centrale*	50
2. *S. papillosum*	52
3. *S. imbricatum*	56
4. *S. magellanicum*	61
Section Acutifolia	65
5. *S. molle*	67
6. *S. subnitens*	71
7. *S. angermanicum*	76
8. *S. subfulvum*	80
9. *S. fuscum*	84
10. *S. quinquefarium*	88
11. *S. capillifolium*	92
11a. *S. capillifolium* var. *capillifolium*	96
11b. *S. capillifolium* var. *rubellum*	96
12. *S. warnstorfii*	98
13. *S. russowii*	102
14. *S. girgensohnii*	107
15. *S. fimbriatum*	111
Section Squarrosa	116
16. *S. teres*	117
17. *S. squarrosum*	122
Section Insulosa	126
18. *S. aongstroemii*	126
Section Polyclada	131
19. *S. wulfianum*	131
Section Hemitheca	136
20. *S. pylaesii*	136

Section Subsecunda ... 140
21. *S. subsecundum* ... 142
   21a. *S. subsecundum* subsp. *inundatum* ... 143
22. *S. auriculatum* ... 150
23. *S. platyphyllum* ... 156
24. *S. contortum* ... 160
Section Cuspidata ... 164
25. *S. cuspidatum* ... 166
26. *S. riparium* ... 171
27. *S. obtusum* ... 176
28. *S. flexuosum* ... 180
29. *S. recurvum* ... 185
   29a. *S. recurvum* var. *mucronatum* ... 185
30. *S. angustifolium* ... 190
31. *S. balticum* ... 195
32. *S. annulatum* ... 199
33. *S. jensenii* ... 203
34. *S. majus* ... 207
35. *S. pulchrum* ... 211
36. *S. lindbergii* ... 215
37. *S. lenense* ... 219
Section Mollusca ... 223
38. *S. tenellum* ... 223
Section Rigida ... 228
39. *S. strictum* ... 230
40. *S. compactum* ... 234
GLOSSARY ... 239
INDEX OF SECTIONS, VARIETIES AND SYNONYMS ... 244
BIBLIOGRAPHY ... 248

PREFACE

Sphagnum is a notoriously difficult genus for the novice because, although it is fairly distinct, the individual component species are very similar in gross features. More detailed features may change according to habitat conditions, and this morphological plasticity can cause problems of identification until the observer knows which characters are important taxonomically, and until he/she has gained a 'feel' for the different species. The early taxonomists tended to assign specific names to many of the forms produced in different habitats and, in many cases, the same specific or varietal epithet was used by different authorities for different forms of the same species. This taxonomic confusion is also somewhat daunting for the beginner.

No key or guide can be infallible because of variability in the interpretation of the author's words or in the plant material itself. If such a guide to the identification of species is not to become too cumbersome for practical use, it cannot attempt to cover the whole range of variation that could be encountered in all situations. Different habitats produce slightly different forms of the plant, and occasionally aberrant forms are found resulting from particularly unusual conditions or genetic accidents. The guide must concentrate on the more typical forms whilst, at the same time, indicating the type of variation likely to be found. The key and descriptions in the present work, and the illustrations accompanying them, are designed to enable reliable identification of fairly typical samples of each species. The 'Detailed descriptions' and 'Additional notes' are intended to help reinforce the key determinations and to indicate the variations likely to be found in less typical specimens.

In any written material, the author has only one chance of making himself understood: there can be no dialogue between author and reader to clarify dubious points. This lack of dialogue produces another possible source of error, one of interpretation. Colour, for example, is difficult to describe in exact terms. Red or green are wide concepts, and 'bright red' or 'dull green' are phrases which mean slightly different shades to many people. We have used the terms as we understand them, so that it may be that some experience of using the guide will be necessary before the user becomes fully familiar, and confident, with it.

We have tried to give a range of aids to identification, and we hope that the combination of field and microscopic characters in the key, detailed species descriptions, habit photographs and line drawings, and ecological notes will allow the majority of specimens to be assigned to a species with a measure of confidence.

A series of distribution maps is included within the text, not to define rigidly the boundaries of distribution, but as a guide to the areas in which particular species are to be found, always bearing in mind the ecological conditions required or preferred by those species. It is also hoped that these maps will provide a stimulus for the more accurate recording and plotting of species distributions.

The ecological notes are, similarly, provided for guidance and comment rather than as definitive statements to be regarded as inviolable, although it is hoped that they give an accurate picture of the ecology of the species found in Europe.

Finally, we hope that, by producing this Handbook, we have filled a gap which we consider existed in the literature and that it will help to stimulate interest in further study of the genus *Sphagnum*, as well as clarifying some of the problems and confusions in the identification of its species.

ACKNOWLEDGEMENTS

We would like to express our thanks and gratitude to a number of people who have helped us in the preparation of this Handbook. We are grateful to Dr H J B Birks, Dr G J Duckett, Dr K Dierssen and T Lindholm for allowing us to reproduce photographs. Professor H Sjors and Dr P Isoviita read parts of an earlier draft and we are indebted to them for their valuable comments. We would like to thank a number of colleagues and friends, mainly from the British Museum (Natural History) and ITE Furzebrook, who tried out alternative versions of the keys. Thanks are also due to Dr M G Morris and Prof F T Last for their comments.

Special thanks are due to the late Dr S W Greene for his encouragement in the early stages of this project: without his valuable guidance, this publication could not have been produced.

INTRODUCTION

Bryophytes (mosses and liverworts) may be considered as a successful plant group because they have evolved into a wide variety of forms, including over 24 000 species. Nevertheless, being small and often ephemeral, they are dominant components of vegetation only under special conditions.

The genus *Sphagnum* ('bog' or 'peat' moss) is exceptional as it covers, often as the dominant or co-dominant plant, large areas of land in the temperate northern hemisphere. Their unique and highly characteristic anatomy gives *Sphagnum* plants special properties which make many of them important ecologically and economically.

The anatomy and growth form of *Sphagnum* plants show adaptations to retain large quantities of water: both living and dead plants have large water-holding capacities. This feature is important to individual plants as they do not have roots for absorbing soil moisture or internal conducting tissues for the transport of water. On a larger scale, the ability to retain water is important as *Sphagnum*, and peat derived from its dead remains, plays a significant role in controlling runoff from wide tracts of land. Commercially, the excellent absorbent qualities of dried *Sphagnum* have made it useful for surgical dressings, whilst wet *Sphagnum* is valuable for packing horticultural and agricultural products for shipping. More recently, *Sphagnum* peat has been used to absorb accidental spillages of oil (Ekman 1969; D'Hennezel & Coupal 1972).

Sphagnum species have a pronounced capacity to exchange hydrogen ions for mineral cations, a feature of particular advantage in habitats where minerals needed for nutrition are in short supply. However, this same capability renders some species particularly susceptible to even low concentrations of some pollutants. Because of its ion exchange ability, *Sphagnum* has proved useful in the purification of waste materials (Belkevich *et al.* 1976).

The accumulation of peat, often consisting of large proportions of *Sphagnum*, represents a considerable store of organic material and energy. Peat is used as a fuel in a number of countries. In areas with extensive peat deposits, it has been traditional to cut blocks for domestic use whilst, more recently, peat has been employed as fuel in electricity generating power stations.

These properties have made *Sphagnum*, or *Sphagnum* peat, particularly valuable for horticulture. It retains moisture and can store minerals for use by growing plants, and it is a source of organic matter and energy for micro-organisms. Many of the traditional horticultural areas of western Europe are on drained peatland. In recent years, the industry has diversified to provide peat or peat-based products, eg peat plant pots. It is no exaggeration to say that horticultural industries with a considerable annual turnover owe a large proportion of their commerce to the accumulation of deposits of *Sphagnum* peat over several thousand years.

Taxonomic relationships

The Class Sphagnopsida contains a single order, the Sphagnales, within which there is one monogeneric family, the Sphagnaceae. *Sphagnum* appears to have no close relatives among other living bryophytes and there is no fossil evidence through which this genus can be related to other mosses or liverworts. Comparisons only serve to emphasize the isolation of the Sphagnopsida from the rest of the Bryophyta.

The actual number of species within the genus *Sphagnum* depends upon the interpretation of 'species'. Different authorities come under different influences and develop different concepts of what constitutes a species, a subspecies or a variety. Some (eg Flatberg, Andrus) follow a tradition that tends to recognize separate subspecies, or even species, on morphological criteria that others, including ourselves, would regard as examples of within-taxon variation reflecting either environmental effects or small-scale differences in genetic composition and warranting, at most, varietal recognition. However, within the genus, there are a number of distinct subgeneric groups with character complexes which, in other groups of plants, may be regarded as sufficient to distinguish them at the generic level. In common with general practice, these are retained as subdivisions of the genus. Although these taxa may be recognized by most workers with an equal degree of distinction (except that, for example, *S. tenellum* is usually placed in the section Cuspidata in most modern treatments, rather than in a section of its own as in the current publication), there is some disagreement as to the status that should be accorded to each of these groups. Some authorities distinguish between subgenera, as represented by the subgroup *Sphagnum*, and sections of the genus such as the section Cuspidata. In the present treatment, we follow the basic example of Isoviita and refer to each of the groups as sections.

The following sections are represented in the *Sphagnum* flora of Europe:
> Section Sphagnum
> Section Acutifolia
> Section Squarrosa
> Section Insulosa
> Section Polyclada
> Section Hemitheca
> Section Subsecunda
> Section Cuspidata
> Section Mollusca
> Section Rigida

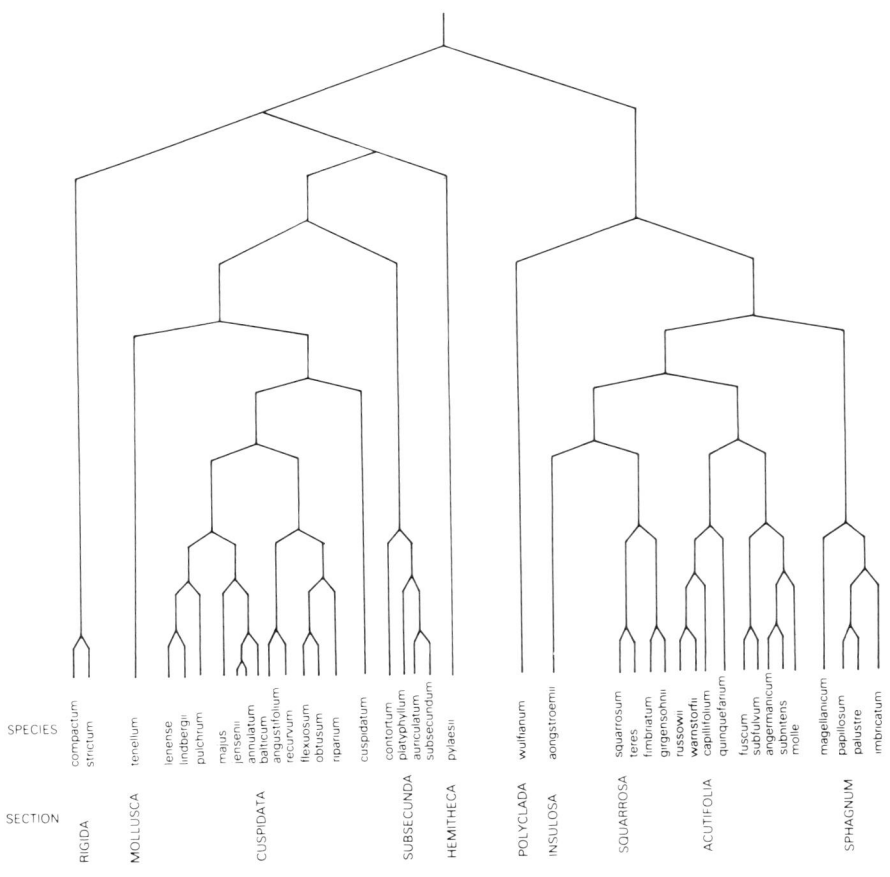

Figure 1. Diagram indicating the possible phylogenetic relationships of European *Sphagnum* species.

Figure 2. Explanatory illustrations of anatomical features used in the key and descriptions
A. Upper part of plant with capitulum (cap) and fascicles of branches
B. A branch fascicle showing spreading branches (sb), pendent branches (pb), and a portion of stem with stem leaves (sl)
C. Cross-section of stem showing thin-walled cortex (hyaloderm) and internal cylinder. The segments demonstrate some of the variation in stem form
D. Lengths of defoliated branch with: left, monomorphic cortical cells of *S. palustre* type; right, dimorphic cortex with imperforate and retort cells (ret)
E. Branch fascicles of dimorphic type (left) and monomorphic type (right)
F. Branch leaf outlines: ovate (left), lanceolate (centre) and linear (right)
G. Margins of branch leaves, with resorption furrow (arrowed left) and intact (right)
H. Areolation of *Sphagnum*: fibrillose and porose hyaline cell and narrow chlorophyllose cells (black): also present are ringed commissural pores (rp), and unringed commissural pore (up), an unringed, free, central (median) pore (cp), a resorption gap (res), a pseudopore (ps) and serial pores (ser)
I. Variation in chlorocyst (chlorophyllose cell) morphology: from the top; oval, completely immersed (eg *S. magellanicum*); triangular, partly immersed (eg *S. recurvum*); trapezoid, fully exposed (*S. cuspidatum*); barrel-shaped (eg *S. subsecundum*)
J. Stem leaf forms: left to right, triangular, oval-triangular, lingulate-triangular, lingulate, spatulate, fimbriate
K. *Sphagnum* spore in polar view
L. Cross-section of leaf of *S. papillosum*, showing papillose internal commissural walls of adjacent hyaline cells
M. Fascicle with female bud (left); fascicle with mature female shoot (centre); fascicle with male branches, detail enlarged to show antheridia (right)

Figure 2.

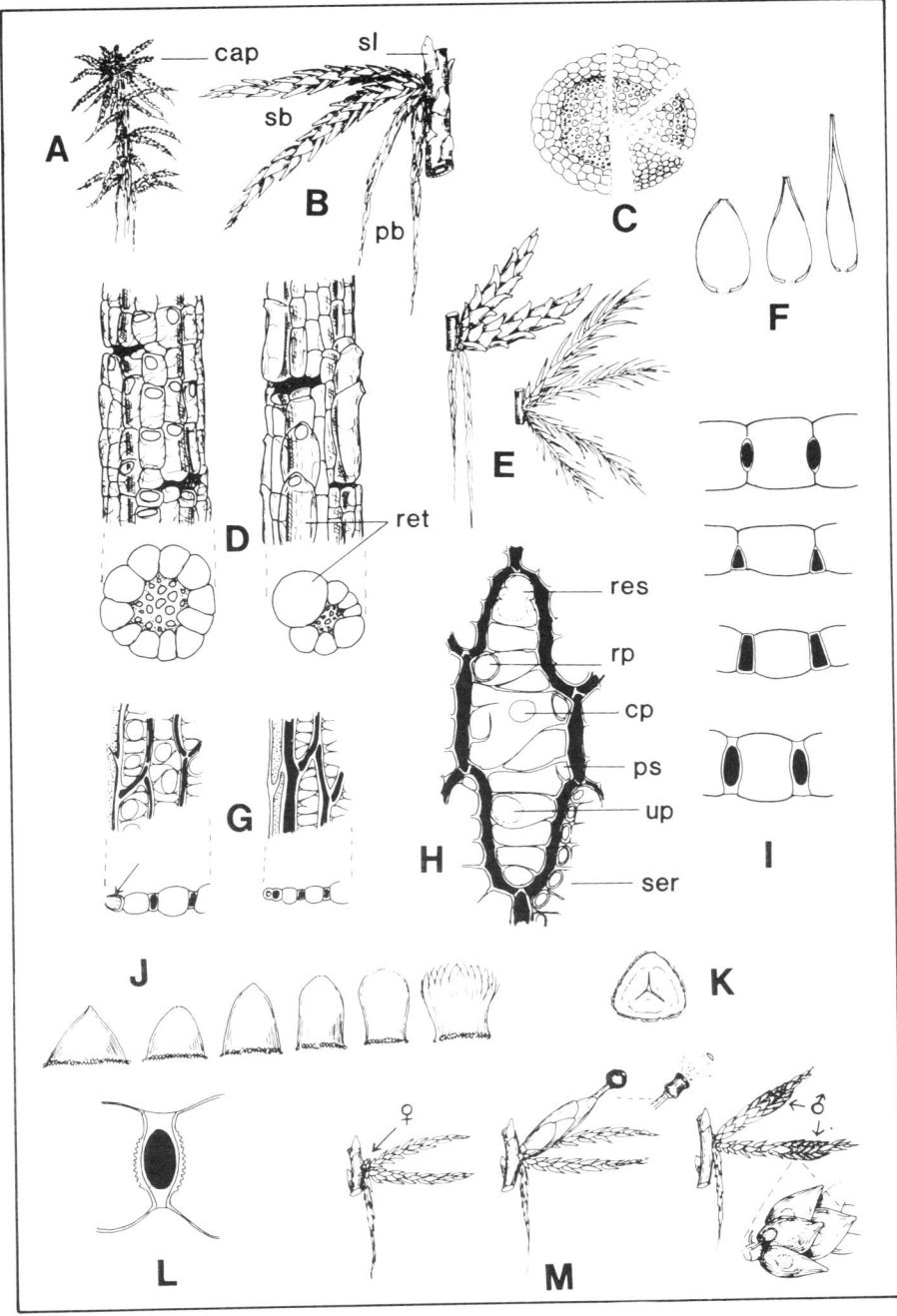

SPHAGNUM L. (Sp. Pl., **2**, 1106. 1753)

Morphology and life history

Although the Sphagnopsida are only distantly related to other classes of bryophytes and are morphologically distinct, their mode of growth and life cycles have similarities with those of other mosses. In common with other bryophytes, there is a distinct alternation of generations, the obvious, leafy plant being the gametophyte, and the sporophyte being much reduced. Also, as in other mosses, spores germinate to produce preliminary, transient protonemata from which the gametophytes arise.

Mature spores of *Sphagnum* have a variable period of dormancy. How long spores can remain viable is not known, but those of *S. compactum* have shown 15% viability after 13 months' storage at room temperature, whilst some spores of *S. capillifolium* germinated after 2 years' storage at 2°C. Spores of various species have been germinated successfully *in vitro* using different techniques. Those of *S. palustre, S. fimbriatum* and *S. subnitens*, for example, have been germinated on agar or moist filter paper at 10°–18°C, over a period of 3–30 days.

Protonemata are seldom noticed in the field, though they are not especially rare, but may occasionally be found on bare, wet peat in the neighbourhood of fruiting plants. A protonema consists of an irregular plate or ribbon of green cells in a single layer which is seldom more than a few millimetres long. Commonly, it produces a single *Sphagnum* plant from its dorsal (upper) surface, though occasionally more than one such leafy gametophyte may be produced. Rhizoids occur only on the underside of the protonema and on very young plants. Unlike most other mosses, the stems of mature gametophytes do not develop rhizoids.

With few exceptions, adult plants are morphologically remarkably uniform: they consist of a primary, erect stem with characteristic clusters (fascicles) of branches borne at more or less regular intervals. Growth of stems and branches is a result of successive divisions of tetrahedral apical cells, but, whereas stems can grow indefinitely, branches are strictly limited in length. Some species, particularly those which grow in pools, continue to grow more or less 'normally' when their stems are prostrate, but, in other species, the apical dominance of the original capitulum (the dense cluster of young branches around the stem apex) is lost, with a tendency to produce numerous 'lateral' stems from the points of insertion of branch fascicles. In certain conditions, branch growth may be reduced, or totally absent, as in *S. pylaesii* and the American *S. cyclophyllum*. Vegetative increase is mainly by pseudo-dichotomy, the visible signs of which are the 'splitting' of the capitula. The dichotomy is not the result of the stem apical cell dividing to give 2 daughter apical cells, but occurs when one of the very young branches begins to assume the characteristics of a stem rather than a branch and continues to develop as a stem. Less commonly, adventitious buds on the stem near the bases of fascicles may become active and produce rapidly growing 'stolons' which may remain branchless until they reach the surface of the *Sphagnum* hummock, or carpet, in which they are growing. Such 'stolons' may be numerous in certain species, eg *S. compactum*, and may be produced in most, if not all, species following damage to the exposed upper parts of the plant (eg as a result of fire or after an extended period of drought).

Although some species remain basically green, the majority produce secondary pigments, unless they are growing in deep shade. Pigmentation may be pronounced only in antheridial branches, though most *Sphagna* have at least tinges of yellow, brown, orange or red, and it is not uncommon to find plants in which the green of chlorophyll is completely masked. The presence of particular pigments is often a very useful aid to identification, as their colours are relatively stable, persisting for many decades in herbarium material stored in normal conditions. Perhaps surprisingly, comparatively little is known about secondary pigments which, being mostly locked in cell walls of photosynthetic cells, are difficult to extract for analysis (see Martensson & Nilsson 1974). Some colours react with weak alkalis (eg saturated Na_2CO_3): the red *S. capillifolium,* for example, will turn violet-blue; colour changes of this sort are normally fully reversible.

In transverse section, a typical *Sphagnum* stem consists of 3 concentric zones (see Figure 2C). The innermost 'pith' is composed of relatively thin-walled, colourless parenchymatous cells, often rich in oil droplets or other storage products. Surrounding this central tissue and grading into it (often rather abruptly) is an 'internal cylinder', consisting of cells which become progressively narrower and thicker towards the periphery: the outer series are commonly deeply pigmented brown or red. The outer cortex consists of enlarged, empty, hyaline cells in 1–4 layers according to species. The walls of these cortical cells may be entire, have one or more large pores or, in the section Sphagnum, have spiral fibrils.

Apart from colour, internal stem tissues have few features of taxonomic value. In contrast, the cortex has a number of features which may be useful in the recognition of subgeneric groups and of significance in phylogenetic studies. Occasionally, cortical characters (usually the presence or absence of pores in cell walls) are useful in making distinctions at the species level. Pores may be obvious in unstained material (eg section Sphagnum), but unringed pores in very thin cell walls can only be discerned after rather intense staining. In fully developed plants of a few species, cortical cells may be virtually indistinguishable from those of the internal cylinder.

The internal cylinder appears to play an insignificant role in the conduction of water and solutes, this function being attributed to the system of hyaline cells in stems, branches and leaves.

Although growth of both stems and branches is from tetrahedral apical cells which have 3 free faces, torsion results in a 5-ranked arrangement of leaves and fascicles. Three-ranked leaves are rare, and then confined to weak, poorly developed plants or the distal parts of branches: a few species have unranked branch leaves.

Fascicles may be close together or well spaced along the stem. Each is composed of a number of branches (commonly 3–5) which are usually of 2 types (dimorphic): (i) unmodified branches spreading out from stems, and (ii) more or less modified slender *pendent* branches, which hang appressed to stems. At the stem apex, developing fascicles are densely packed to form a head (capitulum) which commonly conceals the stem growing point. Branches are normally simple, but occasionally they produce adventitious branches, as a result either of exceptional vigour or of injury.

Fascicle characters, including the length and degree of branch dimorphism, are variable in response to different ecological conditions so that they can only rarely be used as aids in identification. However, where the full potential development of such characters is known, they can provide useful clues to phylogeny. Photosynthetic activity in the pendent branches of highly dimorphic fascicles is almost absent and the main function of such branches is the translocation of water and nutrients.

The anatomy of branches is basically similar to that of stems, but the cortex is always single-layered (some duplication may occur at the bases of branches). The internal cylinders of branches are thinner than those of stems, and the cortical cells are always sharply differentiated from them. Cortical cells may be more or less uniform, as in the section Sphagnum (in which they are also fibrillose), or dimorphic with one or more cells immediately above a branch leaf insertion enlarged and perforated at their upper ends. In their extreme form (as in *S. tenellum*), the pores of these enlarged cells are on protuberances and hence, because of their likeness to the glass retorts formerly used in laboratories, are called 'retort cells' (Figure 2D). Where distinct retort cells are present, the remaining unmodified cortical cells are normally imperforate.

No satisfactory explanation has so far been given for the development of retort cells, though the fact that they occur in all species except those in sections Sphagnum and Rigida suggests that they have a significant role. Their structure and position suggest that they may be involved in translocation to, and from, branch leaves. If so, it is something of a puzzle that they may be found in some species but be absent from others which grow in, superficially at least, similar habitats (eg they are abundant in *S. tenellum* but absent from *S. compactum*).

Branch leaves vary in size and density but, with few exceptions, tend to be similar in superficial appearance. In most species, they are 5-ranked (at least in young branches), often conspicuously so, but are occasionally unranked ('spiral'). Usually they are more or less erect to erecto-patent in attitude, although they may be widely spreading (patent) or, in a few cases, squarrose. Branch leaves vary in length from about 0.7 mm to over 3.0 mm and are ovate to lanceolate, rarely narrowly lanceolate, though they may appear to be so because of inrolled margins in some species. Around the leaf margin is a border of one to several series of narrow cells. In some species, the outer edge of the border is resorbed to form a 'resorption furrow' (Figure 2G), and the oblique cross-walls of the border cells may project to form distinct teeth. Leaf apices are typically truncated and dentate, but may be modified by resorption to a more or less rounded-eroded tip (the normal state in section Sphagnum): rarely, the apices are intact and tapered to more or less acute points (eg *S. riparium*).

The size, arrangement, colour and apices of branch leaves are the characters mainly responsible for the overall appearance of *Sphagnum* plants. With experience, well-grown plants in good light can usually be identified in the field with some degree of confidence, provided that sufficient attention is also paid to the habitat from which the sample was taken. Nevertheless, extreme conditions (eg poor light or immersion of species not usually found in standing water) may affect these superficial characters to an extent that precludes accurate field identification. Fortunately, microscopic characters are less readily distorted.

The degree of divergence in form between pendent and spreading-branch leaves depends upon the degree of dimorphism shown by the branches bearing them. Pendent branches of monomorphic species tend to have leaves which are identical to those of the spreading branches, but, in highly dimorphic species, the pendent branch leaves may be highly modified and bear little resemblance to those of spreading branches. The features most strongly differentiated in pendent branch leaves are: leaf shape (leaves tending to be narrower and often longer and more attenuated); leaf border (which tends to be reduced and, in extreme cases, partially replaced by fibrillose hyaline cells); hyaline cells (which may be proportionately wider and usually have more numerous and larger pores or larger resorption gaps); and photosynthetic cells (which are usually much narrower than in spreading branches). The degree of modification of pendent branch leaves becomes progressively more marked from branch insertion to branch tip; all branches usually have one or more vestigial basal leaves.

Leaves are single-layered, but composed of 2 kinds of cell at maturity: elongated, narrow, living cells containing chloroplasts form a network in which the meshes are occupied by empty hyaline cells (Figure 2H). The hyaline cells of all European species, except where modified in stem leaves, are slightly, to considerably, larger than the photosynthetic cells and have annular or helical strengthening bands (fibrils). The walls of hyaline cells exposed on leaf surfaces are usually perforated by small to large pores (from about 2.0 μm to over 20.0 μm). These pores are more or less circular in outline, though they may appear elliptical or 'half-elliptical' in surface view because of cell wall curvature; they are either simple or surrounded by a ring of thickening. Where the walls of hyaline cells abut the lateral walls of photosynthetic cells (the inner commissural walls), their internal surfaces may be smooth, papillose or have longitudinal to transverse crests or ridges (comb fibrils) according to species. Photosynthetic cells vary in cross-sectional shape and in their position relative to the upper and lower leaf surfaces; their walls may be variously thickened. Photosynthetic cell morphology and the size, number and arrangement of hyaline cell pores provide important taxonomic characters. In a few exotic taxa, branch leaf hyaline cells may be devoid of pores (eg the Malaysian *S. sericeum* C. Mull. and the American *S. macrophyllum* Bernh. ex Brid.).

The hyaline cells in *Sphagnum* provide mechanical support for photosynthetic leaf cells and act as water-conducting systems. In hummock-forming species of drier habitats, the hyaline cells are larger than in those species occupying wetter habitats. Branch dimorphism is also more marked in species of drier situations; in these plants, the pendent branches have almost lost their photosynthetic ability and, instead, they adopt a role in supplementing the water-conducting functions of cortical hyaline cells.

Stem leaves vary in size, shape (Figure 2J), attitude and density. In the more primitive species, they resemble branch leaves in shape and in more detailed anatomy (isophyllous plants), except in details of leaf insertion. In other species, they may be partly differentiated (hemi-isophyllous) or completely different (heterophyllous). Heterophylly results from the differential modification of leaf tissue, and the most common difference between branch and stem leaves is the partial or complete loss of hyaline cell fibrils in the latter, usually accompanied by extensive resorption on one or other leaf surface. In the upper part of the leaf, the border may be lost and the margin become thin, whilst drastic changes may occur in the leaf apex. *S. fimbriatum* shows an extreme example of this apical modification in which the whole of the widely rounded apex has become extensively resorbed and fringed with the remains of photosynthetic cells. A common feature of stem leaves is the apparent expansion of the leaf borders, near the point of insertion, into more or less broad patches of narrow, elongated (prosenchymatous) cells. Photosynthetic cells in such leaves are usually thick-walled with narrow lumina and have a much reduced photosynthetic capability. The insertion line of stem leaves is wider than that of branch leaves and has a different composition, with one or more rows of more or less uniform isodiametric, usually thick-walled, cells that may be conspicuously brown. On one or both sides of the insertion line, there may be, except in isophyllous plants, small projections (sometimes conspicuous) composed of one to a few hyaline cells. These 'auricles' vary even on a single stem but, typically, their cells contain fibrils and pores.

The majority of *Sphagnum* species are unisexual (dioecious), though a few are monoecious (autoecious) with male and female sex organs on different branches of the same plant. Male organs (antheridia) are borne singly in the axils of bracts which strongly resemble branch leaves, except that they are usually more concave and more highly coloured than normal leaves: in some species, fibrils may be lost from the lower hyaline cells. The branch zones (usually the distal $\frac{1}{3}-\frac{1}{2}$ of a branch) that bear antheridia are usually conspicuous because of the denser imbrication (Figure 2M) and strong colour of the perigonial bracts. Antheridia are normally confined to spreading branches, except in the section Rigida where they are more or less confined to pendent branches. They are delicate ovoid bodies (200–330 μm) held on thin multicellular stalks. At maturity, the thin walls, which are one cell thick, dehisce, releasing the biflagellate antherozoids through a number of apical ruptures.

Before fertilization, female 'branches' are inconspicuous, being small and generally colourless. They resemble undeveloped branch buds and appear on the upper side of the point of insertion of some branch fascicles. These female 'buds' contain 1–5 long-necked archegonia at the tip of the branch, though only one will develop after fertilization. After fertilization, the leaves of the female bud (perichaetial bracts) become greatly enlarged to form a protective sheath around the developing sporophyte within the perichaetium (Figure 2M).

Although the female shoots are referred to as 'branches', they are structurally more akin to stems. Female 'branches' arise in addition to the normal branch complement, whereas male branches are always part of the normal complement. Perichaetial bracts, at least in their distal parts, are anatomically similar to stem leaves, except in those species whose stem leaves have undergone extreme evolutionary reduction (eg *S. compactum*). The pseudopodium is an example of a female shoot which has undergone reduction and, in *S. palustre*, although it is more delicate than a normal stem (having a temporary, in contrast to a semi-permanent, role), it tends to have cortical cells in more than one layer and with more than one pore per cell.

Fertilization seems to occur mainly between late autumn and early spring, with the capsules maturing from spring to mid-summer according to species and locality (the earliest is *S. tenellum*). Meiosis takes place while capsules are still green and within the perichaetia. At this stage, a capsule yields an opaque white fluid when squashed and is said to be in the 'milk stage'.

Mature capsules are more or less spherical, with small, shallowly convex lids; they are enclosed in thin, transparent calyptras. With the exception of *S. pylaesii*, capsules are uniform in size and structure throughout the genus. Stomata are poorly developed and are supposedly non-functional, and as a result are often designated 'pseudostomata'. Ripe capsules have airtight walls and pressures build up within them until, in dry conditions, their lids are explosively ejected along with most of the spores. In persistently humid conditions, or when the capsules are submerged, normal dehiscence fails and the capsule walls eventually disintegrate, reportedly without loss of spore viability. The true seta consists of little more than a 'foot' which anchors the capsule to a false stalk (pseudopodium) of gametophyte tissue. It is the pseudopodium that elongates and lifts the ripening capsule clear of the perichaetium: it grows to between about 3.0 mm and 15.0 mm.

Spores develop within a dome-shaped archesporium and the tetrads separate at a comparatively late stage so that the individual spores retain a distinctly tetrahedral shape at maturity. Each ripe spore has 3, more or less smooth, plane walls which form a low pyramid with a characteristic triradiate mark at the apex, and a domed wall (the base of the pyramid) which is variously smooth to strongly papillose, according to species. Spores vary from about 22 μm to about 42 μm in diameter, according to species, and there is no detectable difference between male and female in dioecious species.

The usual haploid complement of chromosomes in *Sphagnum* is $n = 19 + 2$ m. Nineteen normal, large, chromosomes are accompanied by 2 very much smaller ones which stain more weakly and usually dissociate before the larger chromosomes during cell division. In some species, doubling of the chromosomes has occurred so that the normal haploid complement is $38 + 4$ m, as in *S. palustre* and *S. russowii*.

Actively dividing (mitotic) cells of *Sphagnum* are most numerous in and around the stem apex, so that students having cytological interests may have to depend upon tissues buried within the capitula for cells showing stages in division. *In vitro,* decapitated stems laid horizontally in a growth medium will normally produce secondary growth centres, which may provide greater chances of success in the search for mitotic stages.

Sphagnum has not developed obvious special means of vegetative propagation comparable with the gemmae or similar structures found among other bryophytes. However, comparatively small fragments of stems and branches can regenerate new plants, provided that they do not become completely desiccated. Sometimes, small groups of cells may be found forming protuberances on branches but, although their presence has been attributed to nematode attacks, they are often found where there is no evidence of infestation. *In vitro,* adventitious plants have been induced from these cell clusters, although this phenomenon has rarely been observed in the field.

Established plants of many species, especially hummock-formers, may be very resistant to periods of drought, *S. imbricatum* being particularly tolerant to desiccation (Green 1968). This may, in part, be due to the growth form of hummock-forming species, but the secretion of mucilaginous substances by specialized cells in the leaf bases in the capitulum (which may be analogous to the 'slime papillae' of some hepatics) also retards desiccation. Complete drying out for long periods is, however, fatal.

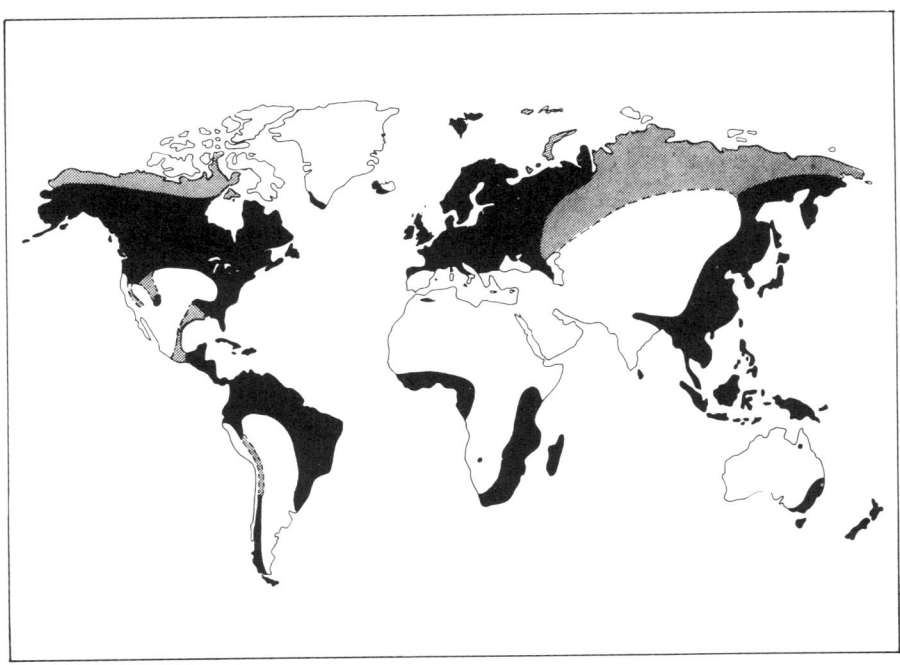

Figure 3. World distribution of *Sphagnum*. Within the half-tone areas, the limits of distribution are less clearly defined

Distribution

The genus *Sphagnum* contains a large number of species and its members are very widespread, being found from the Arctic to the sub-Antarctic (Figure 3). Isoviita (1966) suggests that there are between 150 and 200 species. A high proportion of these are common, many being widespread throughout a range of climatic zones. All the European species are found in North America and some extend to other continents; for example, *S. magellanicum* is fairly abundant in Europe, Asia, North America, tropical Central America, and extends down the Andes as far as Tierra del Fuego, as well as occurring in Malagasy (Madagascar).

European species may be grouped by their distributional tendencies.

1. *A widespread group.* Species in this group are found more or less throughout Eurasia as well as in North America, extending as far south as the sub-tropics or even the tropics: some members of this group are also present in the southern hemisphere. The group contains:

S. magellanicum	extends to tropics and southern hemisphere
S. subsecundum	to the tropics in SE Asia
S. palustre	south to the sub-tropics
S. papillosum	south to the sub-tropics
S. compactum	south to the sub-tropics
S. subnitens	south to the sub-tropics
S. capillifolium	south to the sub-tropics
S. girgensohnii	south to the sub-tropics
S. fimbriatum	in temperate zones of northern and southern hemispheres
S. squarrosum	south to warm temperate zone
S. teres	south to warm temperate zone.

The boreal and sub-arctic zones together form the main centre of distribution of *Sphagnum* in Europe. However, different species have different ranges, some being restricted to particular climatic areas, whilst others are more widespread over a broad latitudinal band and may extend northwards into the Arctic or further south into sub-tropical areas. Some species are restricted to oceanic or sub-oceanic areas which do not experience the extremes of climate found in continental interior regions, whilst others are more abundant in continental areas.

2. *An oceanic or sub-oceanic group.* Species within this group are most abundant close to Atlantic coastal areas, though some may be found a considerable distance north-eastwards in mainland Europe. They are also found in eastern or north-eastern America and may have their main centre of distribution there, eg *S. angermanicum*. A few, eg *S. imbricatum*, may also be found on the Pacific coast of North America.

This group includes:

S. imbricatum	reaching the tropics in West Indies, also inland in the Himalayas
S. molle	extends to sub-tropics in North America
S. subsecundum var. *inundatum*	extends to sub-tropics in North America
S. strictum	extends to sub-tropics in North America*
S. tenellum	extends to sub-tropics and South America
S. pylaesii	main distribution NE North America
S. angermanicum	main distribution NE North America
S. auriculatum	also in west Pacific (Japan) and North Africa (Atlas Mountains)
S. cuspidatum	also in west Pacific (Japan)
S. pulchrum	also in west Pacific (Japan)
S. quinquefarium	also in west Pacific (Japan)
S. recurvum	circumpolar but tending to sub-oceanic in Europe
S. flexuosum	circumpolar but tending to sub-oceanic in Europe

3. *A north-eastern group.* Members of this group are all circumpolar in the boreal and sub-arctic regions of the northern hemisphere, although in some instances gaps occur; eg *S. lenense* is absent from northern Scandinavia. Some species extend into the Arctic. The following species comprise this group:

S. contortum	boreal
S. obtusum	boreal, continental
S. wulfianum	boreal, continental
S. majus	somewhat boreal in Europe: more oceanic in North America
S. centrale	extends to sub-arctic
S. platyphyllum	extends to sub-arctic
S. jensenii	extends to sub-arctic
S. balticum	extends to sub-arctic, slightly continental
S. angustifolium	extends to sub-arctic, slightly continental
S. riparium	extends to sub-arctic; more oceanic in North America
S. fuscum	extends to sub-arctic, slightly continental; in Britain somewhat oceanic

---

* This refers to subspecies *strictum:* subspecies *pappeanum* is found in tropical Central America, southern and eastern Africa and tropical Asia (Indonesia and New Guinea).

S. lindbergii	boreal to arctic
S. lenense	boreal to arctic
S. warnstorfii	boreal to arctic
S. russowii	boreal to arctic
S. subfulvum	boreal to sub-arctic

Although these broad geographical groups may be identified, the more detailed patterns of distribution are dependent upon local climatic, hydrological and chemical conditions.

Ecology

Sphagnum is essentially a genus of wetland habitats, though individual species are adapted to withstand different degrees of desiccation. All, or nearly all, species are probably able to survive short periods of complete drying out.

The morphological structure of the leaves and the presence of hanging branches enable plants to absorb and retain large quantities of water, thus helping to avoid rapid drying out when humidity decreases or water levels fall. In addition to retaining water, hyaline leaf cells may also play a further role in protecting the plant from undue water loss. The presence of these empty cells causes the leaf to become whitish on partial drying which, by increasing reflectivity of heat and light, reduces the absorption of radiant energy and, in turn, reduces evaporation.

Growth habit varies with water level. Plants growing with their apices well above standing water differ in gross morphology from those growing submerged or with their apices close to water level. These differences may be seen by comparing the submerged form of *S. cuspidatum* with a hummock-building species such as *S. imbricatum,* which builds a dome up to one metre above the general water level. In the former instance, differentiation between hanging and divergent branches is very poorly marked and the lax plant may become extremely plumose, with long branches bearing leaves which stand out from the branch axis. By comparison, *S. imbricatum* (and other hummock-forming species) has clearly differentiated hanging and divergent branches, the latter being short and carrying leaves which overlap and lie closely along the branch axis. Dense hummocks, or cushions, remain wet for longer periods than loose mats, even when they stand well above the surrounding water table.

Differences may also be found in the same species growing in different habitats, though the range of these differences is somewhat smaller than that between species of distinctly wet or comparatively dry areas. For example, a lax form of *S. imbricatum* is found more rarely in Britain at present than formerly (though it is still more common elsewhere in Europe): this form is found in situations where the water level is higher than that around the more dense, hummock-producing form. Other species may also have 'wet' forms, eg *S. compactum*, though less marked variation occurs in those plants which normally occupy an intermediate position along the wet–dry gradient.

Whilst the form of some species changes in response to varying water tables, there is a general zonation of species from one type of hydrological condition to another. Hydrology is a major factor in determining the distribution of species; one group, whose members are not necessarily closely related taxonomically, is found in comparatively dry conditions, a second group occupies habitats that have permanently high water tables, and a third group (in addition to any overlap by members of the first 2 groups) grows in intermediate conditions. The distribution of species in relation to general water level in peat is largely a reflection of their differing abilities to withstand desiccation.

A second major factor influencing the distribution of *Sphagnum* species is the chemical status of peat and the water contained within it, particularly the degree of acidity and the quantities of dissolved ions. These 2 aspects are largely complementary, an increase in acidity normally being accompanied by decreased concentrations of mineral nutrients, particularly calcium, available to plants.

The terms 'eutrophic', 'mesotrophic' and 'oligotrophic' are frequently used to define different parts of the range between nutrient-rich waters with a near neutral reaction ('eutrophic') and nutrient-poor waters with a distinctly acid reaction ('oligotrophic'): 'mesotrophic' waters are intermediate between these extremes. In the present context, they are used in a comparative form, although a number of authors have defined them on the basis of pH or calcium concentration (Gorham & Pearsall 1956; Ratcliffe 1964). It is dangerous to define trophic status in absolute terms, particularly when derived from water chemistry, because conditions may vary according to other factors, especially time and the effects of atmospheric or ground water flushing. Sites influenced by late snow melt may have waters with very small concentrations of electrolytes but, in terms of pH, these waters may be regarded as mesotrophic. Similarly, where water flow is fairly rapid and locally concentrated to produce a flush, the vegetation may be comparable with that of more eutrophic situations with little water movement, although the water in the flush may be quite oligotrophic with respect to the content of electrolytes or pH.

Although water level and degree of eutrophy or oligotrophy are important factors in determining *Sphagnum* distribution, it is difficult to relate habitat or geographical occurrences to any single influence. Site conditions are determined by a variety of factors. Where water flow is locally canalized, producing a flush, the vegetation may be that of a eutrophic situation, despite the oligotrophic nature of the water itself.

Habitat preferences of European species

Eutrophic habitats

Although *Sphagna* are found almost throughout the range of north temperate peatlands, and some species extend into the Arctic, they are uncommon, in numbers of both species and individuals, in strongly eutrophic habitats. In eutrophic locations, their place tends to be taken by other moss species, mostly of genera such as *Acrocladium, Campylium, Cratoneuron* and *Drepanocladus* together with *Mnium* or *Philonotis* species. *S. contortum* and *S. squarrosum* are among the few species of *Sphagnum* which occur: both may be found in a range of plant communities, from *Phragmites/Carex* swamp and mixed open fen to fen woodland (carr), provided that they are not shaded too much. More common in carr, where they can form wide carpets on solid peat, are *S. fimbriatum* and *S. palustre*, species which also grow along stream banks, in the 'lagg' areas around raised mires, or in tall fen communities. *S. palustre* var. *centrale*, which has a more northerly distribution, is found in similar habitats, though at the northern end of its range it may extend into more open mire communities.

Slightly less eutrophic habitats favour *S. teres* and *S. warnstorfii*, but these species may be found with *S. contortum* or *S. squarrosum* in bryophyte carpets of open fen, where the more northerly distributed *S. warnstorfii*, in particular, forms low hummocks which appear to be the nuclei upon which more characteristic hummock-building species become established. Similar associations may also be found in upland eutrophic flushes to which *S. squarrosum, S. teres* and *S. warnstorfii* are confined at the southern limit of their ranges. In contrast, *S. contortum* is more abundant at lower altitudes towards the southerly end of its distribution.

S. platyphyllum is of predominantly northern distribution, in wet or flushed, moderately eutrophic peatland or mineral habitats, particularly where seasonal flooding occurs. Typically, it is found in soakways within *Carex* communities of mires, but it also occurs in mineral flushes and pools. In Britain, *S. platyphyllum* is a rarity, confined to the north-west, although it extends across most of northern Europe. Another species found commonly across northern Europe, though absent from Britain, is *S. obtusum*. It grows in wet hollows or is submerged in pools in moderately eutrophic fens and fen woodland, though it may be found as an associate of *S. warnstorfii* or *S. teres* and *S. squarrosum*. In more open, *Carex*-dominated communities of mesotrophic or weakly eutrophic fennoscandian mires, *S. subfulvum* may be present near pool margins, though in the far north it also extends into wet *Betula* woodland. In central Fennoscandia, it often grows with *S. subnitens*, which further south replaces it entirely in this habitat.

Mesotrophic habitats

S. riparium forms loose carpets in very wet mesotrophic areas so that, typically, it is found in ditches and along stream margins, in wet willow scrub or marginal areas of larger wooded mire systems. Less often it is also found in more central parts of these mires. This species is commoner in the north than the south, where it becomes progressively restricted to slightly more oligotrophic montane flushes and stream-sides. In Britain, it is confined to this latter type of habitat.

Three other species, absent from Britain, grow in mesotrophic habitats in Fennoscandia, though none is common. *S. lenense* is really a species of the continental sub-arctic regions of the USSR where it forms small tufts in mesotrophic fens, but it has been recorded as far west as the Kola peninsula. The rare *S. angermanicum* (essentially a species of north-eastern America) has been found along pool and stream margins from a few localities in central Sweden and southern Norway. *S. aongstroemii* is more widely spread through northern Scandinavia and the northern European parts of the USSR, though it is also present (but rare) in more central parts of Scandinavia. It grows mainly along lake and stream margins or in willow scrub, though it may be found, occasionally, on wet rocks.

In rather drier mesotrophic sites, either on hummocks in fens, on flushed slopes or in woodlands, *S. girgensohnii* forms loose patches or small carpets. It is a frequent species of natural moist coniferous forest (taiga) in, for example, Fennoscandia. Like a number of other species, it extends into a more mixed, general mire community including oligotrophic *Salix* carr in northern Scandinavia, where it may also be accompanied by *S. russowii*. Further south (including Britain, from Wales northwards), this latter species is more characteristic of oligotrophic or weakly mesotrophic fens and woodland, mineral slopes or stream edges. Towards the south, it becomes increasingly rare in the lowlands and near its southern limit it becomes confined to upland areas of northern Italy.

Within the species *S. subsecundum* (used in its broad sense), the different subspecies occupy a variety of mesotrophic habitats. Var. *subsecundum* and var. *inundatum* are found at the more eutrophic end of the range, var. *subsecundum* even extending into, and sharing, fens and flushes typically occupied by *S. contortum*. It may also be found in more open, wet grassland. In Britain, it is the rarest of the 3 taxa (which have all, at times, been included within *S. subsecundum sensu lato*), possibly because of a poor competitive ability compared with other members of the Subsecunda group, *S. contortum* at the eutrophic end of the range and var. *inundatum* at the mesotrophic end. *S. subsecundum* var. *inundatum* itself is a plant of mesotrophic fens, flushes or wet hollows, and the margins of water bodies where periodic flooding occurs. The very rare *S. pylaesii* may be accompanied by *S. subsecundum* var.

inundatum in regularly flooded mesotrophic hollows. *S. auriculatum* is a species of more oligotrophic habitats – bog pools, where it is often submerged, wet flushes, or dripping rocks. In Fennoscandia, where it is more frequent in the western parts, it also grows in non-calcareous springs and is even found underwater in clear streams and lakes.

Members of the *S. recurvum* complex are found in a range of mesotrophic to moderately oligotrophic habitats. *S. angustifolium* is found in the widest range of trophic conditions. It grows either as scattered shoots with other species, including *S. recurvum* var. *mucronatum,* or forms low hummocks and lawns in fens with flowing water: it is common in all types of wooded mire. *S. flexuosum* is also found in moderately wet parts of fens or damp woods, or in flushed grass communities on sloping ground over mineral soils with a high organic fraction. *S. recurvum* is confined to more oligotrophic, usually treeless or, at least, fairly open mires where it forms extensive lawns in wet or damp situations. It is also found in oligotrophic flushes or along pool and stream margins. It may also invade damp pine needle litter, though *S. angustifolium* is usually typical of more densely wooded mires.

Oligotrophic habitats

In oligotrophic habitats, the distinction between very wet locations, pools and pool margins, and drier hummocks is reflected in the *Sphagnum* species found in these micro-habitats, although some species extend over a wide range. In Britain, *S. subnitens,* for example, is an oceanic species found as a component of western blanket mire communities, but also forming tufts in oligotrophic to mesotrophic mires of climatically more continental areas. In westernmost Norway, it grows in similar habitats to those in western Britain but becomes progressively more mesotrophic towards the east. It may also occur in damp woods and on rocky banks.

A number of species occur submerged in pools which are moderately to extremely oligotrophic. Apart from *S. auriculatum, S. cuspidatum* is the member of the genus most commonly found submerged in highly to moderately oligotrophic pools, where it becomes extremely plumose, but it is also present in areas around pool margins or in damp hollows where it may, occasionally, be subjected to drought. The most oligotrophic pools are avoided by *S. majus,* which may grow in association with *S. cuspidatum* where there is a slight nutrient enrichment but, in central to northern Fennoscandania, it gradually replaces *S. cuspidatum* even in bog hollows. This species, extremely rare in Britain, is more widespread on the mainland of NW Europe, particularly in Scandinavia. The northerly distributed *S. jensenii* (and the closely related, but rare, *S. annulatum*), which is absent from Britain, grows in a wide range of pools and hollows from distinctly oligotrophic to moderately mesotrophic. In mesotrophic, not too wet habitats, *S. imbricatum* may be found in the lax form more common in moderately continental climates, in contrast to the compact hummock-building form particularly characteristic of the oceanic NW seaboard of Europe and found as far east as SW Sweden.

S. balticum is a species of wet oligotrophic hollows, and the margins of pools: rarely is it actually submerged. This predominantly northern species is of very restricted distribution in Britain, as is the even more markedly northern *S. lindbergii* which is found in Scotland as broad mats in montane flushes, whereas, in northern Europe, it grows as more extensive bog carpets or occurs, occasionally submerged, beside lakes and streams. In Fennoscandia, *S. compactum* also occurs extensively in wet hollows, eg in the 'flarks' (shallow depressions in northern types of mire) of string mires (aapa mires) and between hummocks of, eg, *S. fuscum* and *S. capillifolium*.

Between pools on the flatter 'lawn' areas and at the bases of hummocks on oligotrophic mires, a number of intermediate species are found, eg *S. recurvum* var. *mucronatum, S. subnitens, S. tenellum* or, in the more oceanic areas, *S. pulchrum*. This last species, which extends as far east as the Vologda district of the USSR, has a disjunct distribution in Britain. It is a component of the mixed bryophyte carpet, with numerous leafy hepatics, of oceanic raised and blanket mires of the north-west, but also forms extensive carpets in the valley mires, within areas of dry heath, on the Isle of Purbeck, Dorset. Its habitat in Dorset is more typical of that in continental Europe and Fennoscandia, where it is more frequently found in soakways in oligotrophic mires, especially at the edges of flarks. However, it does not reach the mires near the Arctic Circle.

S. papillosum and *S. magellanicum* may be components of a lawn community, but they also form low hummocks on open, acid mires. *S. magellanicum* appears to favour slightly wetter conditions than *S. papillosum*, being more extensively distributed in oceanic areas and more susceptible to the effects of burning and drainage, though it also grows in much drier situations in wooded oligotrophic mires, eg the *Pinns* bogs of the north and east of Europe.

Although *S. capillifolium* var. *capillifolium* is a plant of oligotrophic mires, it has a wide ecological amplitude and is also found, especially on the continent of Europe and in Fennoscandia, in damp woodland and heath, extending even into slightly mesotrophic mires. *S. capillifolium* var. *rubellum* (where this can be definitely distinguished from *S. capillifolium* var. *capillifolium*) is more restricted to hummocks on oligotrophic, usually raised, mires. In Fennoscandia, these 2 extremes of the *S. capillifolium* complex differ clearly in habit and occupy distinctly different ecological niches, meeting only in few localities. In Britain the morphological and ecological distinction is less readily made.

In north-west Britain, *S. imbricatum* and *S. fuscum* produce large hummocks, up to 1m high, on wet lowland raised or blanket mires. Around the bases of *S. imbricatum* hummocks, particularly where there is lateral water movement, the lax form of the species is sometimes found. As noted above, this is the form most frequently encountered, as carpets in mesotrophic fens or near lake margins, in parts of continental Europe. In Britain, this form is also found extensively in stratigraphical samples, eg from schwingmoor sites in the English west midlands where it is absent from the surface vegetation Similarly, *S. fuscum* is more widespread in mainland Europe, not only on raised bogs where it is often the dominant *Sphagnum* species, but also forming hummocks or ridges in flushes and oligotrophic mires of various types: it is typical, for example, of frost-heaved ridges and mounds where it often accompanies *Empetrum* and *Betula nana*. However, the very high palsas of the far north are too dry for this species, and usually lack *Sphagna* altogether.

Both *S. tenellum* and *S. compactum* may form low cushions, either separately or in association, in wet heath areas or on areas of bared oligotrophic peat, particularly in hollows. *S. molle* is found in similarly damp locations, usually on shallow peat or upland wet heath. In more oceanic areas, *S. strictum*, which has a very restricted western distribution in the British Isles (it is not so uncommon in western Norway), grows in damp, usually open situations on heath, blanket bog or lake-sides.

Two species of acid woodland are (i) the circumboreal *S. wulfianum*, found only in the eastern part of Europe, and (ii) the more westerly distributed *S. quinquefarium*. *S. wulfianum* occurs on mire margins though, more typically, it is found in *Picea* forest with *Vaccinium myrtillus* where it forms loose carpets, often with *S. capillifolium* var. *capillifolium* and *Dicranum undulatum*. *S. quinquefarium* is unusual in being absent from mires: it grows under *Calluna vulgaris* on upland wet heaths or on well-drained woodland slopes, especially under *Betula*.

These notes are intended as a guide to the main types of habitat occupied by different species of *Sphagnum*. The range of habitats in which a particular species will grow is highly variable, some being strictly confined by particular combinations of envrionmental conditions, while others are more catholic in their requirements.

The accompanying diagrams (Plate 1) summarize the distribution of species in relation to the 2 major environmental variables. They do not give precise limits to chemical conditions or state, definitively, heights above water level. However, they indicate the relative positions occupied by individual species along 2 gradients of change: dry → wet habitats and oligotrophic → eutrophic conditions within the peats or peat waters. For clarity, each diagram contains a limited number of species, these species being, where practicable, members of the same subgeneric group. Each diagram thus shows the range of habitats occupied by closely related species and indicates that there is ecological diversification within each of these groups. They also show that some groups are largely restricted to certain types of habitat; eg, most of the members of the section Sphagnum occupy dry or relatively dry situations, in contrast to members of the section Subsecunda whose species are largely found at or near water level. Members of the section Cuspidata and the section Acutifolia are found in a wide range of both chemical and hydromorphological conditions, though the majority of the former are found in wetter areas. Where species are morphologically poorly separated, they also appear to be largely similar in their ecological requirements, suggesting that there is no great disparity in the rates of evolution of physical characters and physiological processes within the genus *Sphagnum*.

Collection, preservation and examination

Although some of the more tolerant species of *Sphagnum*, for example *S. palustre, S. capillifolium* and *S. recurvum,* may still cover wide areas and be abundant locally, the populations of all species have been reduced drastically by large-scale habitat destruction, mainly by drainage or pollution (especially the drift or inflow of fertilizers on to previously oligotrophic sites). The ultimate survival of the rarer species will depend upon deliberate policies of conservation. As an aid to such policies, it is essential that rare species should not suffer from over-enthusiastic collecting. Rarity is often a matter of geography, and a species which is abundant in one area may be rare elsewhere. Distribution maps, such as those included in the present volume, give some information about likely areas of rarity. However, broad-scale maps cannot give a full picture of local conditions as records may consist of single specimens or groups of individuals on the one hand, or extensive stands on the other.

A single stem of a well-grown plant will supply enough material to make several good reference slides, and it is never necessary to collect more than about 10 stems to make up an adequate herbarium specimen.

Collection is no problem as plants are not firmly anchored: they have neither roots nor other attachment organs. Less harm is done to established hummocks if single stems are extracted from different parts of the clump than if a large handful were taken from one spot. Sampling of this type is likely to ensure that some individual variation is found, so that the possible occurrence of some aberrant form may be compensated by the collection of plants with a

Plate 1. The following diagrams show the ecological ranges of European species of *Sphagnum*. In these illustrations, the horizontal axis shows distribution in relation to main chemical gradients; to the left are oligotrophic, usually acid, habitats and to the right are eutrophic, more base-rich, habitats. The vertical axis places dry habitats at the top and wet ones at the bottom: the dashed line indicates water level, so that species which grow submerged are shown extending below this line. In the key (below), species A is confined to dry, oligotrophic sites and species B to wet, eutrophic areas, where it may be found submerged

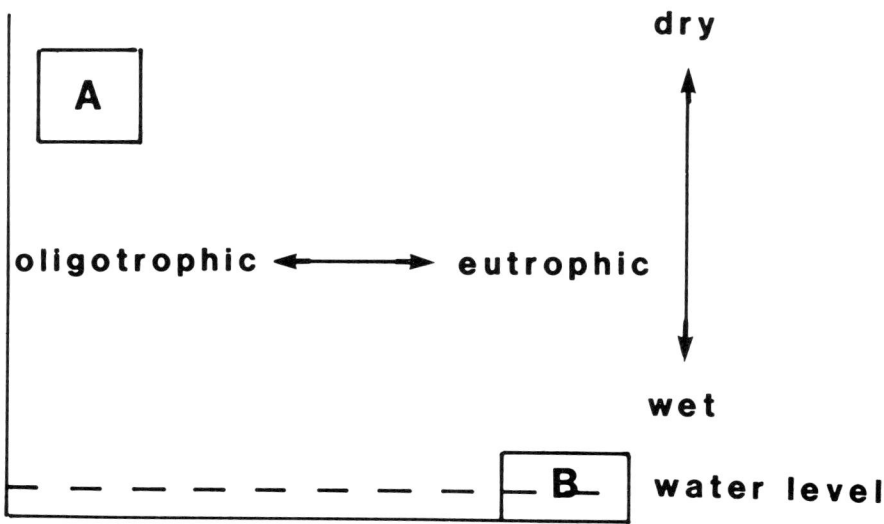

Key. Example of ecological range of European species of *Sphagnum* shown in Plate 1.

Plate 2. *S. palustre* (R E Daniels)

Plate 3. *S. centrale* (K Dierssen) *S. palustre* var. *centrale*

Plate 4. *S. papillosum* (R E Daniels)

Plate 5. *S. imbricatum* (R E Daniels)

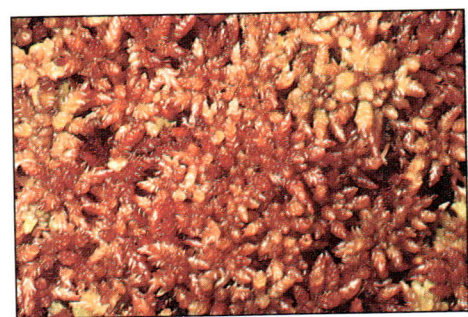

Plate 6. *S. magellanicum* (H J B Birks)

Plate 7. *S. molle* (H J B Birks)

Plate 8. *S. subnitens* (R E Daniels)

Plate 9. *S. angermanicum* (K Dierssen)

Plate 10. *S. subfulvum* (K Dierssen)

Plate 11. *S. fuscum* (H J B Birks)

Plate 12. *S. quinquefarium* (H J B Birks)

Plate 13. *S. capillifolium* var. *capillifolium* (R E Daniels)

Plate 14. *S. capillifolium* var. *rubellum* (+ *S. papillosum*) (R E Daniels)

Plate 15. *S. warnstorfii* (H J B Birks)

Plate 16. *S. russowii* (H J B Birks)

Plate 17. *S. girgensohnii* (H J B Birks)

Plate 18. *S. fimbriatum* (R E Daniels)

Plate 19. *S. teres* (H J B Birks)

Plate 20. *S. squarrosum* (H J B Birks)

Plate 21. *S. aongstroemii* (K Dierssen)

Plate 1i

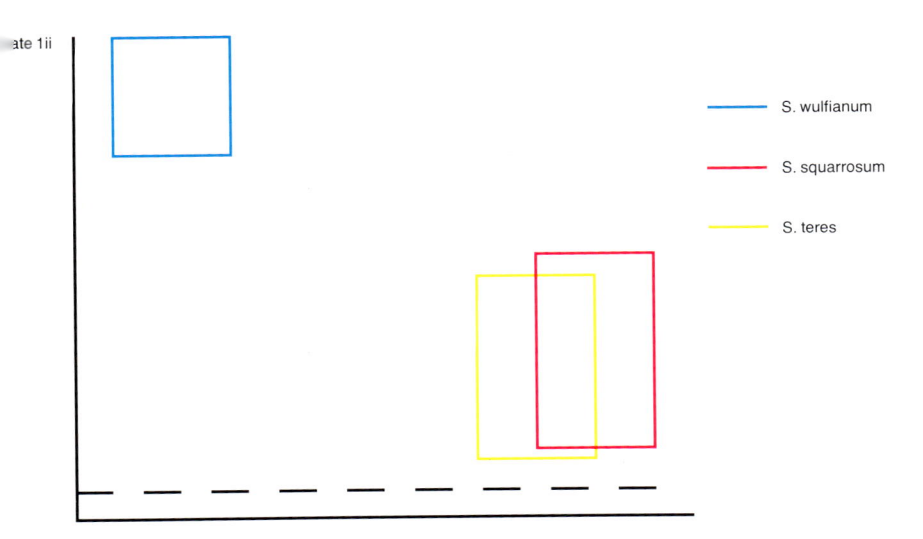

- S. compactum
- S. strictum
- S. tenellum
- S. angstroemii

Plate 1ii

- S. wulfianum
- S. squarrosum
- S. teres

Plate 1iii

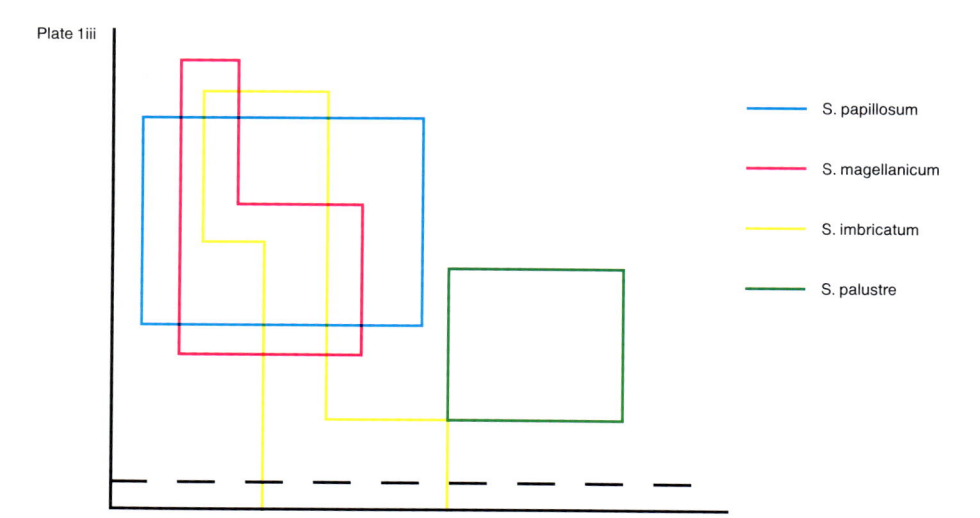

— S. papillosum
— S. magellanicum
— S. imbricatum
— S. palustre

Plate 1iv

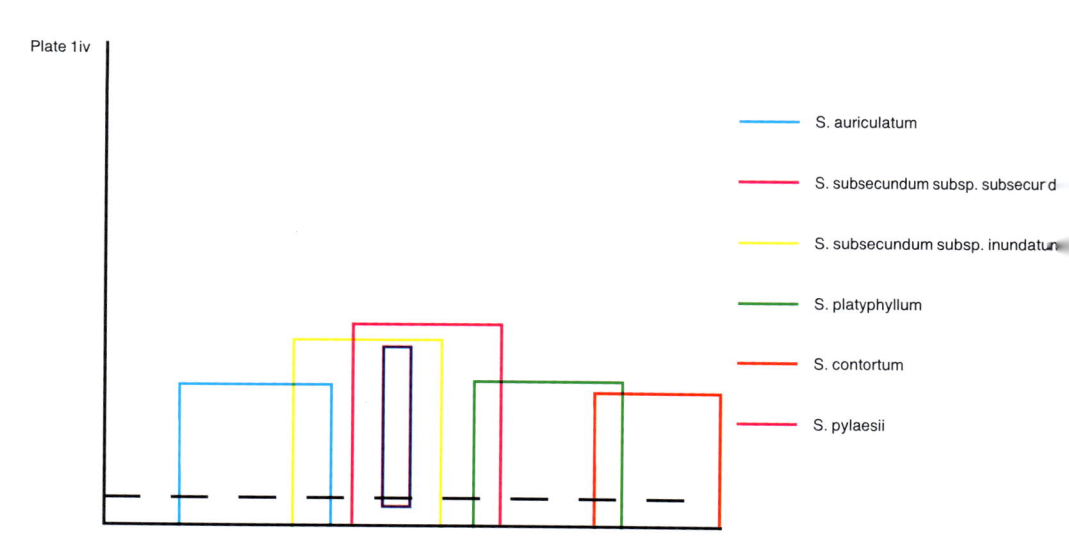

— S. auriculatum
— S. subsecundum subsp. subsecur d
— S. subsecundum subsp. inundatu
— S. platyphyllum
— S. contortum
— S. pylaesii

1v

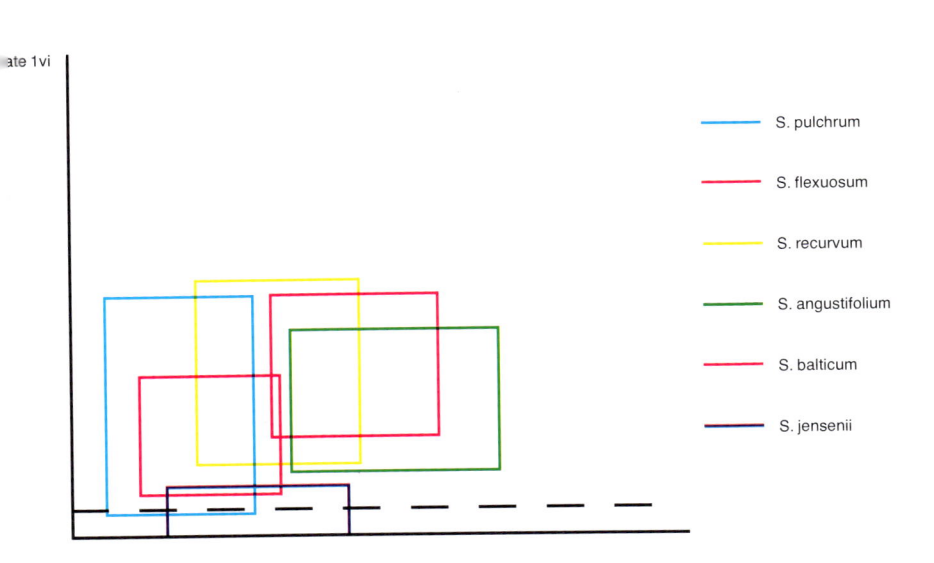

- S. cuspidatum
- S. majus
- S. lindbergii
- S. lenense
- S. riparium
- S. obtusum

1vi

- S. pulchrum
- S. flexuosum
- S. recurvum
- S. angustifolium
- S. balticum
- S. jensenii

Plate 1vii

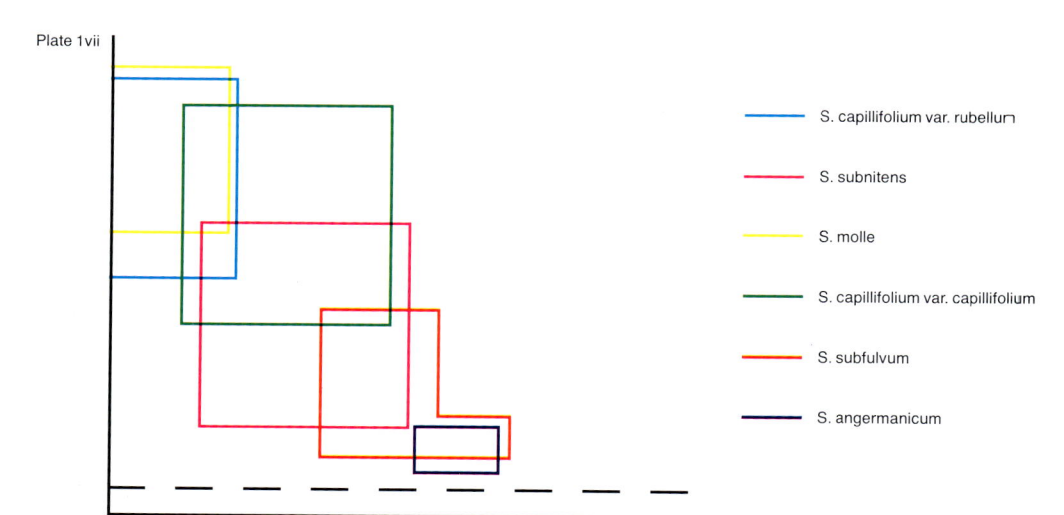

- S. capillifolium var. rubellum
- S. subnitens
- S. molle
- S. capillifolium var. capillifolium
- S. subfulvum
- S. angermanicum

Plate 1viii

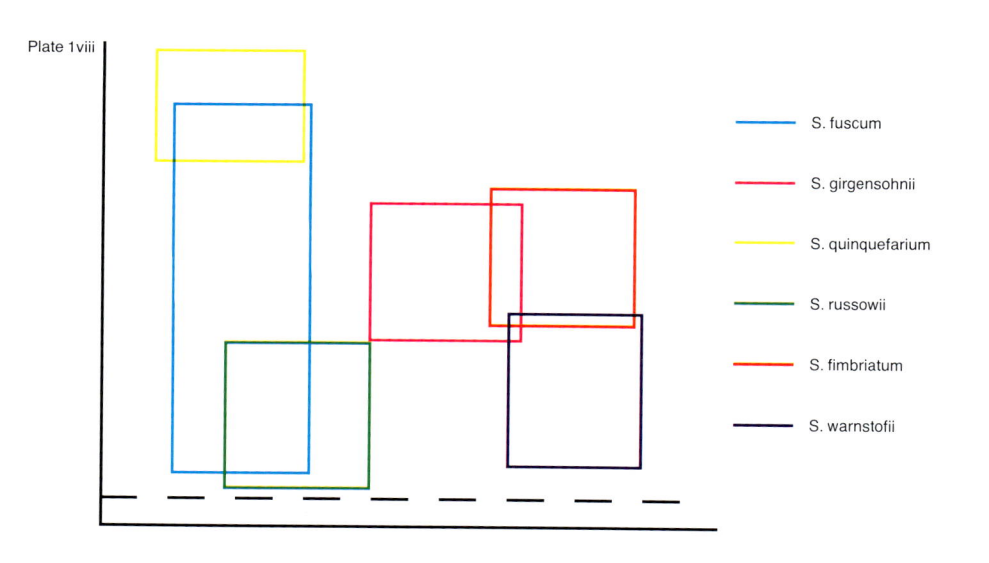

- S. fuscum
- S. girgensohnii
- S. quinquefarium
- S. russowii
- S. fimbriatum
- S. warnstofii

different genotype, or not influenced by some localized environmental factor causing the aberration.

Specimens for the herbarium can simply be left to dry or, after squeezing out excess water by hand, pressed lightly between sheets of absorbent paper. A gentle heat source, eg a central heating radiator, may be used to accelerate drying.

Living material will survive for long periods, sealed in polythene bags and kept away from direct sunlight. Alternatively, the plants can be held more or less dormant in a refrigerator at about ±2°C. Many will survive deep freezing, but it is not advisable to place an actively growing plant directly into a freezer, if future culture work is intended.

With experience, it should be possible to identify almost all of the European species with the aid of a simple hand lens of 10× to 15× magnification. There will always be circumstances, however, when the use of a microscope is essential. Although a 200× magnification may suffice, 400× is better: at 300×, it may not be possible to make out the faint details necessary for the identification of, for example, *S. obtusum*. Access to a dissecting microscope is an enormous advantage when manipulating material (eg cutting leaf or stem sections), but is not essential. Some workers find an eyeglass, a large mounted lens or a reading glass helpful. For dissection, a pair – preferably 2 pairs – of fine-pointed forceps (the best are the watchmaker's variety) and a sharp tool for cutting sections (a safety razor blade is perfectly adequate) are the basic essentials. Optional extras include a fine pipette or dropper, and a small (about size no. 1) artist's brush.

As the hyaline cells of *Sphagnum* are very thin, certain important features, such as unringed pores, may be difficult, or even impossible, to detect without prior staining. Staining techniques are very simple, and virtually any cationic dye will suffice. Alcoholic solutions of Methylene Blue are effective, as is a fairly strong solution of Gentian Violet (=Crystal Violet) – some claim that the latter is better. Many proprietary coloured inks have been found to work quite well. To identify pores, either a drop of stain may be added to material on a microscope slide, or whole plants (or parts of them) may be stained prior to dissection.

Usually leaves can be detached by pulling them with forceps, although stem leaves are easily torn. Auricles are present on stem leaves, but these are never diagnostic, so that small fragments left behind will not usually hinder identification. In the case of branch leaves, these should be detached in the direction of the branch tip, if the intention is to expose the branch cortex: pulling towards the branch base may remove strips of cortex but does usually ensure more complete leaves.

KEY TO EUROPEAN SPECIES OF *SPHAGNUM*

1. Cortical cells or branches (and usually at least the internal one of stems) with spiral fibrils. Apices of branch leaves blunt and hooded, appearing minutely roughened on the convex surface due to projecting, partly resorbed, hyaline cells. Plants usually robust. ....
   .... Section Sphagnum 2

   Cortical cells without fibrils. Branch leaf apices usually acute, truncate or, if hooded, then smooth on the convex surface. Plants robust or small.
   5

2. Internal commissural walls of branch leaf hyaline cells (at least in the lower lateral parts of leaves) with papillae or sparse to dense lamellae or crests (ie appearing as though the photosynthetic cells have papillose or ornamented walls). Plants green, yellow, ochre or brown, never crimson. 3

   Internal commissural walls smooth throughout. Plants, except for capitula, usually green or crimson-tinted. 4

3. Internal commissural walls finely to rather coarsely papillose. Photosynthetic cells in TS, urn-shaped with oval lumina and thick adaxial walls. Plants often ochrous: widely distributed and locally abundant on oligotrophic to slightly mesotrophic mires. ....
   .... *S. papillosum* (52)

   Internal commissural walls with scattered dense, oblique to transverse crests or lamellae ('comb fibrils'). Photosynthetic cells in TS, broadly triangular with triangular lumina: thin-walled. Plants dull brown to orange-brown, rarely green or yellowish: local to rare in, mainly lowland, oligotrophic to slightly mesotrophic mires. ....
   .... *S. imbricatum* (56)

4. Plants usually with at least some flecks of crimson, often entirely deep, wine-red. Spreading branches usually blunt. Photosynthetic cells not, or rarely, exposed on the concave leaf surface: in TS, oval and mostly enclosed by hyaline cells. Plants widespread but local (and then often abundant) on oligotrophic mires. ....
   .... *S. magellanicum* (61)

   Plants never red (sometimes pinkish-brown or grey-purple, especially in the capitulum). Spreading branches long-tapering. Photosynthetic cells well exposed on the concave leaf surface: in TS, trapezoid to narrowly oval-triangular. Plant common in mesotrophic mires or parts of mires. ....
   .... *S. palustre* 4a (56)

4a. Photosynthetic cells in TS oval-triangular to trapezoid: thin-walled. Plant widespread and common in the south: becoming rarer further north. ....
....*S. palustre* var. *palustre* (46)
Photosynthetic cells in TS, oval to urn-shaped with strongly thickened adaxial walls. Plant widespread in the north (ie boreal and sub-arctic): confined to montane areas further south. ....
....*S. palustre* var. *centrale* (50)

5. Branch leaves large (more than 1.5 mm) and broad. Hyaline cells of upper mid-leaf romboid, not more than 5 times longer than wide. Stem cortex 2-or more layered. Photosynthetic cells oval in TS, immersed or almost so. 6
Branch leaves narrow or, if broad, then hyaline cells long and narrow, more than 6 times as long as wide, and stem cortex usually 1-layered. Photosynthetic cells (except *S. wulfianum*) obviously exposed on one or both surfaces. 8

6. Branch leaves large. Stem leaves very small (< 1.1mm), triangular: convex surface of hyaline cells intact. Branch cortical cells all alike, most or all with a single large pore. Capitula ± hidden among upwardly directed branches. ....
..................Section Rigida 7
Branch and stem leaves of similar size. Stem leaves lingulate to rectangular: convex surface of hyaline cells resorbed. Branch cortical cells of 2 kinds; groups of 2–3 retort cells (large cells each with a distal, protuberant pore) and smaller, narrower imperforate cells. Capitula not ± hidden.
..................Section Insulosa ....*S. aongstroemii* (126)

7. Branch leaf photosynthetic cells deeply immersed between hyaline cells: in TS, thin-walled, oval. Internal commissural walls smooth. Older stems dark brown. Plants densely tufted: widespread and common, typically on wet heath or along the margins of flarks (large pools in patterned boreal peatlands). ....*S. compactum* (234)
Branch leaf photosynthetic cells narrowly exposed, at least on the convex leaf surface: in TS, with thickened abaxial walls. Internal commissural walls minutely papillose. Stems pale. Plants usually pallid with squarrose branch leaves: local to rare on shallow peat near Atlantic coasts. ....
....*S. strictum* (230)

8. Fascicles of mature plants consisting of at least 7 (usually 8 or more) branches. Stem leaves small (less than 1.2 mm). Photosynthetic cells in TS, oval. Plants rigid with dense, acute branch leaves and conspicuously large capitula. A continental species of moderately wet coniferous forest and, rarely, damp heath. ....
..................Section Polyclada ....*S. wulfianum* (131)

Fascicles never with more than 7 (usually fewer than 6) branches. Stem leaves usually over 1.2 mm long. Photosynthetic cells in TS, triangular, trapezoid or barrel-shaped. Capitula small to well-developed. 9

9. Photosynthetic cells in TS, triangular or trapezoid with widest exposure on convex leaf surface, or ± barrel-shaped and ± equally exposed on both leaf surfaces. Plant, if small-leaved, not red. 10

Photosynthetic cells in TS, triangular or trapezoid with widest exposure on concave leaf surface. Includes small-leaved (ie under 1.6 mm long) red plants. Section Acutifolia 34

10. Stem leaves large, lingulate, never fibrillose in mature plants; border not expanded below. Plants often robust, with large branch leaves that may be abruptly narrowed and reflexed at mid-leaf. Never red. ....Section Squarrosa 11

Stem leaves various, if lingulate and efibrose, then with borders markedly expanded below. Branch leaves rarely squarrose. Plants sometimes red, then with very large (>2 mm long) branch leaves. 12

11. Branch leaves small (less than 2.2 mm long), usually with apices erect or only slightly spreading (if squarrose, then branches long and attenuated) so that branches appear cylindrical and tumid. Plants green or, usually, brownish. A local to frequent species (common in the Arctic) of eutrophic mires. ....*S. teres*   (117)

Branch leaves large (usually over 2.3 mm long), usually distinctly squarrose, giving the plant a prickly appearance. Plants green or yellowish, rarely brownish, robust. Common in mesotrophic mires. ....*S. squarrosum*   (122)

12. Mature plants without branches or with, at most, 1–2 short, small-leaved branches per fascicle. Stem leaves much larger than branch leaves, overlapping and concealing stems. Leaves sometimes with resorption gaps but never with clearly defined pores. Plants purple to blackish or dark brown, rarely dull olive-green: ± aquatic: very rare in NW France and NW Spain. Section Hemitheca ....*S. pylaesii*   (136)

Mature plants with well-developed fascicles, rarely of fewer than 3 branches (if fewer, then stem leaves not conspicuously larger than branch leaves, not concealing stems, or at least some branch leaves with clearly defined pores). 13

13. Branch leaf hyaline cells short and proportionately broad (less than 6 times as long as broad). Stem leaves almost as large as branch leaves,

Plate 22. *S. wulfianum* (K Dierssen)

Plate 23. *S. pylaesii* (R E Daniels)

Plate 24. *S. subsecundum* (R E Daniels)

Plate 25. *S. subsecundum* var. *inundatum* (H J B Birks)

Plate 26. *S. auriculatum* (R E Daniels)

Plate 27. *S. contortum* (H J B Birks)

Plate 28. *S. cuspidatum* (R E Daniels)

Plate 29. *S. riparium* (H J B Birks)

Plate 30. *S. obtusum* (K Dierssen)

Plate 31. *S. flexuosum* (H J B Birks)

Plate 32. *S. recurvum* var. *mucronatum* (R E Daniels)

Plate 33. *S. angustifolium* (R E Daniels)

Plate 34. *S. balticum* (J G Duckett)

Plate 35. *S. jensenii* (T Lindholm)

Plate 36. *S. majus* (+ *S. jensenii*) (T Lindholm)

Plate 37. *S. pulchrum* (H J B Birks)

Plate 38. *S. lindbergii* (H J B Birks)

Plate 39. *S. tenellum* (J G Duckett)

Plate 40. *S. strictum* (H J B Birks)

Plate 41. *S. compactum* (H J B Birks)

both ovate and concave. Branches appearing 'beaded' because of widely spaced, concave branch leaves. Plants small, delicate and pale green or yellowish. A common species of damp oligotrophic mires and wet heath. ....
.................Section Mollusca ....*S. tenellum* (223)

Branch leaf hyaline cells long and narrow (at least 6 times as long as broad). Stem leaves various but, if similar to branch leaves, then plants not delicate. Branches not 'beaded'. Plants mostly of medium size or robust. 14

14. Leaves from middle of branch, ovate, less than twice as long as wide, with broad, ± hooded apices (if narrower, then stem cortex with pores). Branch leaf hyaline cells with few to numerous small, normally thick-ringed, pores along the commissures on the convex or both leaf surfaces (very rarely only on the concave). Photosynthetic cells of branch leaves almost equally exposed on both leaf surfaces. ....
.................Section Subsecunda 15

Leaves from middle of branch, lanceolate, at least twice as long as wide, never with hooded apices (apices often apparently acute due to inrolled margins). Stem cortex never with pores. Branch leaf hyaline cells with pores various but rarely numerous on convex leaf surface, and then not normally along the commissures. Photosynthetic cells with much wider exposure on the convex leaf surface, often not reaching the concave surface. ....Section Cuspidata 20

15. Stem leaves ± as large as branch leaves, strongly fibrillose almost or quite to insertion. Stems pale, greenish, yellowish or pale brown, never dark brown or blackish. Plants medium-sized to robust, ± flaccid. 16

Stem leaves much smaller than the largest branch leaves, sometimes minute, fibrillose in the upper two-thirds at most. Stems pale or dark. Plants sometimes small. 17

16. Branch leaves short and strongly concave throughout. Branches blunt. Stem cortex 2-layered (1- or 3-layered in parts): outer cells occasionally with a large pore or thinning. Plants dull green or olive, sometimes purplish but never red. Local to rare in mesotrophic locations which are, at least periodically, flooded. ....*S. platyphyllum* (156)

Upper branch leaves longer and narrower than the lower. Branches blunt or, commonly, somewhat acute and tapering. Stem cortex 1-layered, rarely with occasional pores. Plants sometimes distinctly red. Common and widespread in oligotrophic to mesotrophic pools and ditches or hollows; also on dripping rocks. ....*S. auriculatum* (150)

17. Stem cortex of 2 or more layers. Internal cylinder of stems always pale, yellowish to pale brown, never dark brown or blackish. Plants small-leaved. Rare to locally frequent in eutrophic mires. ....
....*S. contortum* (160)

    Stem cortex always single-layered. Internal cylinder of stem commonly dark brown or blackish, at least in part. Plants small- or large-leaved. Frequent to common in oligotrophic to mesotrophic mires or pools, on wet rocks or along seepage lines. 18

18. Stem leaves small (less than 1.2 mm long): fibrillose only near the apex (fibrils often incomplete, rarely absent). Branch leaves small, the lower usually curved, asymmetric and secund. Fascicles of fully developed plants with 5–6 branches (if small-leaved but with 3–4 branches per fascicle and stem leaves fibrillose for more than one third of length, see *S. auriculatum*). .... *S. subsecundum* subsp. *subsecundum* (142)

    Stem leaves at least 1.2 mm long: fibrillose at least in upper quarter and commonly to about two-thirds from apex. Branch leaves rarely under 1.3 mm long, the lower curved or straight. Fascicles of fully developed plants commonly with only 3 or 4 branches. 19

19. Fascicles of 3–4, rarely 5, branches. Branch leaves mostly symmetrical, ± suberect and convolute. Branches ± tumid, often curved and horn-like. Stem leaves lingulate to spatulate: fibrillose at least in upper third and often to below half-way. Hyaline cells of stem leaves with fewer pores on the adaxial surface than on the abaxial: abaxial pores often in regular rows along the commissures. Plants often very robust, sometimes tinged wine-red. Common in oligotrophic or mesotrophic hollows and pools, springs and seepage lines and on wet rocks. ....
    ....*S. auriculatum* (150)

    At least some fascicles with 5 well-developed branches (except in weak plants). Branch leaves often curved and asymmetric near the branch bases, mostly erect-spreading. Branches rarely tumid, never curved and horn-like. Stem leaves triangular-lingulate, narrowed above the insertion: fibrillose one quarter to one third from apex, rarely to mid-leaf. Hyaline cells of stem leaves with more numerous pores on adaxial surface than on abaxial, or both surfaces multiporose. Plant often orange but never red. An uncommon species of mesotrophic mires and stream-sides. .... *S. subsecundum* subsp. *inundatum* (143)

20. Stem leaves shortly lingulate, as wide or wider above than at insertion: fimbriate with a wide zone of apical resorption: hyaline cells enlarged and lacking fibrils. Stems dark brown to almost black. Plants brown. ....
    21

Stem leaves as wide as, or narrower above than, at insertion: subacute to obtuse; if narrowly fimbriate, then stems and plants pale. If plants brown, then stem leaves either ± acute or with fibrils. Stems mostly pale, sometimes dark. 22

21. Plants robust. Stem leaves large (ca 1.0 mm wide): expanded above and fimbriate across the whole upper part. A widespread plant of the boreal to arctic area of N and NE Europe. ....*S. lindbergii* (215)

Plants small. Stem leaves small (less than 0.8 mm wide): not, or only slightly wider at apex than insertion and fimbriate only across the narrowed apex. A rare arctic to sub-arctic species of the extreme NE of Europe. ....*S. lenense* (219)

22. Branch leaves (except for a few basal ones) narrowly lanceolate to linear-lanceolate, more than 3 times (often more than 5 times) as long as wide. Hyaline cells on convex surface of branch leaves without pores, or with a single pore in the apical angle. Stem leaves ± acute; fibrillose above. Plants yellowish green to pale green: common in pools and wet hollows. 23

Branch leaves (especially those of pendent branches) mostly less than 3 times as long as wide or, if longer, then hyaline cells with abundant pores or stem leaves without fibrils. Stem leaves various, mostly with acute apices. Plants various, sometimes brown. 24

23. Photosynthetic cells of branch leaves in TS, trapezoid, widely exposed on both leaf surfaces. Branch leaves commonly more than 4 times as long as wide. Fascicles with poorly differentiated or undifferentiated pendent branches. Plants often pale green: widespread, usually in bog pools or wet hollows. ....*S. cuspidatum* (166)

Photosynthetic cells of branch leaves in TS, triangular, not, or barely, reaching the concave leaf surface. Branch leaves rarely more than 3 times as long as wide. Fascicles with slightly to moderately differentiated pendent branches bearing ovate-lanceolate to lanceolate leaves. Plants usually deep green. .... *S. recurvum* var. *mucronatum* (185)

24. Plants green, yellowish or tinged brown: if overall brown, then stem leaves acutely pointed or apparently mucronate, strongly deflexed; non-fibrillose. 25

Plants brown, except in shade: if paler, then stem leaves ± spreading, obtuse and concave; mostly fibrillose, at least near apex. 30

25. Stem leaves triangular to oval-triangular: apices acute or mucronate due to incurved or 'pinched' margins: if narrowly obtuse, then branch leaves strongly 5-ranked, more than 1.4 mm long and stem cortex ± distinct around whole of circumference. 26

Stem leaves shortly lingulate to triangular-lingulate: apices broadly rounded-obtuse, sometimes notched, torn or fimbriate: if apparently mucronate, then branch leaves mainly less than 1.4 mm long and stem cortex distinct only adjacent to leaf and fascicle insertions. 27

26. Branch leaves widest one quarter to one third above insertion. Stems pale, green or yellowish. Branches long (usually more than 15 mm). Plants green to yellow-orange, occasionally olive-brown. Common and widespread, often forming extensive carpets in mesotrophic mires. ......
.... *S. recurvum* var. *mucronatum* (185)

Branch leaves widest at, or just below, half-way. Stems brownish, rarely pale. Branches short (usually less than 15 mm). Plants gold, golden brown or rich brown. Local in lowlands, mainly towards south and west.
.... *S. pulchrum*   (211)

27. Stem leaves deeply notched or bifid, appearing as if partially torn down the middle. Branch leaf apices narrow, slightly spreading when dry, ± acute and composed of undifferentiated prosenchymatous cells (except in weak plants). Plants robust: shade-tolerant in wet oligotrophic to mesotrophic mires. A sub-arctic and boreal species, common in the north but more local or rare in the south. .... *S. riparium*   (171)

Stem leaves not deeply notched or bifid, sometimes minutely notched. Branch leaf apices narrowly truncated and composed of both hyaline and photosynthetic cells. 28

28. Stem leaves shortly triangular to oval-triangular, ± equilateral: apices obtuse, often slightly concave, rarely pinched to give a mucronate appearance. Pendent and spreading branches strongly dimorphic. Hyaline cells of pendent branch leaves distinctly wider at apical end. Leaves of spreading branches rarely more than 1.5 mm long. ....
.... *S. angustifolium*   (190)

Stem leaves triangular-lingulate: apices rounded or truncate. Pendent and spreading branches weakly dimorphic. Hyaline cells of pendent branch leaves not distinctly wider at apical end. Leaves of spreading branches rarely less than 1.5 mm long. 29

29. Photosynthetic cells in TS, triangular, not, or rarely, exposed on concave leaf surface. Hyaline cells of lower lateral parts of branch leaves with few to many small, faint pores remote from the commissures (strong staining required). Branch leaves 5-ranked, ± uniform along branch. Plants robust, green or yellowish, in wet mesotrophic to eutrophic mires: often periodically submerged. Scattered to locally frequent, rare in south and extreme north. .... *S. obtusum*   (176)

Photosynthetic cells in TS, trapezoid, exposed on concave leaf surface. Hyaline cells of lower lateral parts of branch leaves without small, faint pores. Branch leaves not consistently 5-ranked: distal leaves of spreading branches often linear. Branch leaf apices with photosynthetic cells as wide as hyaline cells. Plants medium-sized to rather robust, green or ochre, in wet mesotrophic mires. Widespread but less common in the north and confined to montane areas in the south. ....
....*S. flexuosum* (180)

30. Plants small. Fascicles often of only 3 branches. Branch leaves small (up to 1.6 mm), often curved and secund. Branch leaf hyaline cells with pores and pseudopores confined to commissures on convex leaf surface. Stem leaves proportionately large, spreading, concave: apices rounded-obtuse. Pendent branch leaves, at least in lower lateral parts, with conspicious, large resorption gaps in the apical angles of hyaline cells. A species of wet oligotrophic mires, often forming extensive 'lawns'. Common in the north and east: more scattered further south and west. ....*S. balticum* (195)

    Plants medium-sized to robust. Fascicles rarely of fewer than 4 branches. Branch leaves rarely less than 1.5 mm long: seldom secund. Branch leaf hyaline cells with few to many pores, remote from commissures on the convex leaf surface (if lacking such pores, then stem leaves ± triangular and strongly deflexed). Apical resorption gaps, when present, not conspicuously large (ie less than 12.0 μm). 31

31. Stem leaves triangular to oval-triangular, strongly deflexed: apices narrowly obtuse to, apparently, mucronate. Branch leaf hyaline cells in upper mid-part of leaf on convex surface without pores, or with 1–2 pores confined to apical and upper lateral angles. A lowland species, scattered in the south and west. ....*S. pulchrum* (211)

    Stem leaves short-lingulate, concave, ± spreading (sometimes weakly deflexed): apices broadly obtuse. Branch leaf hyaline cells in upper mid-part of leaf on convex surface with few to numerous pores not confined to cell angles. Northern or alpine species. 32

32. Hyaline cells of branch leaves with numerous pores on convex leaf surface; pores absent or few on concave surface. Photosynthetic cells of branch leaves widely exposed on the concave leaf surface. Plants lax: long-leaved. ....*S. majus* (207)

    Hyaline cells of branch leaves with numerous small pores on both leaf surfaces. Photosynthetic cells mostly immersed on concave leaf surface, or only narrowly exposed. Plants lax or firm: not notably long-leaved. 33

33. Branch leaf hyaline cells with numerous, small pores in one or 2 rows, remote from commissures, on both leaf surfaces. A widespread but uncommon boreal to arctic species of oligotrophic mires. ....
....*S. jensenii* (203)

    Branch leaf hyaline cells with pores usually along the commissures; seldom numerous or in regular rows on the concave surface. Rare boreal or sub-arctic species. ....*S. annulatum* (199)

34. Outer edge of branch leaf border resorbed, forming a furrow (seen as a notch in TS). Stem leaves large: strongly fibrillose and, at least in upper halves, ± identical in structure to branch leaves. Plants medium-sized but low-growing: usually pale, tinged with pink. Locally frequent in the west on shallow peat of wet heaths. ....*S. molle* (67)

    Outer edge of branch leaf border intact. Stem leaves various, without fibrils or weakly fibrillose (strongly fibrillose in *S. angermanicum*), not identical to branch leaves in upper part. 35

35. Stem leaves lingulate to spatulate (wider above the middle than at insertion), with patches of enlarged hyaline cells above base and also just below apex: hyaline cells never fibrillose: apices rather narrowly to widely fimbriate. Stem cortex with large, distinct pores. Stem bud conical and ± projecting from capitulum. Branch leaves never 5-ranked. Plants pale green to pale ochre, never red or dark brown....................... 36

    Stem leaves various; if wider above the middle, then hyaline cells at least partly fibrillose, branch leaves 5-ranked, or plants red or brown: cells above base not markedly enlarged. Stem cortex without pores, or pores indistinct (sometimes pores more distinct in *S. russowii*). Stem buds rarely projecting from capitulum. Plants often with at least some red or brown coloration (sometimes confined to internal cylinder of stems and branches). 37

36. Stem leaves expanded above and resorbed-fimbriate around the whole upper part. Stem buds conspicuous, projecting. Plants often tall and thin. Common, especially in the south, in mesotrophic mires, frequently in shade. Fruit common. ....*S. fimbriatum* (111)

    Stem leaves not widely expanded above, fimbriate only across the apices. Stem buds slightly, but inconspicuously, projecting. Plants not usually attenuated-looking. Local in south, common in north in mesotrophic mires and wet woodlands. Fruit rare. *S. girgensohnii* (107)

37. Stem leaves wider at mid-leaf than at insertion: hyaline cells strongly fibrillose. Branch leaves easily flattened: apices broad, not inrolled,

strongly dentate. Capitula with projecting stem buds. Plants flaccid, often rather pale: rare in weakly mesotrophic mires. .... *S. angermanicum* (76)

Stem leaves at mid-leaf as wide as, or narrower than, at insertion: hyaline cells rarely strongly fibrillose. Branch leaves difficult or impossible to flatten because of inrolled margins towards apices (at least in lower leaves of branches). Capitula without projecting stem buds. 38

38. Stems brown. Plants usually brown (at least in part), rarely green, never red. Stem leaves mostly without fibrils. Branch leaves never 5-ranked. 39

    Stems green, red or violet. Plants green, red or violet (at least in part): if entirely green, then stem leaves distinctly fibrillose near apex. Branch leaves 5-ranked or not. 40

39. Dried plants irridescent when viewed with a lens. Branch leaf hyaline cells on convex leaf surface mostly with a large (12.0–15.0 µm) pore in the apical angle. Plants small to medium-sized in somewhat mesotrophic habitats Rare to local: boreal. ....*S. subfulvum* (80)

    Dried plants matt, lacking irridescence. Branch leaf hyaline cells on convex surface mostly lacking apical pores: if present, then less than 12.0 µm. Plants small, in oligotrophic mires. Common and widespread in the north: rare to locally frequent towards the south. .... *S. fuscum* (84)

40. At least some fascicles with 3 spreading branches. Stem leaves triangular to triangular-lingulate: usually less than 1.4 mm long. Branch leaves conspicuously 5-ranked. Stem cortex with occasional to frequent faint pores (strong staining required). Plants tall, usually variegated pale green and red. Widespread but mainly southern in rather dry, often shaded, habitats: virtually absent from mires. .... *S. quinquefarium* (88)

    Fascicles normally with 2 spreading branches (rarely with an additional branch at a stem bifurcation). Stem leaves lingulate or, if triangular, then over 1.4 mm long and branch leaves not 5-ranked. Stem cortex (except *S. russowii*) without pores. 41

41. Stem leaves triangular: apices markedly pointed, because of inrolled margins: hyaline cells lacking fibrils. Plants irridescent when dry, often rather loose, with large capitula. ....*S. subnitens* (71)

    Stem leaves lingulate: apices obtuse or ± acute: hyaline cells mostly fibrillose, at least towards the apex (fibrils sometimes weak). Dried plants not irridescent. 42

42. Branch leaves distinctly 5-ranked, especially near the capitulum. Stem leaf hyaline cells without fibrils, or fibrils weak. 43

    Branch leaves not distinctly 5-ranked. Stem leaf hyaline cells strongly fibrillose in upper part. .... *S. capillifolium* var. *capillifolium* (96)

43. Stem leaf lingulate: apex broadly rounded or truncate, often fimbriate. Cells of stem cortex with occasional pores or thinnings. Branch leaf hyaline cells with large, circular pores throughout on concave leaf surface. Plants medium-sized to rather small: widely distributed in oligotrophic mires and paludified woods. Species with northern tendencies. ....*S. russowii*   (102)

    Stem leaf lingulate to lingulate-triangular: apex narrowly rounded to subacute, not fimbriate. Outer cells of stem cortex without large pores or thinnings. Branch leaf hyaline cells in mid-leaf on concave leaf surface without pores. 44

44. Branch leaf hyaline cells, in upper half of leaf, with small, circular, thick-ringed pores on the convex leaf surface (pores less than 5.0 μm, including ring). Plants usually deep red, occasionally green with red flecks: in eutrophic habitats. Abundant in the north, but towards the south increasingly confined to montane areas. ....
    ....*S. warnstorfii*   (98)

    Branch leaf hyaline cells with larger (more than 5.0 μm), apparently half-elliptical pores against the commissures. Plants of mesotrophic and oligotrophic habitats. ....*S. capillifolium*   44a (92)

44a. Capitula ± hemispherical. Branch leaves not consistently 5-ranked. Plants often compact and dense. Branch leaf hyaline cells with pores 10.0–15.0 μm on convex surface at mid-leaf. ....
    .... *S. capillifolium* var. *capillifolium* (96)

    Capitula ± flat. Branch leaves 5-ranked. Plants usually lax. Branch leaf hyaline cells with pores 6.0–12.0 μm on convex surface at mid-leaf. ....
    .... *S. capillifolium* var. *rubellum* (96)

SPECIES DESCRIPTIONS

In the following descriptions of the subgeneric groups and of individual species, a standard format is followed. The name in current use is followed by a list of the commoner synonyms and the authorities to whom they are attributed. Next follows a formal description of the plants, their organs and cell structure — features of particular taxonomic value are shown in italic type — followed by notes on habitat preferences and distributions throughout the world, in Europe and in Britain. The supplementary notes are designed to give less formal clues to identification and to point out where confusions may arise. Accompanying the text are line drawings showing salient features and maps showing distributions in Europe.

Abbreviations used in Figures

asl	apex of stem leaf
abl	apex of branch leaf
b	branch
bl	branch leaf
f	fascicle
ld	dorsal (=convex or abaxial) surface of branch leaf
lv	ventral (=concave or adaxial) surface of branch leaf
pbl	leaf of pendent branch
pld	convex surface of leaf of pendent branch
plv	concave surface of leaf of pendent branch
sl	stem leaf
sld	convex surface of stem leaf
slv	concave surface of stem leaf
tsl	transverse section of branch leaf
tss	transverse section of stem

SECTION SPHAGNUM

Sphagnum sect. *Cymbifolia* Schimp. (*Syn. musc. eur.* 2nd ed., 847. 1876)
Sphagnum [sect.] I. *Inophloea* Russ. (*Schr. NaturfGes. Univ. Dorpat,* **3,** 27. 1887)

PLANTS: Normally large, with distinct capitula, stiff stems and somewhat turgid branches. **Fascicles:** Usually with 3–5 strongly dimorphic branches: spreading branches blunt or tapering distally, pale to variously coloured; pendent branches thin and colourless, variable in length, appressed to the stem. **Stems:** Relatively stout, usually with a dark internal cylinder and well-developed 3–4-layered cortex (taking up more than half of the stem diameter); cortical cells large and hyaline, nearly always (at least in the interior layers) *with distinct spiral fibrils*; external face of outer cortical cells with *(1–) 2–6 or more large, thin-ringed circular pores*. **Branch anatomy:** A single layer of large, cylindrical hyaline cells in which *spiral fibrils are numerous* (often better developed than in the stem); cortical cells uniform, mostly with a large pore at the distal end. **Stem leaves:** Relatively large, up to, or exceeding, 2 mm in length, lingulate or spatulate, widely rounded at apex; apex bordered by 1–3 series of thin hyaline cells, so appearing fimbriate; adaxial face of hyaline cells intact and, in the upper part of the leaf, often with fibrils; abaxial face wholly or partly resorbed. **Branch leaves:** Large, uniform or modified in the distal portions of the branch, mostly 1.4–2.0 mm long, distinctly concave and often strongly overlapping to give branches a turgid appearance; *apex strongly hooded, minutely scabrid at the back because of projecting remains of half resorbed hyaline cells*; border apparently one cell wide, *with ill-defined outer edge due to the presence of a resorption furrow*. **Hyaline cells** of branch leaves large (70–120 × 20–40 μm in upper mid-leaf); adaxial surface in mid-leaf without, or with one (rarely more) large pore, usually towards the upper end; abaxial surface with few to several large, ringed pores adjacent to the commissures and sometimes with a large, free, ringed or unringed pore near the upper end; at the point of convergence of a basal and 2 lateral angles, a triple pore (the site of a pseudolacuna, see Figure 2) is usually present; internal commissural walls sometimes ornamented. Photosynthetic cells completely immersed (*S. magellanicum*) or emergent with widest exposure on the adaxial leaf surface. **Fertile plants:** Dioecious (or with male and female inflorescences on separate stems). Antheridial bracts not markedly different from vegetative branch leaves; perichaetial bracts numerous, resembling stem leaves in shape, the lowermost also similar in size; bracts graded upwards to large spatulate, convolute inner ones, which may exceed 5 mm; tissue more or less uniformly prosenchymatous near the base, further upwards with a zone in which fibrils are lacking but hyaline and photosynthetic cells remain distinct, and in the upper regions with areolation more or less the same as that of branch leaves; border as in stem leaves. Paraphyses present at the insertion of inner bracts, composed of filaments of, usually, 3 cells, the terminal of which is enlarged. Capsule and spores typical of the genus. Geographical range almost that of the genus.

In the field, the section is often referred to as the 'Cymbifolia' or the 'Inophloea', although, strictly speaking, these terms are nomenclaturally incorrect. It is, perhaps, the first group to be recognized by the novice, being composed of rather coarse, fat-looking species with large, somewhat rounded and strongly concave branch leaves whose apices are distinctly hooded (cucullate). A number of essential characters may be observed with the naked eye, but a hand lens (×10 is adequate, but ×20 is better) will reveal the minutely roughened, cucullate, branch leaf apices. Under the microscope, the presence of spiral fibrils in the branch cortex is a distinctive feature.

1. SPHAGNUM PALUSTRE

Sphagnum palustre L. (*Sp·Plant.*, **2**, 1106. 1753)
S. cymbifolium Hedw. (*Fund. musc. frond.*, 86. 1782)

PLANTS: Robust, normally pale green or yellow-brown with a more strongly coloured capitulum varying from red-brown to straw to pinkish, but *never deep red*, occasionally the whole plant green (in wet habitats in deep shade); capitulum less strongly coloured in shaded plants. **Fascicles:** Distant or rather crowded; of 3–6 dimorphic branches; spreading branches 2–3, long, tapering distally; pendent branches 1–4, pale and thin, as long as or longer than the spreading. **Stem:** typical of the section; outer cortical cells nearly always with spiral fibrils, the inner ones always so; outer faces of outer cortical cells mostly with 2–5 fairly large pores; internal cylinder dark brown to almost black, paler or green in shaded plants. **Branch anatomy:** Branches 15–25 mm long; cortical cells uniformly fibrillose, most with a single large pore at the distal end; internal cylinder pale brown, yellow-brown or almost concolorous with leaves. **Stem leaves:** Erect spreading or hanging; spatulate to rectangular; hyaline cell walls resorbed on abaxial surface but usually fibrillose (weakly so in comparison with branch leaves) near the apex on the adaxial surface, but frequently lacking fibrils. **Branch leaves:** Typical of the section, large (1.5–2.5 mm, though smaller in stunted forms and sometimes over 3.0 mm in shade forms), ovate or broadly ovate, strongly concave and with cucullate apices. **Hyaline cells:** Variable in size but always relatively wide (ca 20–30 µm wide in upper mid-leaf and up to 45 µm below); adaxial face in mid-leaf entire or with a single (rarely more), large, ringed or unringed circular pore near the upper or upper-lateral angle; abaxial face with numerous ringed pores, mainly along the commissures, and so appearing elliptical; triple pores, typical of the section, present; 4–5 lower marginal series of cells with numerous large pores on both faces; *internal commissural walls smooth.*
Leaf TS (transverse section): Hyaline cells highly inflated on the abaxial leaf surface, slightly so on the adaxial surface; photosynthetic cells relatively small, narrowly triangular or trapezoid with straight or slightly curved sides; exposure wider on adaxial surface via thin or slightly thickened wall (wall strongly thickened in var. *centrale,* see below). Photosynthetic cells with colourless walls, sometimes yellow-brown, rarely deep yellow-brown in fresh specimens (old, dried specimens may darken considerably). **Fertile plants:** Dioecious or male and female inflorescences on separate stems. Young antheridial branches morphologically similar to vegetative branches, but often more strongly coloured: lower perichaetial bracts similar to stem leaves; upper bracts large (to 4.5 mm or more), spatulate and forming a more or less conspicuous sheath; apices widely rounded, truncate or, occasionally,

Figure 4i. Distribution of *S. palustre* var. *palustre*

Figure 4ii. Distribution of *S. palustre* var. *centrale*

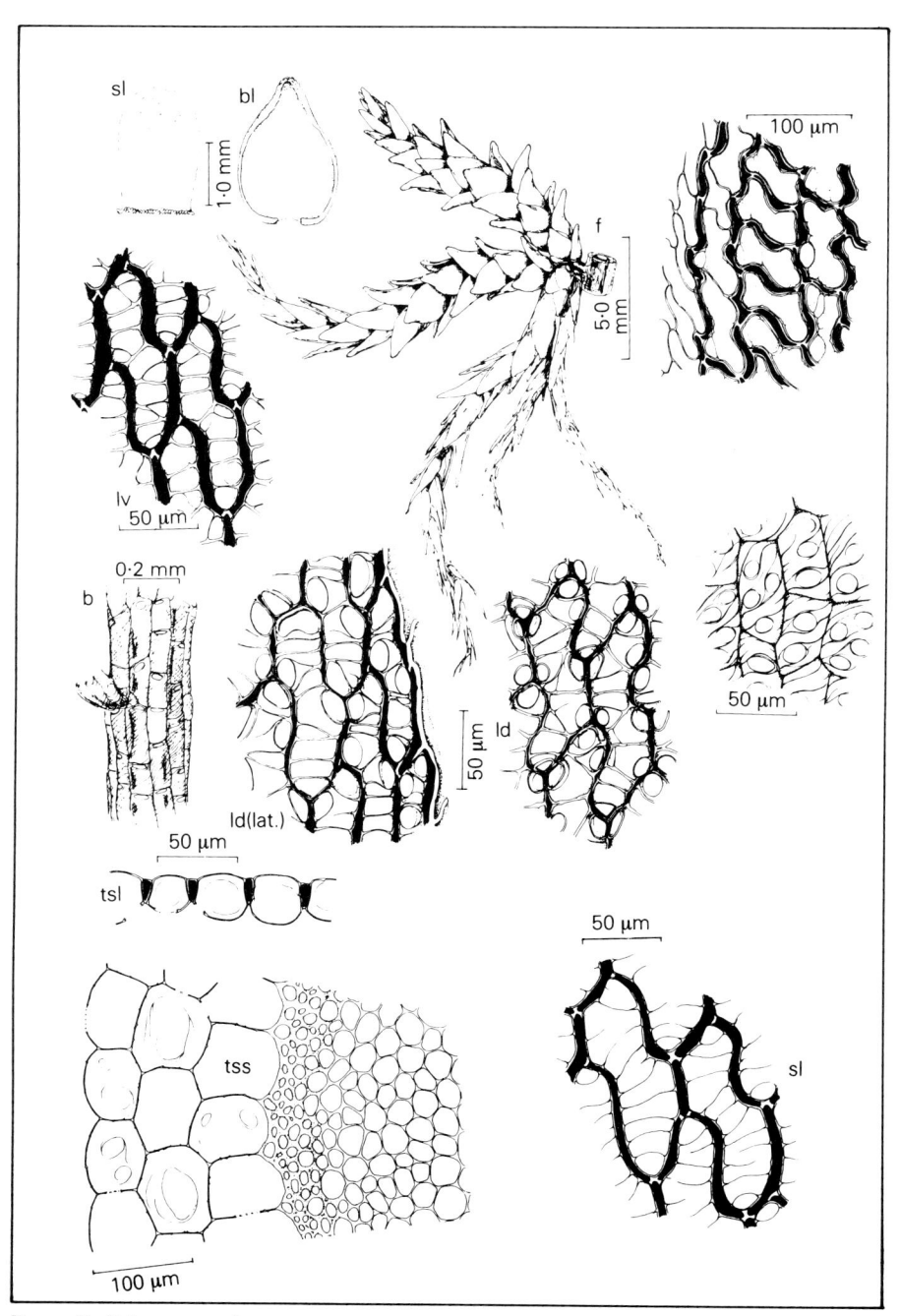

Figure 5i. *Sphagnum palustre*

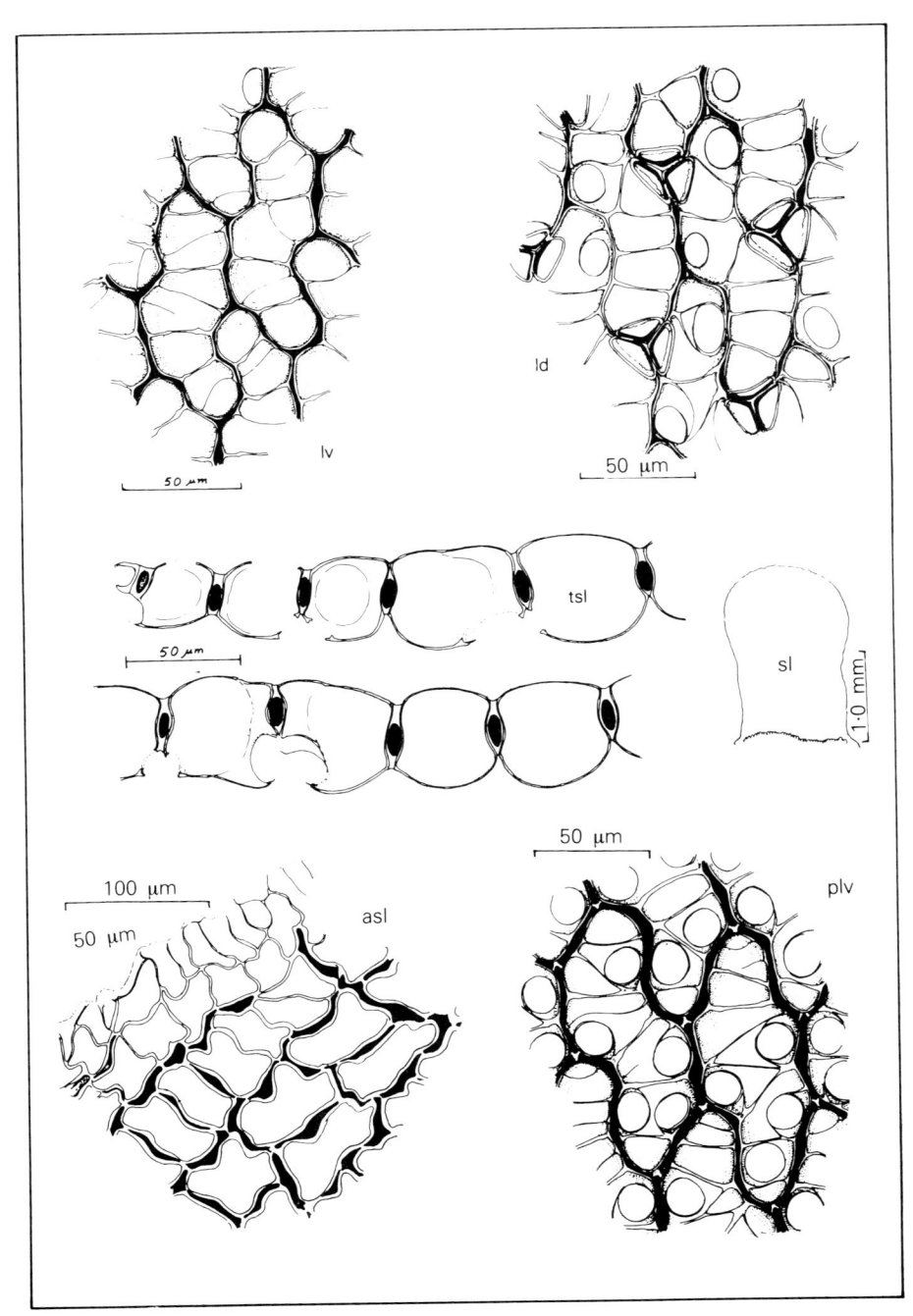

Figure 5ii. *Sphagnum palustre* var. *centrale*

somewhat retuse; tissue at bract base more or less uniformly prosenchymatous, but progressively differentiated towards the upper tissue which is similar to that of branch leaves. Capsules produced frequently: spores yellow-brown, strongly papillose, 26–32 µm diameter.

HABITAT: A common and widespread species in a wide range of mesotrophic peatland habitats, but absent from both highly calcareous and strongly acid locations. It appears to be one of the more shade-tolerant species and will form loose carpets or tussocks in wet fen woodland. In more open situations, it is also found in ditches, along stream and lake margins, on flushed hillsides and in mesotrophic fens.

DISTRIBUTION: Discontinuously circumboreal, with oceanic tendencies: Europe, East Asia (including Japan), Pacific and Atlantic coasts of North America as far south as Mexico. Widespread in north, west and central Europe, but absent from extreme north and east (but see *S. palustre* var. *centrale,* below). Common throughout most of Britain and Ireland.

1a. *S. palustre* var. *centrale*

S. palustre var. *centrale* (C. Jens.) A. Eddy, comb. nov.
S. centrale C. Jens. (Arnell & Jensen, *Bih. K. Svenska Vetensk-Akad. Handl.,* ser. 3, **21,** 34, 1896)

Over large areas of North America and in north-east Europe, a plant occurs which is virtually indistinguishable in the field from *S. palustre.* It is identified by North American and Scandinavian botanists as *S. centrale.* Under the microscope, it differs from typical *S. palustre* in the form of its photosynthetic cells which, in TS, are small with *oval* lumina and *strongly thickened adaxial walls.* Supporting characters based on fascicle structure (Nyholm 1969), stem leaf anatomy and stem cortex features (Lid 1925) are neither consistent in *S. centrale* nor always absent in unequivocal *S. palustre.* Furthermore, there has been considerable confusion between *S. centrale* and green forms of *S. magellanicum* and, in some regions, 'normal' *S. palustre* has been consistently identified as *S. centrale.* It is doubtful whether, without recourse to sophisticated analysis, poorly developed states of *S. centrale* could ever be identified with certainty. In the circumstances, we have treated *S. centrale* as a variety of *S. palustre.* In some parts of North America and east Europe (particularly more continental areas), it is the commoner form. Its occurrence in Britain is uncertain, as it has been largely overlooked or ignored by recent workers. Investigations in the herbarium and the field suggest that var. *centrale* is a British rarity, although it has been collected recently in Perthshire, Angus, Roxburghshire and Northumberland.

HABITAT: Ecologically there is little distinction from normal *S. palustre*, except that var. *centrale* does have a more northern and continental distribution.

DISTRIBUTION: Discontinuously circumboreal with continental tendencies: inland in Europe, north Asia and North America: central and north-east Europe, becoming less common towards the south and west, but recorded from as far south as Bulgaria (there confined to montane areas). Rare in Britain, although possibly under-recorded, being included with *S. palustre* ss.

2. SPHAGNUM PAPILLOSUM

Sphagnum papillosum Lindb. (*Acta Soc. Sci. fenn.*, **10,** 280. 1872)

PLANTS: Robust and tumid (similar to *S. palustre* and sometimes difficult to distinguish from that species in the field); typically rather short, *pale ochre-brown throughout,* though commonly green or yellowish; capitulum often brown but *never red*. **Fascicles:** Seldom with more than 4 branches, 2 spreading and 2 *rather short* pendent branches; spreading branches *short and blunt, not long tapering* (except in some shade forms). **Stem:** Typical of the section; outer surface of cortical cells often only weakly fibrillose, mostly with 1–4 pores; internal cylinder dark brown to almost black, sometimes green. **Branch anatomy:** Cortex typical of the section, strongly fibrillose. **Stem leaves:** Erect, hanging or spreading; rectangular or spatulate; abaxial surface resorbed, fibrils on adaxial surface sparse or inconspicuous. **Branch leaves:** Of uniform size (ca 1.4–1.9 mm), typical of the section, ovate or broadly ovate, strongly concave with cucullate apices. **Hyaline cells:** Variable in size but comparatively wide; adaxial face in mid-leaf entire or with a single (rarely more), large, ringed or unringed circular pore near the upper or upper lateral angle; abaxial face with numerous ringed pores, mainly along the commissures, and so appearing elliptical; triple pores, typical of the section, present; lower marginal series of cells with numerous large pores on both faces; internal commissural walls rough with *projecting papillae* (at least in the upper part of the leaf). Photosynthetic cells: usually *yellowish to pale brown, seldom green*. **Leaf TS:** Hyaline cells highly inflated on abaxial leaf surface, slightly so on the adaxial; internal commissural walls papillose. Photosynthetic cells relatively small, oval to barrel-shaped with *more or less oval lumina;* moderately or barely exposed on adaxial surface via a *much thickened cell wall*. **Fertile plants:** Dioecious. Antheridial branches similar to vegetative branches, but usually ochre and closely imbricated; archegonial branches typical of the section. Capsules produced moderately frequently: spores finely papillose, 26–30 μm diameter.

HABITAT: A common and widely distributed species of open acid peatlands, growing in distinct hummocks or more extensive carpets and often the principal peat former. In upland areas, it may also form hummocks beside streams or in flushes, whilst, in the lowlands, it can also be present in transitional mires with, eg, *Juncus effusus* or *Carex rostrata,* but usually only where the peat surface has been raised above direct groundwater influence: it is distinctly less tolerant of base-rich conditions than *S. palustre,* though, towards the northern part of its range, it may be found in more mesotrophic situations; in Finland, it is regarded as an indicator of mineral water influence.

DISTRIBUTION: Circumboreal in Europe, Asia and North America and extending as far south as the Himalayas and Azores, but with somewhat oceanic tendencies. Common in much of western and northern Europe but absent from much of eastern Europe. Common to abundant in suitable habitats in Britain and Ireland.

S. papillosum can normally be recognized easily in the field by its short, blunt branches and its brown-ochre coloration. Young growth may be greener, but there is usually some brownish colour present: in late summer and autumn, the whole plant may become brown. Stunted plants on the tops of hummocks may, at times, be difficult to distinguish from similarly dense states of *S. imbricatum*. Confusion may also arise with green forms of *S. magellanicum,* but the latter usually has at least some red pigment. Under the microscope, the papillae are diagnostic, this being the only species of the section to have them.

Occasionally, plants may be found in which not all the internal commissures are papillose, some near the apex and base of the leaf being almost, or quite, smooth. Such forms have been named, in the past, var. *sublaeve* Limpr. However, there is no clear distinction between these forms and the usual forms with papillae: there may also be considerable variation even within a single specimen. As there are no ecological or distributional trends associated with degree of papillation, there seems little to be gained by upholding the variety. Some shade forms lack the characteristic thickening of the photosynthetic cell walls and may also lack the brownish colour of typical plants. This variant has, at times, been identified as *S. hakkodense,* but appears to be no more than a habitat-induced form of *S. papillosum*. Extreme cases have been reported in which papillae are completely absent (var. *laeve* Warnst.). A few such plants from northern Europe have been seen but, of the British specimens examined, either some degree of papillation is present or the specimen has turned out to be *S. palustre.*

Figure 6. Distribution of *S. papillosum*

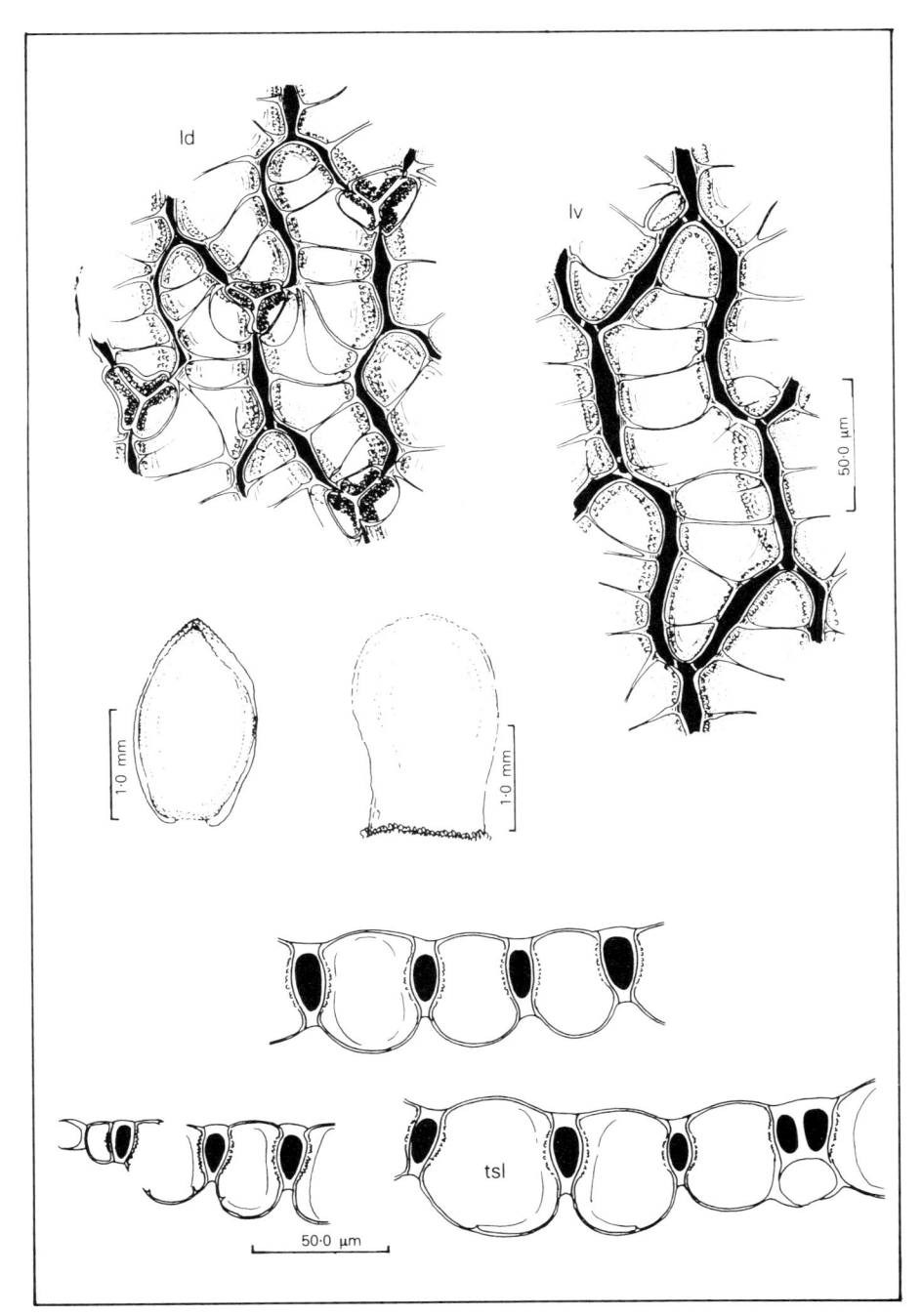

Figure 7. *Sphagnum papillosum*

3. SPHAGNUM IMBRICATUM

Sphagnum imbricatum Russ. (*Beitr. Kennt. Torfm.*, 21. 1865)
S. austinii Sull. ex Aust. (*Musci appalach.*, 3. 1870)

PLANTS: With habit typical of the section, lax or, more usually, rather dense (then resembling smaller states of *S. palustre* or *S. papillosum*), occasionally compact so that the stem is completely hidden by the closely packed spreading branches; dull green to yellow-brown or chestnut, compact forms often very dark with some purple-brown coloration (then resembling depauperate or unhealthy states of *S. magellanicum*). **Fascicles:** Typically of 4, sometimes only 3, branches; 2 blunt or distally tapered spreading branches and 1–2 pendent branches which are shorter, weaker and usually unequal in length. **Stem:** Typical of the section, inner layers of cortex strongly fibrillose, the outer layers weakly so (exposed face often lacking fibrils); outer face of cortical cells with 2–3 pores; internal cylinder dark red-brown, seldom pale, yellow-brown. **Branch anatomy:** Spreading branches ca 15 mm, blunt or shortly tapered; pendent branches rarely exceeding 12 mm; cortex fibrillose and porose; hyaline cortical cells, where they lie against cells of the internal cylinder, *densely fibrillose so that wall appears obliquely and densely striate;* internal cylinder red-brown or yellow-brown. **Stem leaves:** Erect, spreading or hanging; shortly lingulate to spatulate, rarely exceeding 1.5 mm in length; abaxial surface resorbed; adaxial surface entire; fibrils mostly absent but numerous septa present; abrupt transition from upper tissue with short, wide hyaline cells to lower tissue with much narrower ones; hyaline leaf border often well developed. **Branch leaves:** Typical of the section, concave, broadly ovate (sub-orbicular when flattened), with cucullate apices, rarely exceeding 1.9 mm; resorption of the abaxial surface of hyaline cells often extending well below the leaf apex, so that much of the leaf may appear rough under a lens. **Hyaline cells:** Comparatively wide; adaxial face in mid-leaf entire or with a single (rarely more), large, ringed or unringed circular pore near the upper or upper lateral angle; abaxial face with numerous ringed pores, mainly along the commissures, and so appearing elliptical; triple pores, typical of the section, present; lower marginal series of cells with numerous large pores on both faces; internal commissural walls, at least towards the leaf base, ornamented with numerous to very abundant transverse lamellae which, in surface view of the leaf, appear pectinate or comb-like (these are the so-called 'comb fibrils'), photosynthetic cells nearly always *strongly coloured pale to rich brown*, relatively wide (6–8 μm). **Leaf TS:** Hyaline cells with strongly inflated abaxial face; adaxial face slightly inflated; internal commissural walls with local thickenings (an optical effect caused by the presence of comb fibrils) or projections where the section cuts through the end region of a photosynthetic cell, where the fibrils are oblique. Photosynthetic cells *broadly trapezoid or isosceles to equilateral triangular* with rather thin walls and more or less angular lumina; reaching both leaf surfaces but broadly exposed only on the adaxial surface. **Fertile plants:** Dioecious, rare. Antheridial branches scarcely distinguishable from vegetative; female inflorescences more or less typical of the section, but usually with only a small region of fibrillose cells below the bract apices, sometimes none on the innermost bracts. Capsules rare: spores yellow-brown, papillose, 26–30 μm diameter.

HABITAT: Scattered in relatively undisturbed, wet, lowland, oligotrophic mires in areas of strong maritime influence, eg NW Scotland, SW Sweden. In these areas, it forms large, dense, usually brown, hummocks. It also occurs as carpets in wet mesotrophic fens or near lake margins in slightly more continental locations. In this type of habitat, and sometimes around the bases of hummocks, the growth form is more lax.

Bands of *S. imbricatum,* characteristically a rich, fulvous brown, are abundant in peat of many sites where the living plant is no longer found. It was clearly much more common in former times and its drastic reduction is possibly a reflection of climatic change. Recent extinctions, however, are undoubtedly the result of habitat modifications by techniques such as moor burning, to which this species seems particularly sensitive.

DISTRIBUTION: Discontinuously circumboreal with strong oceanic tendencies, in Europe, east Asia, Atlantic and Pacific coasts of North America, extending southwards to the Himalayas, West Indies and Chile. Restricted to north-western part of Europe from central Sweden and south-eastern Baltic coastal areas to Holland and Belgium. In Britain, confined to the north and west where it occurs infrequently.

In the field, typical *S. imbricatum* may be recognized by its comparatively small size and compact habit compared with other members of the section, the usual presence of at least some brown coloration, and the small, closely set branch leaves which tend to give the branches a cigar shape. The only other species in Europe which forms large brown hummocks on oligotrophic mires is *S. fuscum,* but this is a much more delicate plant than the present species. It may be confused with small forms of *S. papillosum,* but the usually tapered branches and stronger coloration of *S. imbricatum* are often sufficient to confirm identification. Microscopically, the comb fibrils in the branch leaf hyaline cells are distinctive. Only one other species possesses this feature, *S. portoricense,* but this is a more robust, dull green plant, distinguished by protruding end walls in its branch cortical cells: it is also confined to tropical and sub-tropical America. In the very rare cases when comb fibrils are virtually absent or difficult to detect, the closely striate inner walls of the branch cortical cells and the wide, triangular photosynthetic cell section will confirm the plant's identity.

As with *S. papillosum,* a number of varieties of *S. imbricatum* have been described, mainly based on the height and extent of the comb fibrils. Thus, forms with tall and conspicuous lamellae throughout the leaves have been named var. *cristatum* Warnst, forms in which the upper portions of leaves lack comb fibrils

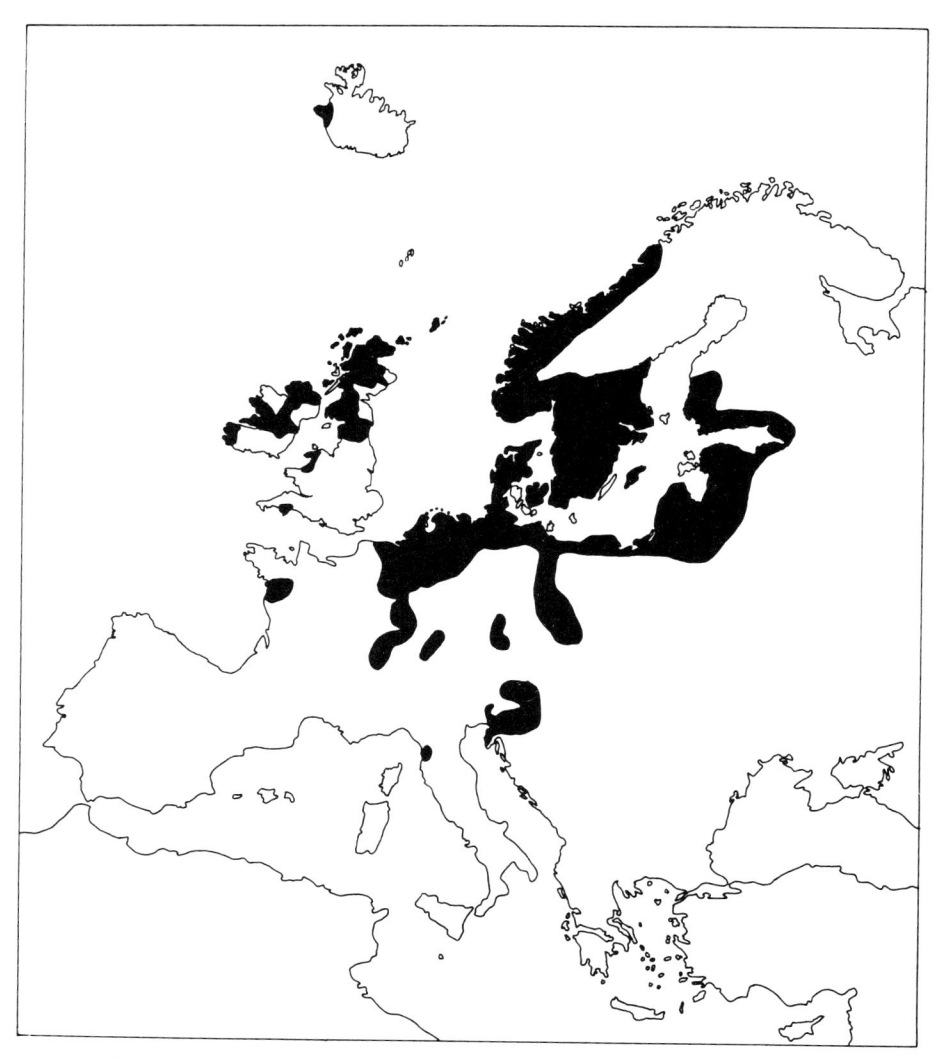

Figure 8. Distribution of *S. imbricatum*

have been called var. *sublaeve* Warnst, whilst plants which lack them altogether have been designated as var. *affine* Warnst. If such forms do occur in Europe, they must be very rare. The var. *affine* has been reported but appears to be otherwise confined to regions adjacent to north-eastern and north-western Pacific coasts. There is a continuous range of states between the '*cristatum*' and '*sublaeve*' conditions and any genetically determined predisposition towards one

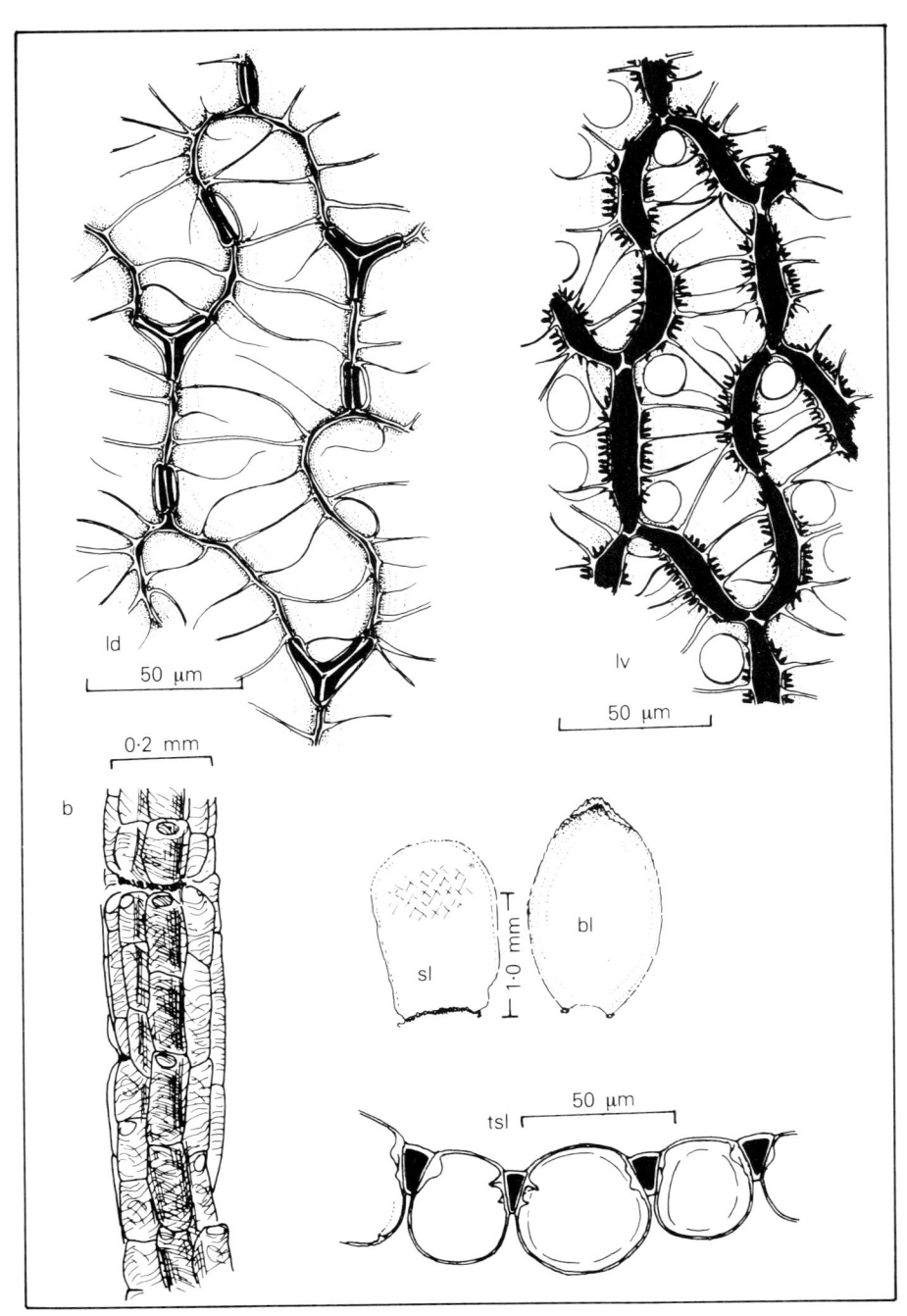

Figure 9. *Sphagnum imbricatum*

or the other is likely to be masked by habitat-induced variation, so that it seems impossible to apply either epithet with any real meaning. The comb fibrils themselves are regarded by some authors (eg Hill 1978) as homologues of ordinary fibrils. The function of the latter is to give mechanical support to the delicate hyaline cells, and it is difficult to reconcile such a function of the comb fibrils with their distribution in the plant. It would seem more likely that they might, in some way, assist in the absorption of mineral ions, by increasing surface area. This would explain why they are confined to hyaline cell walls which lie against photosynthetic cells in the leaves, and against the internal tissues of the branches. Perhaps, in this respect, comb fibrils have the same function as the papillae in *S. papillosum*.

Recently Flatberg has suggested that *S. imbricatum* consists of a complex of taxa, and he recognizes a number of morphological and ecological differences by which they may be distinguished. Under his proposals, 2 of the 3 subspecies he defines (elevated to species status by Andrus, together with the arctic form, *S. steerii* Andrus) occur in Europe: *S. imbricatum* subsp. *imbricatum* is confined to eastern Asia.

Hill (1988) presents convincing support for Flatberg's (1984) thesis that the two distinct forms, *S. imbricatum* subsp. *austinii* (Sull.) Flatberg and *S. imbricatum* subsp. *affine* (Ren. & Card.) Flatberg, are found in Europe. Subspecies *austinii* is typically found as a hummock-forming component of bogs, though occasionally forming lawns in some areas. It is characterised by usually having only one pendent branch per fascicle, comb fibrils in the stem leaves and the absence of dense fibrillae (comb lamellae) on the inner surface of the stem cortex. Subspecies *affine,* according to Hill, is found in more minerotrophic sites, in association with species such as *Sphagnum recurvum* and *Carex rostrata*. It differs from subsp. *austinii* in being paler, usually with 2 pendent branches per fascicle, in having more septate stem-leaf hyaline cells and having well-developed fibrillae in the inner surface of the stem cortex.

A reappraisal of herbarium material by one of the present authors (Eddy) was hampered by the paucity of information attached to the majority of specimens and by the scarcity of plants that could be assigned unequivocally to subsp. *affine*. Although the morphological division between the subspecies seemed less sharply defined than implied by Hill, with some specimens showing a mix of characters, it was generally possible to assign plants to one taxon or the other, without undue difficulty. In the British Isles, at least, subspecies *affine* is either scarcer than subsp. *austinii*, or has often been overlooked.

We have taken a conservative approach, but would welcome a genecological study to help resolve the relationships between the taxa proposed.

4. SPHAGNUM MAGELLANICUM

Sphagnum magellanicum Brid. (*Musc. recent.*, **2,** 24, 223.1798)

PLANTS: Robust but, typically, rather short (in size and habit rather similar to *S. papillosum*) but not tumid; pale green (though nearly always with at least some flecks of red or pink) to wine-red throughout. **Fascicles:** Most commonly with 4, occasionally 5, branches; spreading branches usually 2, blunt; pendent branches thin and pale. **Stem:** Typical of the section, but cortex often lacking fibrils or only faintly fibrillose; outer cortex with (1–)2–4 pores per cell, occasionally more; internal cylinder dark purple-brown or deep red. **Branch anatomy:** Branches mostly 15–20 mm long; cortex fibrillose and porose; internal cylinder pale to deep red, rarely green. **Stem leaves:** Erect, hanging or spreading; rectangular to lingulate; hyaline cell walls resorbed on abaxial surface; on adaxial surface with or without fibrils in the upper half of the leaf. **Branch leaves:** Typical of the section, of uniform size (1.5–2.4 mm), broadly ovate, concave with cucullate apices. **Hyaline cells:** Often somewhat angular; adaxial face, except for about 4 marginal rows, usually without pores; abaxial face with large, ringed pores along the commissures and well-defined triple pores; internal commissural walls smooth. **Leaf TS:** Hyaline cells slightly inflated on both surfaces, internal commissural walls smooth. Photosynthetic cells *oval, thin-walled, and mostly completely enclosed by hyaline cells*, rarely narrowly exposed on one or both surfaces via thickened end walls. **Fertile plants:** Dioecious. Antheridial branches almost impossible to distinguish, in the field, from vegetative ones, densely imbricated; archegonial branches typical of the section. Capsules rare in Europe (but frequent in parts of South America); spores papillose, 26–30 μm diameter.

HABITAT: Widespread, but often localized, in oligotrophic mires where it grows, usually in association with *S. papillosum* or *S. capillifolium,* as broad carpets or low hummocks. It is typical of wet raised bog sites, but may also be found on wet blanket bog. In oligotrophic, lowland valley mires, it can form loose patches on an underlying carpet of other *Sphagna* which isolate it from direct contact with minerotrophic groundwater. It is more susceptible to burning and drainage than *S. papillosum,* and appears to favour wetter or more oceanic situations than that species, at least in Britain. It is, however, also found in drier, wooded, oligotrophic sites in north and east Europe.

DISTRIBUTION: Circumboreal in Europe, Asia and North America and extending as far south as the Himalayas, through Central and South America to Tierra del Fuego and the Malagasy Republic (Madagascar). Common throughout most of Europe but more scattered towards the north and absent

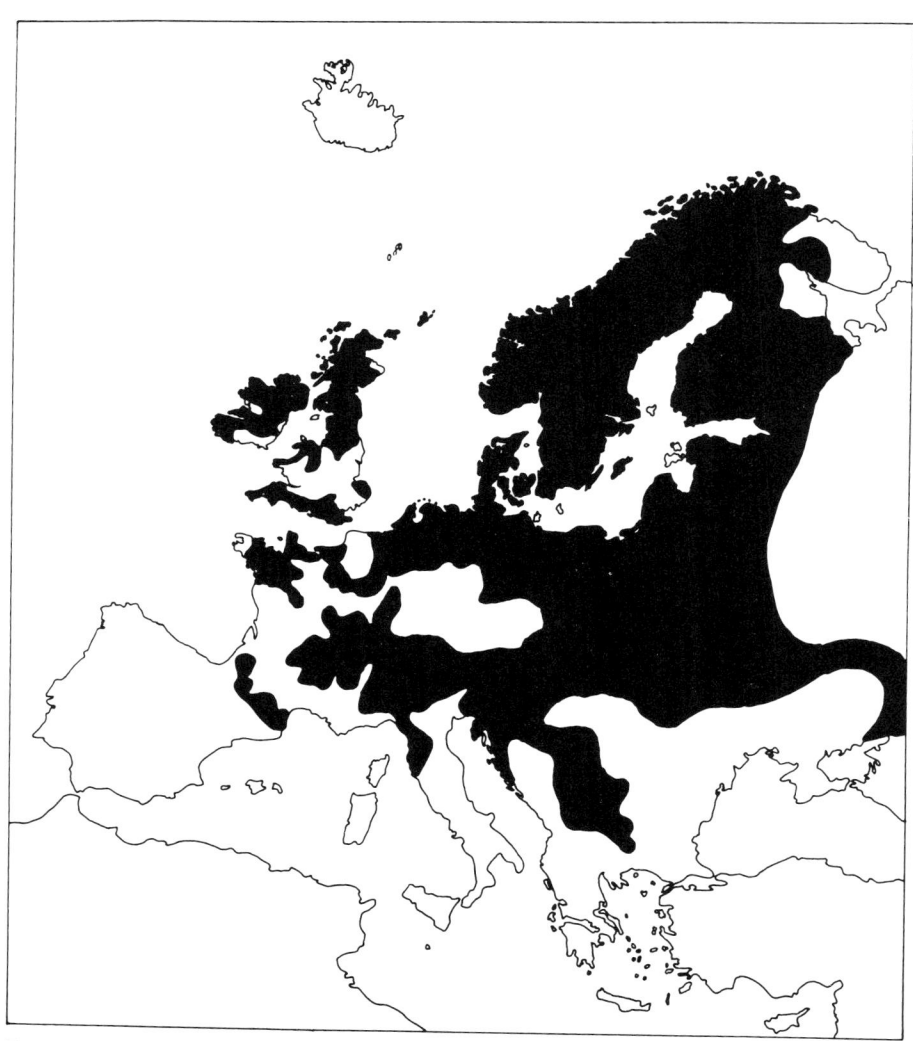

Figure 10. Distribution of *S. magellanicum*

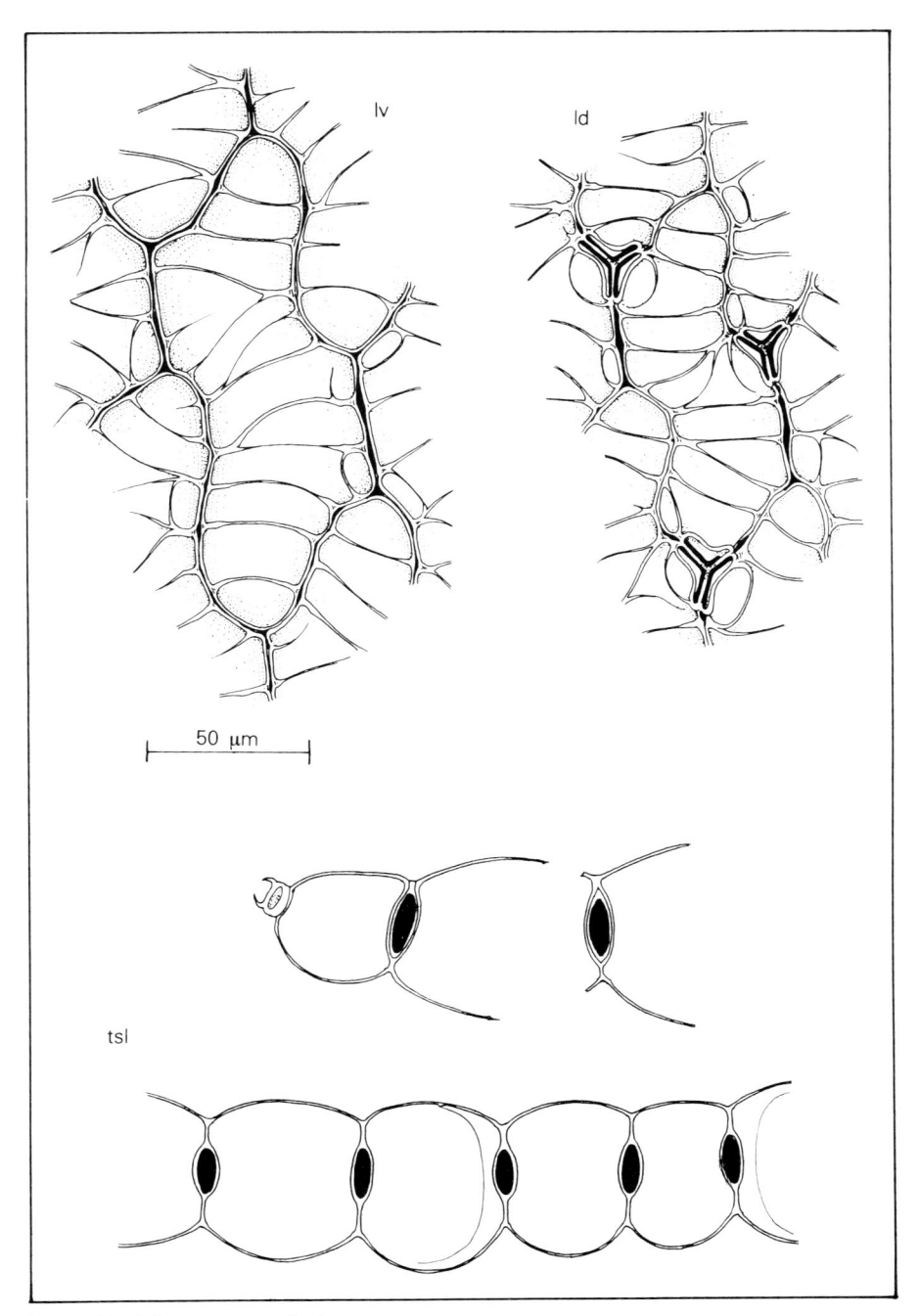

Figure 11. *Sphagnum magellanicum*

from the extreme north and east. In Britain, more common in the north and west, but also present along the south coast in suitable habitats.

S. magellanicum can nearly always be recognized in the field by its '*cymbifolium*' habit and red colour. The intensity of this colour can vary with season and habitat and may be almost absent under dense shade, but usually, even under these conditions, some stems have flecks of red or rose-pink. Defoliated stems held up to the light have a pale to deep wine-red or purple colour not possessed by other European members of the section. Some green forms exist in which the absence of secondary pigmentation is genetically determined, and positive recognition of these can only be achieved by microscopic examination of photosynthetic cell characters. Another rare form combines a lack of pigmentation with some thickening of the abaxial and adaxial fused commissural walls of the hyaline cells. This form may be confused with *S. palustre* var. *centrale,* but lacks the long, tapered branches of that taxon and is also distinct in the symmetrical, not adaxially displaced, photosynthetic cells.

SECTION ACUTIFOLIA

Sphagnum sect. *Acutifolia* Wils. (*Bryol. brit.,* 20. 1855) Excl. parte

PLANTS: Small to medium-sized, rarely robust; mostly with well-developed capitula; usually (except in *S. fimbriatum* and *S. girgensohnii*) with red or, occasionally, brown pigments, especially in the antheridial branches. **Fascicles:** Well-developed; nearly always distinctly, and often strongly, dimorphic. **Stem:** With a *well-developed cortex of at least 2, usually 3–4 layers of highly inflated hyaline cells,* the outer surface with or without a large pore; internal cylinder well-developed, usually (except in *S. fimbriatum* and *S. girgensohnii*) at least partly coloured brown, red, purple or violet. **Branch anatomy:** Retort cells distinct, sometimes in linear pairs or threes, but *commonly solitary;* internal cylinder usually red, purple or brown. **Stem leaves:** Erect and appressed to spreading, rarely strongly deflexed and hanging; relatively large and more or less equalling or exceeding branch leaves in length, rarely less than 1.2 mm; variable in shape, hemi-isophyllous to strongly heterophyllous, ovate to lingulate to broadly spatulate and fimbriate; border mostly distinct and often widened markedly in the lower half of the leaf; hyaline cells with or without fibrils, often abundantly septate, typically with extensive resorption of the adaxial surface. **Branch leaves:** Small to medium-sized, commonly between 1.2 and 1.9 mm long; erect to erect spreading, very rarely squarrose; commonly 5-ranked; more or less lanceolate and tapered to truncate-dentate apices; apices rarely somewhat eroded and somewhat hooded (forms of *S. capillifolium*); border narrow, of 1–2 cells, with a resorption furrow only in *S. molle.* **Hyaline cells:** Near the leaf apex, often narrow, but otherwise broad, and highly inflated abaxially, often those of the lower parts of the leaf many times larger than those near the apex (similar to section Squarrosa). Abaxial surface almost always porose, pores large (except some in *S. warnstorfii*) and mostly distinctly ringed. Adaxial surface without, or with one or a few, large circular pores. Two to several series of hyaline cells in the lower lateral parts of leaves with large (15.0 μm or more) ringed pores on both surfaces. **Leaf TS:** Hyaline cells strongly inflated on abaxial surface, almost plane or shallowly inflated on the adaxial; internal commissural walls smooth. Photosynthetic cells *triangular to trapezoid with widest exposure on the adaxial leaf surface.* **Fertile plants:** Dioecious or monoecious. Antheridial bracts similar to branch leaves, but often contrasting in colour. Inner perichaetial bracts variable; apices obtuse and minutely retuse, largely prosenchymatous; hyaline cells differentiated and intact in upper part of bract and, in most species, more or less lacking fibrils.

The section includes strongly minerotrophic (*S. warnstorfii*) and strongly oligotrophic species which, for the genus as a whole, occupy relatively dry sites. Its members are widely distributed, but absent from Australasia (with the exception of a dubious record of *S. fimbriatum*) and generally poorly represented in the southern hemisphere. The maximum diversity of species in this section is reached in the temperate northern hemisphere, particularly the more oceanic regions of Europe and North America.

Students usually recognize this section initially by the distinct, and sometimes very pronounced, red colour of most of its commoner species. Even in the absence of such coloration, the well-developed stem cortex, strongly dimorphic branches and relatively large, more or less erect, stem leaves (which can be seen with a hand lens) should prevent confusion with the section Cuspidata. The most important microscopic features are the photosynthetic cells, which are widely exposed on the adaxial leaf surface, and the highly inflated hyaline cells with at least some large (over 12.0 μm) ringed pores on their abaxial faces. The hyaline cells of branch leaves of members of this section are very similar to those of members of the section Sphagnum, but the dimorphic branch cortical cells, lacking fibrils, clearly identify the group. If *S. fimbriatum* were excluded, the section would form a sharply defined natural group, at least in the northern temperate zone. Unfortunately, that species, which differs widely from the Acutifolia norm in its distribution and ecology, as well as in some morphological characters, connects to the more typical taxa via *S. girgensohnii*. A relationship between section Acutifolia and section Squarrosa is clearly indicated in these 2 species. There are also other evolutionary links with the (more primitive) section Subsecunda and the (more advanced?) section Sphagnum, although the evidence for such links is derived mainly from tropical taxa.

5. SPHAGNUM MOLLE

Sphagnum molle Sull. (*Musci alleghan.*, 205. 1845)

PLANTS: Normally low-growing, unless etiolated, and often rather dense, forming low, pale hummocks (most closely resembling under-developed *S. subnitens* or diminutive *S. palustre* or, when compact, *S. compactum*); tufts usually containing numerous under-developed stems intermingled with normal ones, the former lacking, or with much reduced, branches; pale green, yellowish with distinct flecks of pale violet-red or pink, *never strongly coloured*. **Fascicles:** Closely set and usually concealing the stem; of 3–4 dimorphic branches (the degree of dimorphism may vary); spreading branches 2, rather short (6.0–9.0 mm) but often attenuated distally and then up to 12.0 mm or more, horizontal or, in compact forms, directed upwards as in *S. compactum*; pendent branches 1–2, varying in length but always deflexed and more or less appressed to the stem. **Stem:** Rather thin, up to 0.8 mm diameter; cortex of 2–3 layers of inflated hyaline cells, mostly without pores, occasionally (particularly in cells adjacent to a stem leaf insertion) with a large pore; internal cylinder yellowish to pale brown, *never dark brown or violet*. **Branch anatomy:** *Retort cells in groups of 3 or more* (not solitary as in other members of the section Acutifolia), varying in size but usually 1–3 times larger than other cortical cells (some of these ordinary cortical cells may also be perforated); internal cylinder pale yellowish or brownish. **Stem leaves:** Large, up to 2.0 × 1.3 mm, ovate-lingulate to ovate-spatulate, *wider at mid-leaf than at insertion*; apex rounded-truncate; border narrow, rather indistinct above, *not, or scarcely, widened below*. Hyaline cells mostly without fibrils or a few cells fibrillose to varying degrees (leaves of weak stems may be particularly strongly fibrillose); septa numerous. **Branch leaves:** Suberect, laxly imbricate; relatively *large*, 1.6–2.2 mm long; ovate and widest at about mid-leaf, *not abruptly narrowed above to an apparently acuminate apex*; border narrow, mainly 2 cells wide, *the outer wall of the outer marginal series resorbed to form a resorption furrow* (as in, eg, *S. palustre*), the remaining cross-walls projecting as remote teeth (this species is sometimes described as having denticulate branch leaves); apex truncate-dentate. Pendent branch leaves similar to those of the spreading branches; the lower short and wide, the distal narrower. **Hyaline cells:** Large to very large, 160–180 × 25–45 μm, of rather uniform size throughout the leaf. Abaxial surface strongly inflated (often sufficiently so to almost obscure the pores); pores numerous, large (12–22 μm diameter), ringed, those of adjacent hyaline cells often in opposed pairs; pseudolacunae (triple pores) often fairly well developed; irregular resorption gaps frequently present in apical angles. Adaxial surface without pores, except in 1–3 marginal rows of

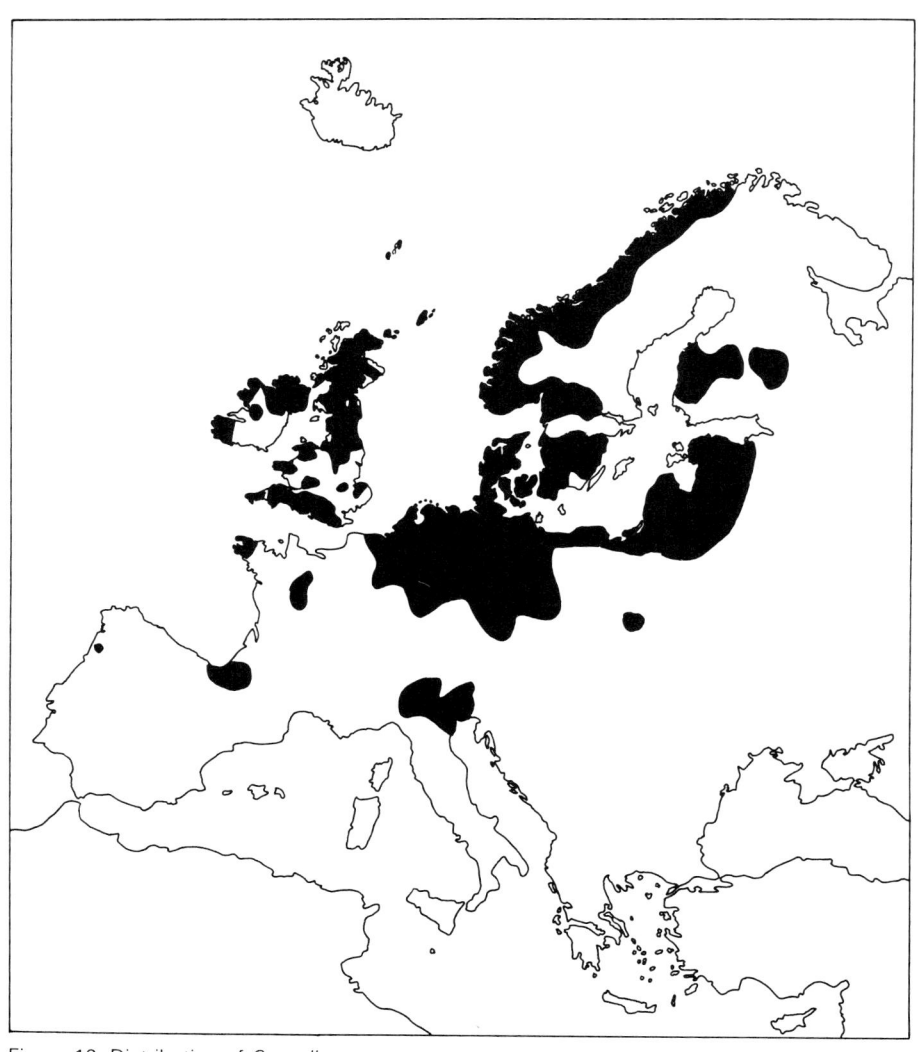

Figure 12. Distribution of *S. molle*

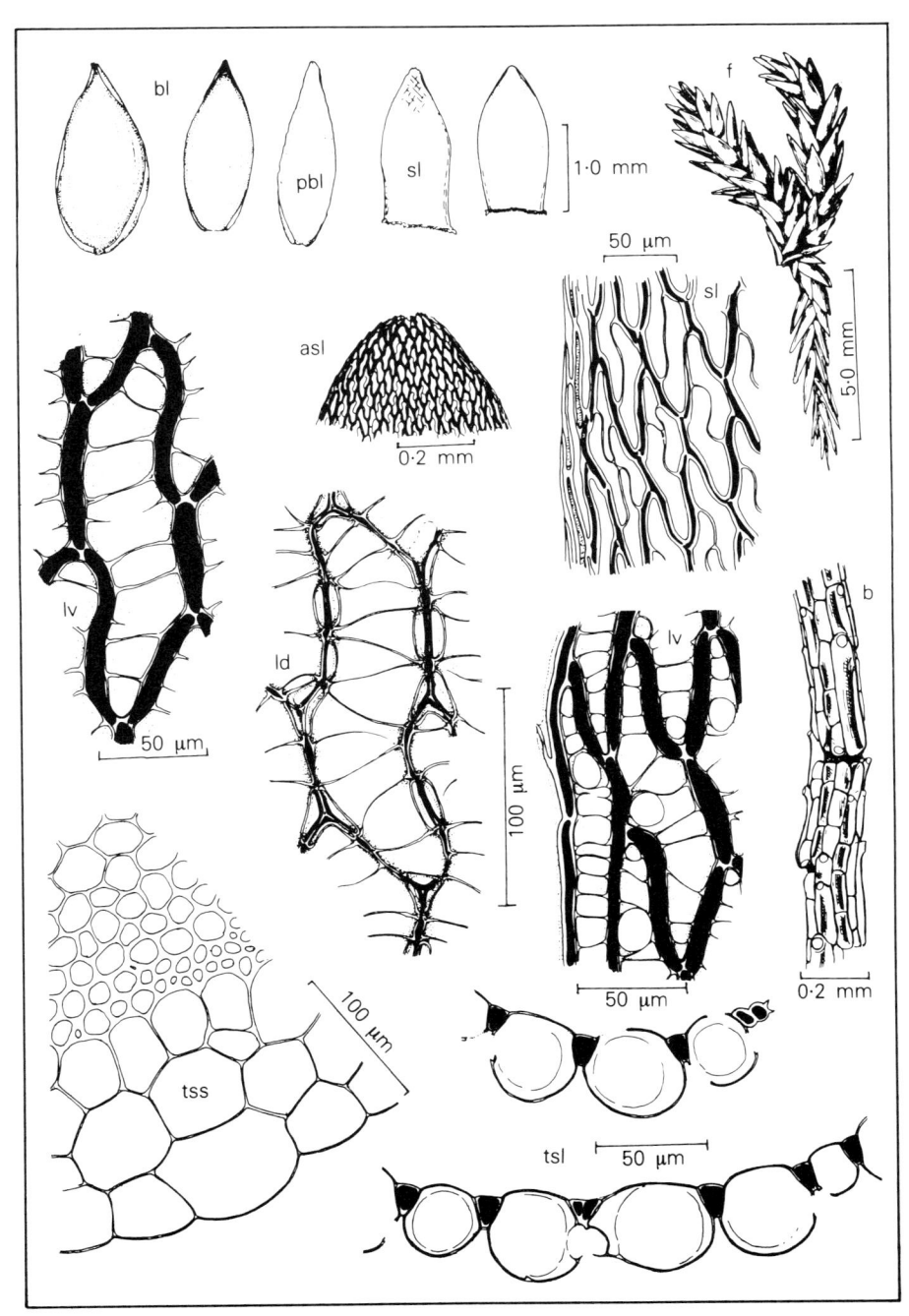

Figure 13. *Sphagnum molle*

cells. **Leaf TS:** Hyaline cells strongly convex on the abaxial leaf surface, shallowly convex or almost plane on the adaxial. Photosynthetic cells triangular, occasionally trapezoid, thin-walled with more or less angular lumina; widely exposed on adaxial surface, not, or hardly, exposed on the abaxial. **Fertile plants:** Monoecious. Antheridial branches similar to spreading sterile branches, concolorous or, frequently, distinctly red-violet. Perichaetial bracts ovate, tapering from above the middle to a *somewhat acuminate apex*; hyaline cells more or less differentiated throughout, intact and lacking fibrils, or a few cells with fibrils and pores on the abaxial surface. Capsules frequent; spores yellow-green, more or less roughened, 28.0–32.0 μm diameter.

HABITAT: A plant of damp, open oligotrophic situations. Because it cannot compete successfully with rank vegetation, it is confined to open areas. In this, it resembles *S. compactum* and *S. tenellum* and, to some extent, *S. subnitens*: indeed, these are its most frequent associates. *S. molle* is found on damp areas of western raised and blanket bog, along the wet heath margins of valley mires or, occasionally, along the margins of oligotrophic lakes.

DISTRIBUTION: An oceanic species of coastal areas on both sides of the Atlantic, being found only in Europe and along the eastern seaboard of North America from Labrador to Florida. In Europe, it is found mainly in the lowlands from south-west Scandinavia (although it does extend further northwards along the Norwegian coast) to the Pyrenees and northern Italy. Rare and scattered in much of Britain, but a little more common in the north-west.

S. molle is often an inconspicuous plant, especially when mixed with some other species, and could easily be dismissed as weak forms of them. However, in the more oceanic districts of the west coast of the British Isles and Scandinavia, it often forms quite large, pure stands which are a pink-flecked, pale green. These may resemble pale forms of *S. subnitens*, but the plants of *S. molle* have broader, laxer, branch leaves. Dense forms may resemble *S. compactum*, but have smaller branch leaves and much larger stem leaves. Under the microscope, the resorption furrow and groups of retort cells distinguish *S. molle* from other members of the section Acutifolia.

6. SPHAGNUM SUBNITENS

Sphagnum subnitens Russ. & Warnst. ex (Warnst.) (*Verh. bot. Ver. Prov. Brandenb.*, **30,** 115. 1888)
S. plumulosum Röll. (*Allg. bot. Z.*, **16,** 70. 1910)

PLANTS: Medium-sized, relatively robust (for the section), smaller in dry or exposed situations, short in open habitats, elongated in shade; capitula rather large; colour variable, green, ochre, brown, red or purple, rarely red throughout. **Fascicles:** Crowded to rather distant (in shade); of 3(–4) dimorphic branches; spreading branches 2, long, up to 25 mm or more, tapering; pendent branches 1–2, shorter or longer than the spreading. **Stem:** Up to 0.9 mm diameter; cortex well developed, of (2–)3–4 layers of highly inflated hyaline cells without pores on the external face; internal cylinder brown or purplish, sometimes distinctly red near the capitulum. **Branch anatomy:** Retort cells distinct, one per leaf axil; internal cylinder pale red. **Stem leaves:** Erect; relatively large, (1.2–)1.4–1.6 mm long, *triangular to triangular-lingulate*; apex acute and sometimes shortly cuspidate, or narrowly rounded truncate; border distinct, of 3–6 rows of cells, strongly widened below. Hyaline cells *without fibrils,* rarely very weakly fibrillose, but mostly septate, at least in upper half of leaf; abaxial surface intact; adaxial surface partially or wholly resorbed. **Branch leaves:** Suberect to somewhat spreading; not 5-ranked in mature plants (sometimes in immature plants the leaves may be weakly 5-ranked); large, 1.3–2.0 mm long, some, at least, exceeding 1.4 mm; ovate or ovate-lanceolate with the upper margins inrolled, commonly narrowed abruptly in the upper half (cf section Squarrosa); apex truncate but often appearing acute due to the inrolled margins; border 1–2 cells wide, without resorption furrow. Pendent branch leaves similar to those of spreading branches, but more delicate and, relatively, narrower. **Hyaline cells:** Varying in size, and often several times larger in the basal part of the leaf than in the apical part; in the upper leaf 100–130 × 18–20 μm, in the lower part longer (to 160 μm or more) and wider (to 40 μm): Abaxial face strongly inflated so that pores may be almost obscured; pores few to several (up to 8 per cell), sometimes more numerous towards the lower lateral parts of the leaf; large (15–30 μm), more or less circular to broadly elliptical, along the commissures, often in triplets or opposing pairs in adjacent cells. Adaxial face mostly without pores, except in a few marginal series of cells, which have a large circular pore on both faces. Pendent branch leaf hyaline cells similar to those of the spreading branches, except that the adaxial pores are more numerous and widespread. **Leaf TS:** Hyaline cells highly inflated abaxially, almost plane or shallowly convex on adaxial face. Photosynthetic cells triangular or trapezoid; rather thin-walled and with angular lumina; not

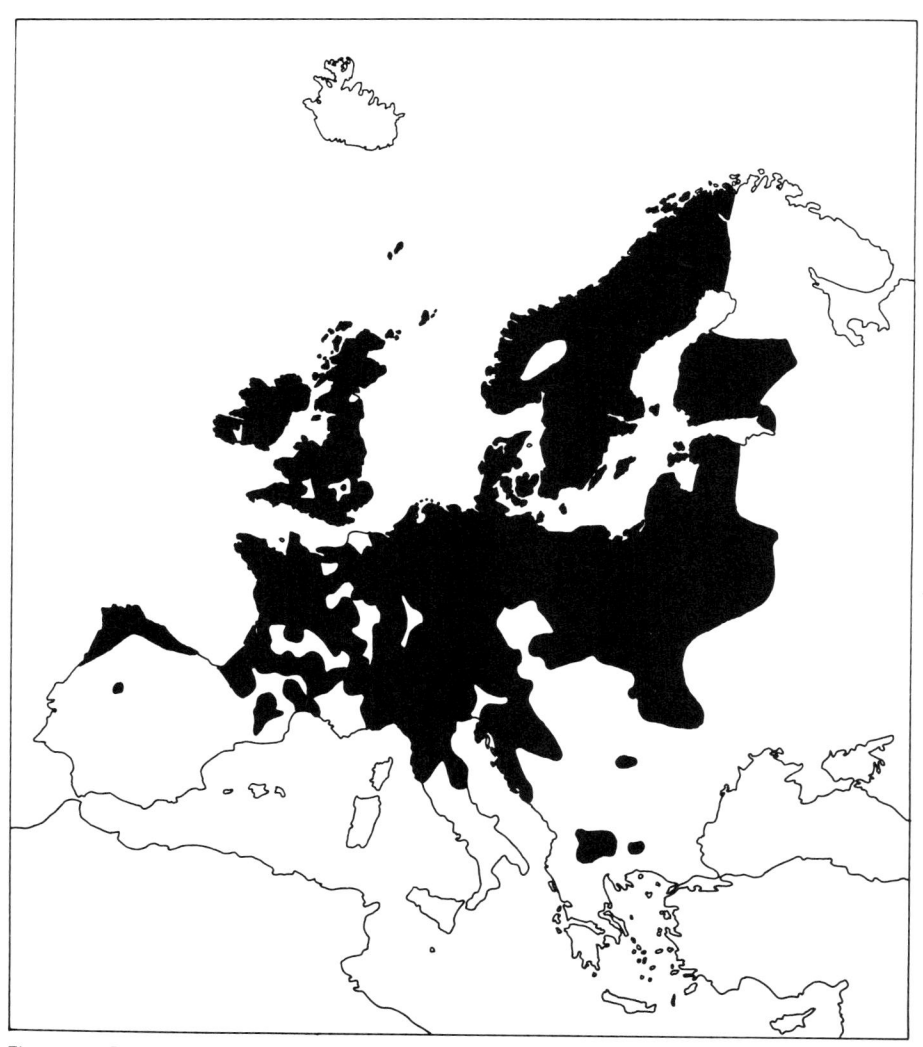

Figure 14. Distribution of *S. subnitens*

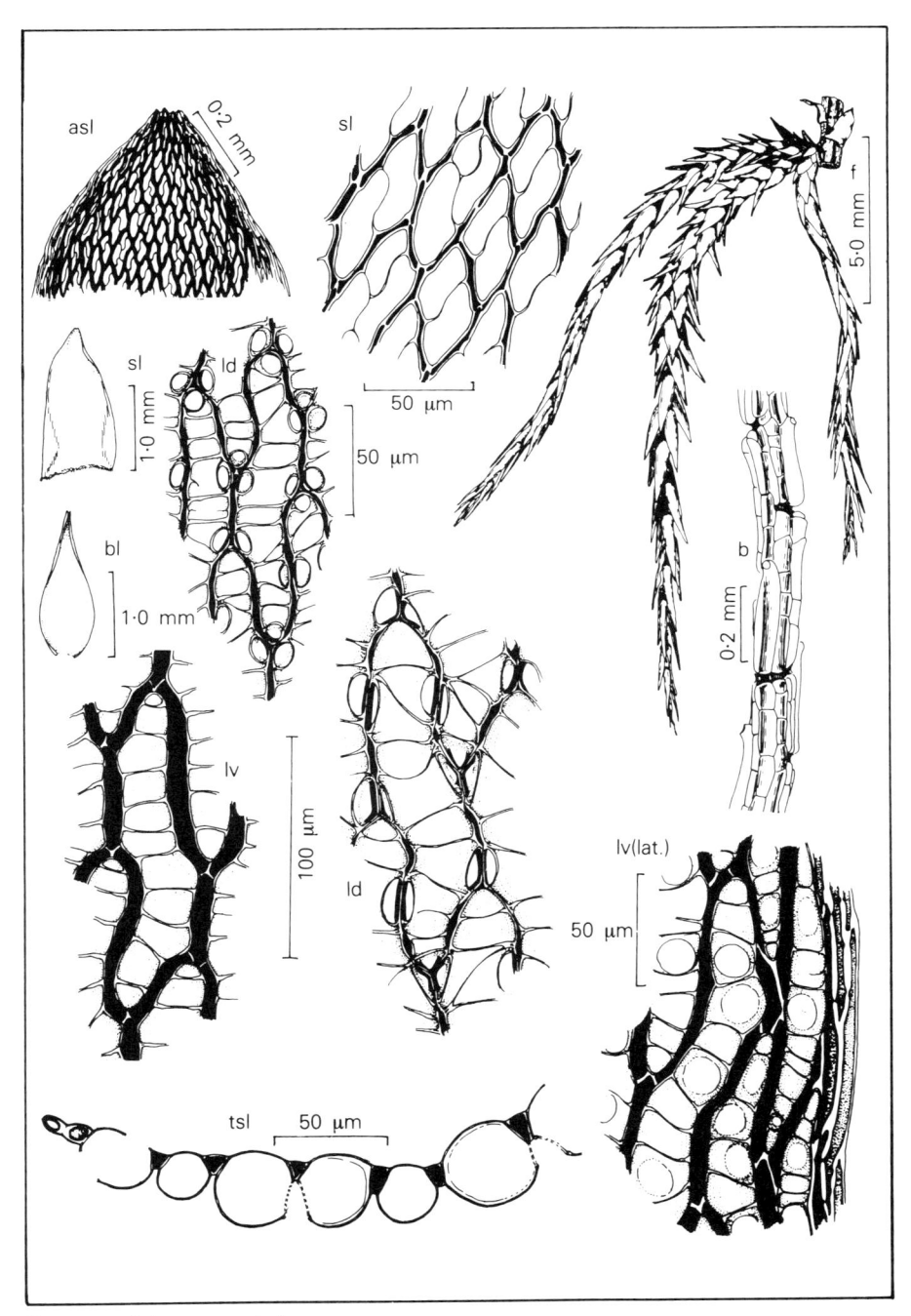

Figure 15. *Sphagnum subnitens*

enclosed on abaxial surface, but often with very narrow exposure there, always much more widely exposed on adaxial surface. **Fertile plants:** Monoecious, though sometimes apparently dioecious (ie only one sex evident in a clone). Antheridial bracts not, or scarcely, distinct from branch leaves. Perichaetial bracts large, the inner up to 3.5 mm long and 1.7 mm wide; apex variable, usually widely obtuse, truncate, or retuse, intact; hyaline cells only weakly differentiated from photosynthetic cells, without fibrils or pores. Capsules common, often abundant; spores yellow-brown, papillose; 26–32 μm diameter.

HABITAT: In a range of oligotrophic to mesotrophic habitats as individual shoots, small cushions or more extensive mats. In common with a number of species, its distribution in relation to chemical gradients varies with geographical location. Consequently, it is found as an ombrotrophic species on wet, oceanic raised and blanket bogs of north-west Britain and western Norway, but further east it forms small, rather oligotrophic tussocks within areas of more mesotrophic mire. It may also occur on wet, rocky banks, on peaty slopes or in damp woods, where it is able to tolerate moderate levels of shade. It is an intermediate species in relation to water level, occurring most commonly slightly above the level of standing water.

DISTRIBUTION: Widespread, with a warm, oceanic tendency, extending from Scandinavia to the Azores, eastern and western North America, along the Andes to Chile, and in the Pacific coastal area of Asia. Increasingly rare towards the north and east of Europe, it is a common plant in the west. It is mainly a lowland species, but reaches 1900 m in southern Europe. Abundant in north-west Britain, but more localized in the south and east.

This is an extremely variable species and many forms occur which are difficult to identify in the field with complete confidence. However, typical plants are usually short and rather 'scruffy', having muted colours, fascicles which hide the stem, and branches that spread in various directions: frequently the capitulum appears to be disproportionately large. Plants are generally greenish, brownish or yellowish, variegated with pink or reddish purple, and often appear greasy when wet. Although some red colouring is normally present (and may be quite conspicuous in some plants), it is never as intense as in other members of the section Acutifolia and is often confined to the internal stem and branch tissues. In dried specimens, the branch leaves usually show a marked irridescence (best seen with a hand lens), which is particularly apparent in the lower spreading branches. Whilst by no means unique among *Sphagna*, this 'metallic' or 'rainbow' sheen is sufficiently pronounced in *S. subnitens* to be an effective aid to identification. Brown forms may be referred to subspecies *ferrugineum* of Flatberg (1986). Some compact forms from montane districts may lack secondary pigments almost completely and resemble forms of *S. capillifolium*. Such forms can usually be identified by their larger branch leaves and more triangular stem leaves with cuspidate points and hyaline cells lacking fibrils. Confusion may also arise with *S. molle* and *S. angermanicum*, but the combination of the sheen and the stem leaf shape should resolve the identification. The closely related *S. subfulvum* may be more difficult to distinguish because that species may also show a sheen and the stem leaf is of a similar shape, although not as acutely pointed as in *S. subnitens*. However, in *S. subfulvum* the stem is always dark brown.

A species of *Sphagnum* new to the European flora was discovered in the Suardal area of Skye by Flatberg in 1987. Although clearly related to *S. subnitens*, it differed from that species in a number of details, and was described as *S. skyense* Flatberg (Flatberg 1988a). Although too divergent from any known European species of section Acutifolia to be a product of recent speciation, such an isolated occurrence of an endemic *Sphagnum* species in the region of its discovery is highly improbable. On the other hand, the plant has much in common with the current concept of *S. junghuhnianum* Dozy & Molk. (1854), a species whose centre of distribution is in tropical Asia. However, it does extend northwards into Taiwan, Japan and China, and was recently found on the Pacific coast of North America (Crum 1984). *Sphagnum junghuhnianum* differs from *S. subnitens* in the following ways: plants are more robust, with branch leaves up to 2 mm long; branch leaf hyaline cells in mid-leaf are 25–30 µm wide (usually 20–25 µm in *S. subnitens*); although stem leaves are variable, they may be longer than those of *S. subnitens* (up to 1.8 mm), are mainly triangular-lingulate and are often fibrillose in the upper part, seldom devoid of fibrils (rarely fibrillose in *S. subnitens*); branch leaves are often distinctly 5-ranked (never 5-ranked in *S. subnitens*). *Sphagnum molle* differs from *S. junghuhnianum* in its larger stem leaves, which are typically wider above than at insertion, and in the presence of a resorption furrow along the branch leaf border.

Climatic conditions of the island off the west coast of Scotland have much in common with those of the coastal regions of western North America and parts of Japan, and a number of bryophytes with disjunct distributions are found in this type of region. In such a context, the occurrence of *S. junghuhnianum*, a species with strong oceanic preferences, in the western islands of Scotland is credible, adding one more species with wide distributional disjunctions to an already extensive list. Although at present known from only the single locality, the species may be present elsewhere along the Atlantic seaboard of Europe, and it should be looked for, particularly, in frost-free, wooded valleys on the western coasts of Scotland and Scandinavia.

7. *SPHAGNUM ANGERMANICUM*

Sphagnum angermanicum Melin (*Svensk bot. Tidskr.*, **13,** 21. 1919)

PLANTS: Medium-sized to fairly robust, rather flaccid (somewhat resembling *S. auriculatum*, *S. teres* or taller forms of *S. molle*); capitula rather small with *projecting terminal stem buds*; pale, white-green or yellow-green with or without pink or purplish flecks (coloration develops more strongly during the growing season and may be absent or indistinct at the beginning). **Fascicles:** Crowded to rather distant; of 3, occasionally 4, more or less irregularly orientated, *not, or slightly, dimorphic branches* which are variable in length, up to about 18 mm, and tapered distally. **Stem:** 0.5–0.7 mm diameter; cortex of 2–3 layers of large hyaline cells, the outer layer without pores; internal cylinder thin, of a few layers of cells with moderately thickened walls, pale, yellowish or faintly brownish. **Branch anatomy:** Retort cells distinct, *mainly in linear pairs,* occasionally in threes; other cortical cells long and narrow; internal cylinder pale brown. **Stem leaves:** Erect or hanging; *large,* up to 2.5 mm or more long and about 1.5 mm wide in mid-leaf; broadly ovate to ovate-lingulate, widest at or above the middle ($1\frac{1}{2}$ times as wide as at insertion); apex more or less flat, rather widely rounded-truncate, eroded and finely fimbriate across the truncate portion; border narrow throughout. Hyaline cells elongated-rhomboid; strongly fibrillose in at least the upper half of the leaf (the upper half of the leaf anatomically similar to branch leaves); adaxial face with large resorption gaps below and large resorption pores above; abaxial surface with numerous large pores and, usually, a resorption gap in the apical angle; septa usually present, especially in the lower part of the leaf. **Branch leaves:** Suberect, laxly imbricate or contiguous; *large* (for the section), 1.7–2.5 mm long; *ovate-lanceolate to more or less lingulate,* tapering slightly to a *very wide, shallowly concave to almost flat, multidentate, apex*; border narrow, 1–2 cells wide, without resorption furrow. **Hyaline cells:** Large and almost uniform throughout the leaf. Abaxial surface strongly inflated; pores numerous, scattered, large and ringed, lying along the commissures; irregular resorption gap frequently present in apical angle. Adaxial surface often lacking pores except in 1–3 marginal rows of cells, sometimes with a single large pore near the apical angle. **Leaf TS:** Hyaline cells convex on adaxial face, very strongly so on the abaxial. Photosynthetic cells triangular to trapezoid, thin-walled; widely exposed on the adaxial leaf surface, not, or hardly, exposed on the abaxial. **Fertile plants:** None were examined.

HABITAT: Forming low hummocks or mats in slightly mesotrophic fens, particularly around the margins of pools or along the edges of streams or runnels. It may occur in sparse, wet woodland with light shade, or in more open peatland communities where it is usually associated with other *Sphagnum* species, eg *S. subnitens, S. papillosum, S. pulchrum, S. capillifolium* var. *capillifolium, S. tenellum* or *S. subfulvum*, or in slight hollows with *S. majus*.

DISTRIBUTION: The main centre of distribution of this species is in eastern North America (the maritime provinces of Canada and the north-eastern USA. In Europe, it is confined to southern Norway and central Sweden.

S. angermanicum can be mistaken in the field for a species from some other section, its subdued colour and relatively broad leaves giving a superficial resemblance to, eg, *S. auriculatum*. Under the microscope, however, it shows the characteristic '*acutifolium*' leaf anatomy, and even under a ×10 lens, the more delicate nature of the leaf tissue can be determined, with experience. The similarity in shape of branch and stem leaves (isophylly) might suggest a juvenile or etiolated form of some other species, eg *S. subnitens*, but in plants with fully developed branch fascicles the differences are always more distinct. The absence of a resorption furrow is a useful means of distinguishing *S. angermanicum* from lax forms of *S. molle,* and neither *S. molle* nor *S. subnitens* have a projecting stem bud in the capitulum.

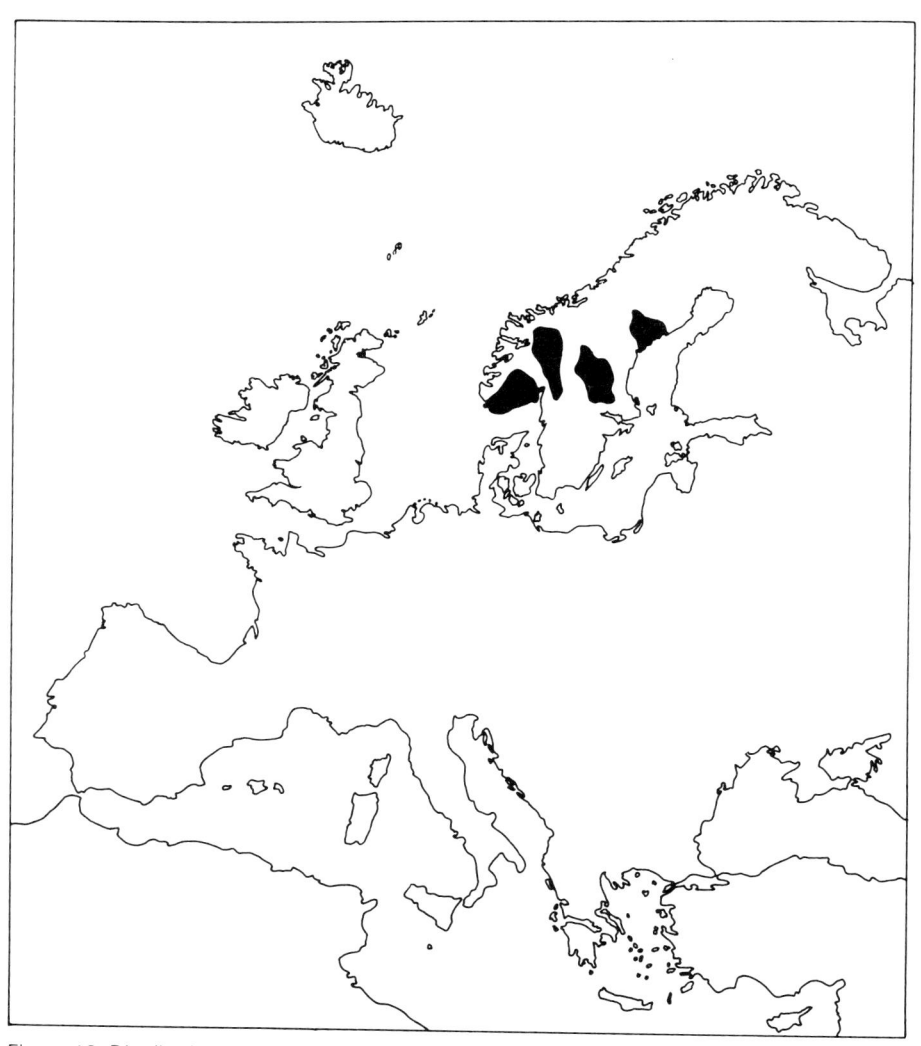

Figure 16. Distribution of *S. angermanicum*

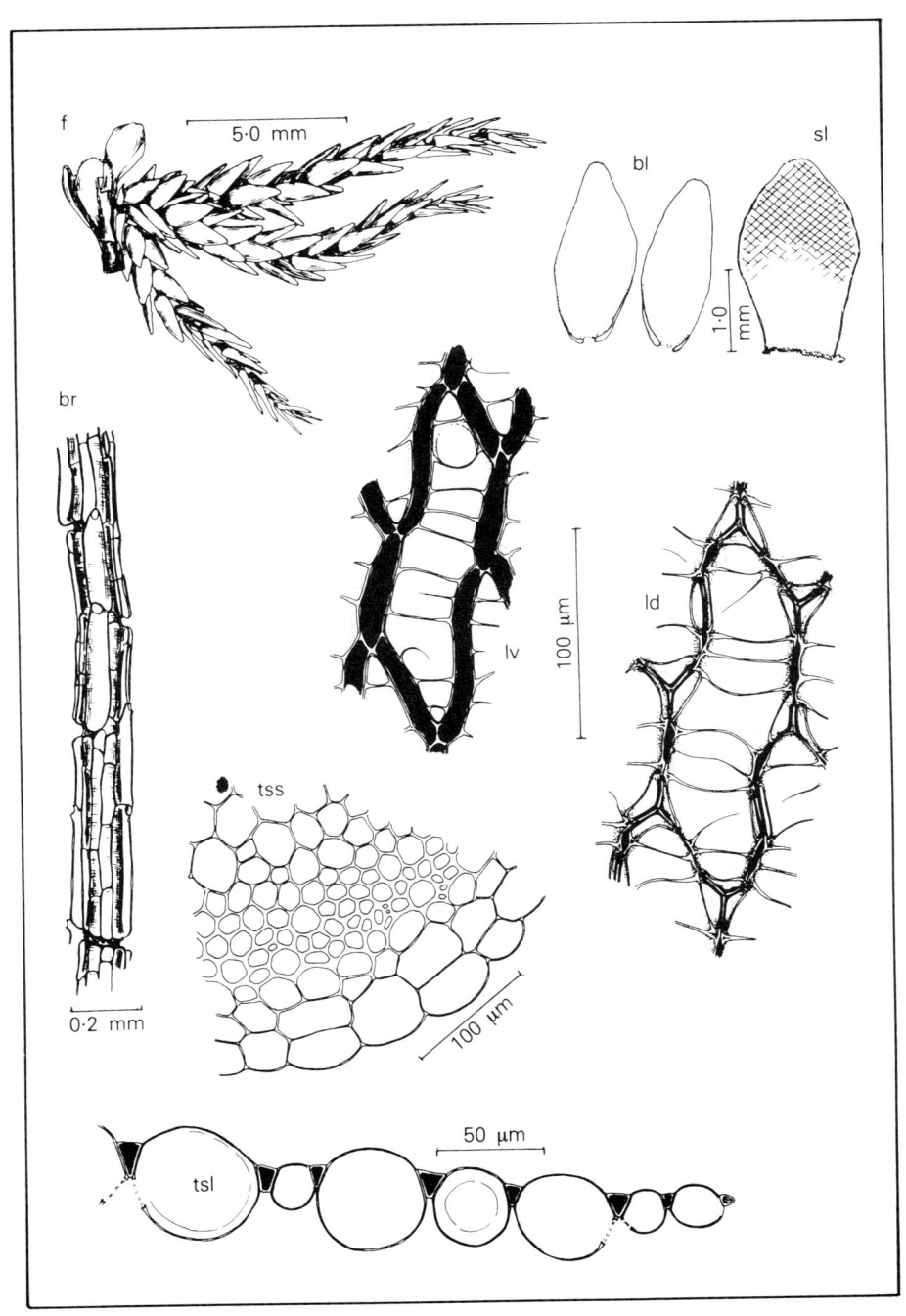

Figure 17. *Sphagnum angermanicum*

8. SPHAGNUM SUBFULVUM

Sphagnum subfulvum Sjörs (*Svensk bot. Tidskr.*, **38,** 404. 1944)

PLANTS: Medium-sized to rather small, rarely robust; capitula rather large; yellowish brown to brown, *never with red or purple*. **Fascicles:** Rather crowded to slightly distant; of 3(–4) dimorphic branches; spreading branches 2, fairly long, tapering; pendent branches 1(–2), shorter or longer than spreading. **Stem:** 0.6–0.9 mm diameter; cortex of 3(–4) layers of hyaline cells without pores in the outer walls; internal cylinder well developed, often with more than 7 series of small cells with strongly thickened walls, dark brown to almost black, never red. **Branch anatomy:** Retort cells distinct, one per leaf axil (similar to those in *S. subnitens* but with *longer necks*); internal cylinder brown, never red. **Stem leaves:** Lingulate to lingulate-triangular; apex broadly rounded or somewhat square and truncate; border of 2–5 rows of cells, *widely expanded below* (more so than in *S. subnitens*) into patches of undifferentiated tissue. Hyaline cells without, or with few, weak, fibrils, but mostly septate, at least in upper half of leaf; abaxial surface intact; adaxial surface partially or wholly resorbed. **Branch leaves:** Suberect to lightly imbricated; not 5-ranked, rarely obscurely 5-ranked in part; fairly large (sometimes as long as those in *S. subnitens,* but usually rather shorter); ovate with the upper margins inrolled; narrowed above half-way; apex truncate and widely acuminate due to inrolled margins (apex broader and acumen less acute than in *S. subnitens*); border 1–2 cells wide without resorption furrow. Pendent branch leaves ovate to ovate-lanceolate and similar to those of the spreading branches. **Hyaline cells:** Varying in size, and often several times larger in the basal part of the leaf than in the apical part; abaxial face strongly inflated. Adaxial surface with few, or no, pores, except in the lower marginal areas of the leaf, where cells have large, circular pores on both faces. Abaxial surface with numerous circular to more or less elliptical pores, mainly along the commissures and with a large resorption gap in the apical angle. **Leaf TS:** Hyaline cells highly inflated on abaxial surface, almost plane or shallowly convex on the adaxial. Photosynthetic cells triangular to trapezoid with rather thin, usually brown, walls; widely exposed on adaxial surface and narrowly exposed on the abaxial. **Fertile plants:** Monoecious. Antheridial branches resembling sterile branches. Perichaetial bracts large (similar to those in *S. subnitens*); apices variable, widely obtuse, truncate or retuse, intact; hyaline cells only differentiated in upper part of bract, without fibrils or pores. Capsules frequent; spores brown, very coarsely granulate, 28–29 μm diameter.

HABITAT: Forming loose mats or low hummocks in mesotrophic to moderately eutrophic mires, usually in *Carex* communities with, eg *C.*

lasiocarpa or *C. rostrata*, and most frequently along pool margins. It is moderately shade-tolerant and in the north of Fennoscandia may grow under *Ledum* in *Betula* woodland, together with *S. warnstorfii, S. fuscum* and *S. papillosum*. It tends to replace the closely related *S. subnitens* in northern mesotrophic mires.

DISTRIBUTION: A mainly sub-arctic to northern boreal species in Europe, north-eastern North America and north-east Asia. In Europe, it is largely confined to Fennoscandia and the Baltic coast of Russia, although it is also recorded from an isolated station in Switzerland. Absent from Great Britain, though formerly recorded in error from north Wales. It has recently been recorded from Ireland.

Sphagnum subfulvum is closely related to *S. subnitens*, and many records, including an old one from Britain, are based on mis-identifications of the latter species. The brown colour, at times as deep as that in *S. fuscum*, is a positive feature by which *S. subfulvum* is most readily recognized in the field. The anatomical differences between this species and *S. subnitens*, although slight, are sufficiently consistent to enable confirmation of normal plants. The metallic irridescence seen in *S. subnitens* is equally, if not more, conspicuous in *S. subfulvum*, so that herbarium specimens of the latter may appear distinctly pink to purplish. This irridescence disappears on soaking. Small forms could be confused in the field with *S. fuscum*, but that species lacks large resorption pores in the branch leaves.

A similar, apparently exclusively North American, taxon, *S. flavicomans* (Card.) Warnst., differs from *S. subfulvum* in a few minor details, such as: generally larger dimensions, more flaccid stem leaves which often have fibrils in the upper half, and more strongly developed pendent branches. It occurs only near the east coast of North America and may possibly be best regarded as a geographical subspecies of *S. subfulvum*. Several authors have shown how these species can be separated systematically from *S. subnitens*, but there seems to be no publication which shows how they can safely be distinguished from each other: see, eg, Osvald (1940), Sjörs (1944), Nyholm (1969), Crum (1976).

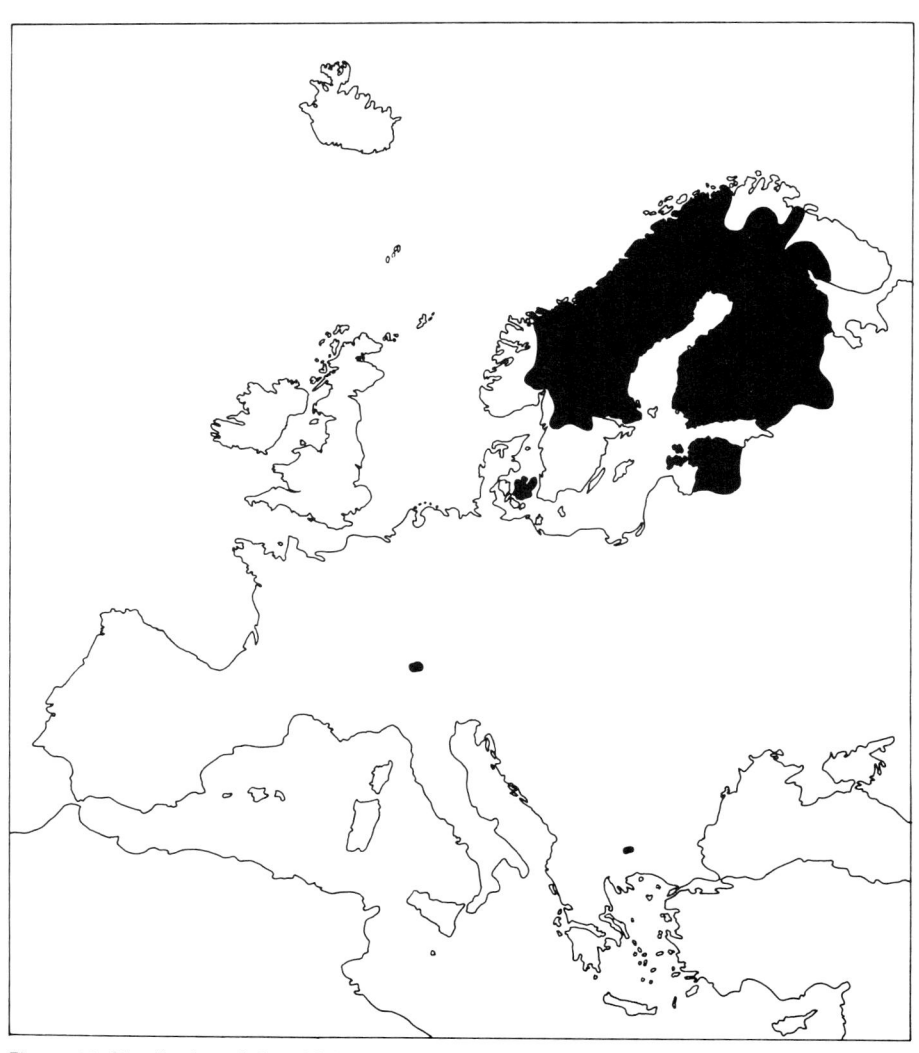

Figure 18. Distribution of S. *subfulvum*

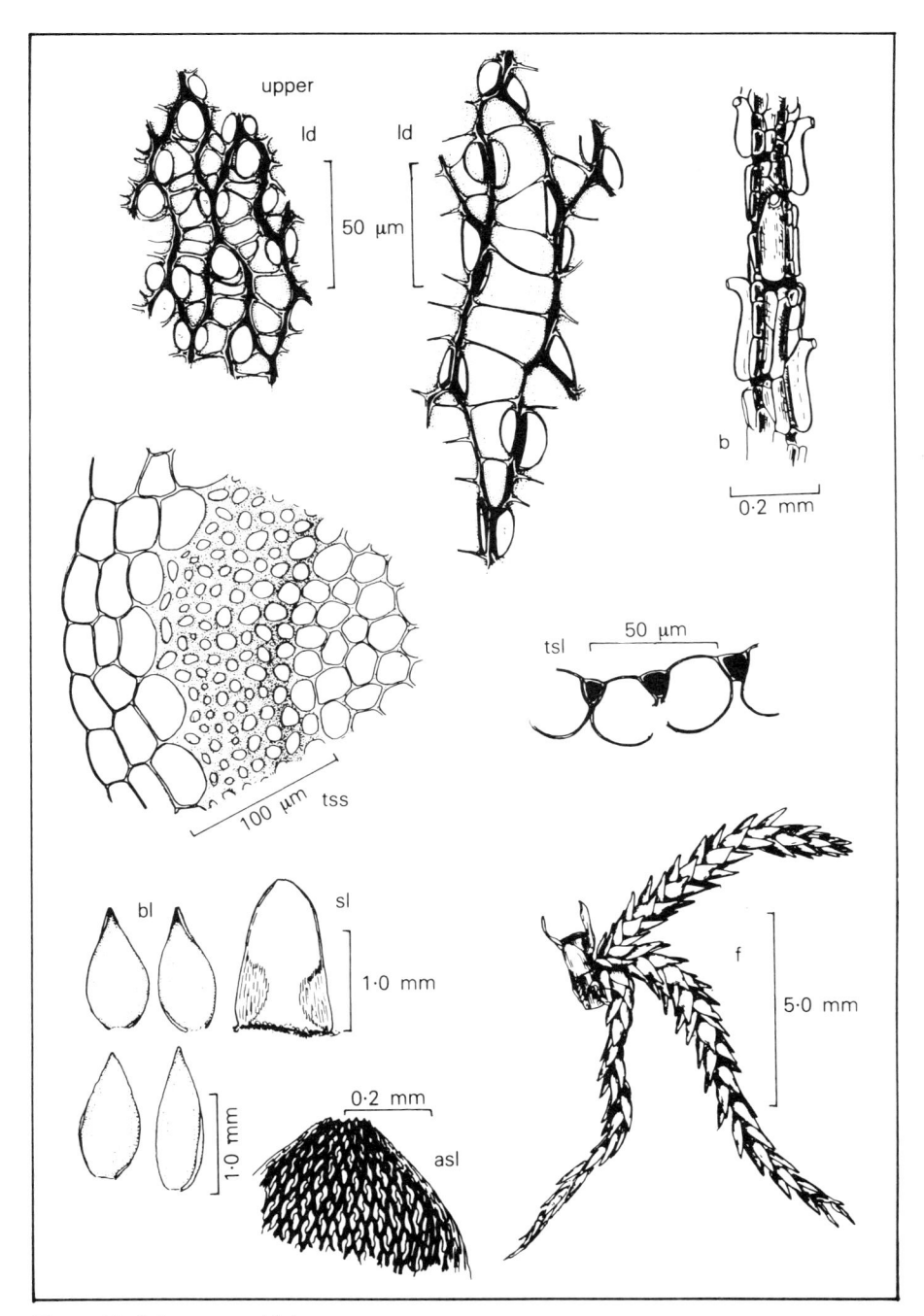

Figure 19. *Sphagnum subfulvum*

9. SPHAGNUM FUSCUM

Sphagnum fuscum (Schimp.) Klinggr. (*Schr. phys.-ökon. Ges. Königsb.,* **13,** 4. 1872)
S. acutifolium var. *fuscum* Schimp. (*Mém. prés. div. Sav. Acad. Sci. Inst. Fr.,* **15,** 64. 1857)

PLANTS: Small, normally compact; capitula small, more or less flat (plants resembling compact forms of *S. capillifolium* var. *capillifolium* except in colour); *mid- to deep brown, never with any trace of red,* very rarely all green. **Fascicles:** Moderately to densely crowded, rarely distant; of 3, very occasionally 4, dimorphic branches: spreading branches 2, short, less than 7.0 mm long (sometimes distally attenuated and up to 10.0 mm) shortly tapered; pendent branches 1(–2), long tapering, usually somewhat longer than the spreading branches but rarely exceeding 18.00 mm; pale and terete. **Stem:** Rather thin, 0.4–0.6 mm diameter; cortex well developed, of 3–4 layers of large hyaline cells without pores on the external surface; internal cylinder well developed, *dark brown to almost black* in surface view (except in etiolated plants). **Branch anatomy:** Retort cells distinct and conspicuous with well-developed rostra, solitary; internal cylinder brown. **Stem leaves:** Erect and appressed to stem; about 1.3 × 0.7 mm; *lingulate to slightly spatulate,* thin textured; *apex broadly rounded with rather wide resorption area across the tip* (cf *S. russowii*); border strong, expanded below to occupy one third to one half or more of leaf base. Hyaline cells *without fibrils,* septa numerous, occurring in nearly every cell. **Branch leaves:** Densely imbricated, especially in the lower half of the branch; not 5-ranked; small, ca 1.1–1.3 mm long; ovate to ovate-lanceolate; border 2 cells wide, without resorption furrow. **Hyaline cells:** Variable in size, ca 90 × 15 µm, near apex, becoming larger downwards to ca 130 × 25 µm. Adaxial surface more or less plane; pores absent, except near leaf margins (there occurring singly and rather large, 10.0–20.0 µm). Abaxial surface strongly convex; pores in upper cells small to medium-sized (5.0–10.0 µm diameter), more or less circular, strongly ringed, mainly in cell angles; in lower cells, pores more uniform, larger (to 20.0 µm or more), appearing half elliptical due to wall convexity; near leaf apex an additional large, ringed or unringed pore is present in the apical angle of hyaline cells. **Leaf TS:** Hyaline cells more or less plane on adaxial face, strongly inflated on the abaxial. Photosynthetic cells triangular, occasionally trapezoid, with rather thin, usually brown, cell walls; widely exposed on the adaxial surface, narrowly exposed on the abaxial. **Fertile plants:** Dioecious. Male plants apparently rare, at least in Britain. Antheridial branches dark brown. Female plants apparently frequent; perichaetial bracts large, the inner up to 4.0 mm long, ca 2.2 mm wide, convolute; apex more or less retuse and intact;

hyaline cells differentiated almost, or quite, to base, without pores or fibrils. Capsules rare; spores yellow-brown, distinctly to strongly papillose, 23–27 μm diameter.

HABITAT: Usually an ombrotrophic species, forming dense, usually low and wide but occasionally higher hummocks in open oligotrophic and mesotrophic mires, in the latter giving rise to oligotrophic 'islands'. Although this species is able to tolerate some shading and may be found in wooded parts of mires, it is more usually found in open communities. It also occurs in upland flushes. In Britain, it shows a slight oceanic tendency, being found mainly on north-western raised and blanket bogs, whereas in Eurasia as a whole it is rather northern and continental: it is most abundant in west Siberia where it is the major peat former over a very extensive area. Although occurring as pure stands in hummocks, *S. fuscum* has a high degree of association with other species of *Sphagnum*, eg *S. russowii*, *S. angustifolium*, *S. capillifolium* var. *capillifolium*, *S. papillosum*, *S. magellanicum* and *S. subnitens* in open mires, whilst on upland slopes over largely mineral ground it may also be found with *S. compactum*.

DISTRIBUTION: Circumboreal with distinctly continental and arctic-alpine tendencies. Common in Fennoscandia, but decreasing in abundance further southwards and becoming more restricted to sub-alpine or alpine areas. In Britain, it is rare and local, mainly in upland areas in the north, though it may be locally frequent in mid-altitude bogs of northern England and Scotland.

Sphagnum fuscum appears to be most closely related to the rare *S. subfulvum*, although it superficially resembles *S. capillifolium* var. *capillifolium*. In the field, it may be recognized by the characteristic brown coloration, sometimes more conspicuous in the interior of hummocks, and the total lack of red pigmentation. Where visible, the broad apices of the stem leaves (similar in shape to those of *S. russowii*) provide useful confirmation, especially in those forms where the brown colour is less distinct or absent. *S. subfulvum* is a more robust plant with larger leaves, which is distinguished microscopically by its larger pores and greater number of hyaline cells having large apical resorption gaps on their abaxial faces. Dried specimens of *S. fuscum* appear distinctly matt, lacking the conspicuous irridescence of *S. subfulvum*. Compact, dull forms of *S. subnitens*, which sometimes have little or no red in them, can be separated by the shape of their stem leaves, more or less triangular and narrow-pointed, not lingulate and rounded above.

In culture, *S. fuscum* appears to dislike temperatures sustained above 25°C, especially when waterlogged, and is much less tolerant of such conditions than either *S. capillifolium* or *S. subnitens*. This may offer a partial explanation of its pattern of distribution.

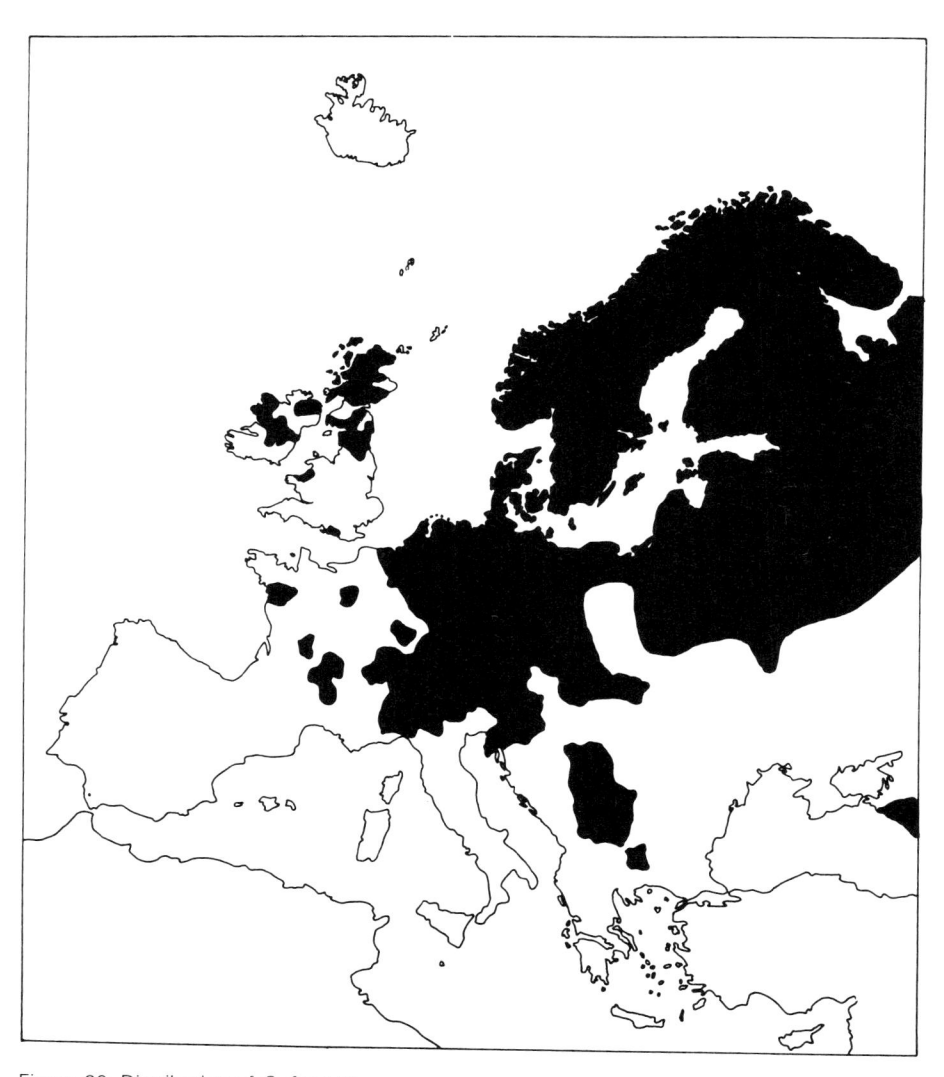

Figure 20. Distribution of *S. fuscum*

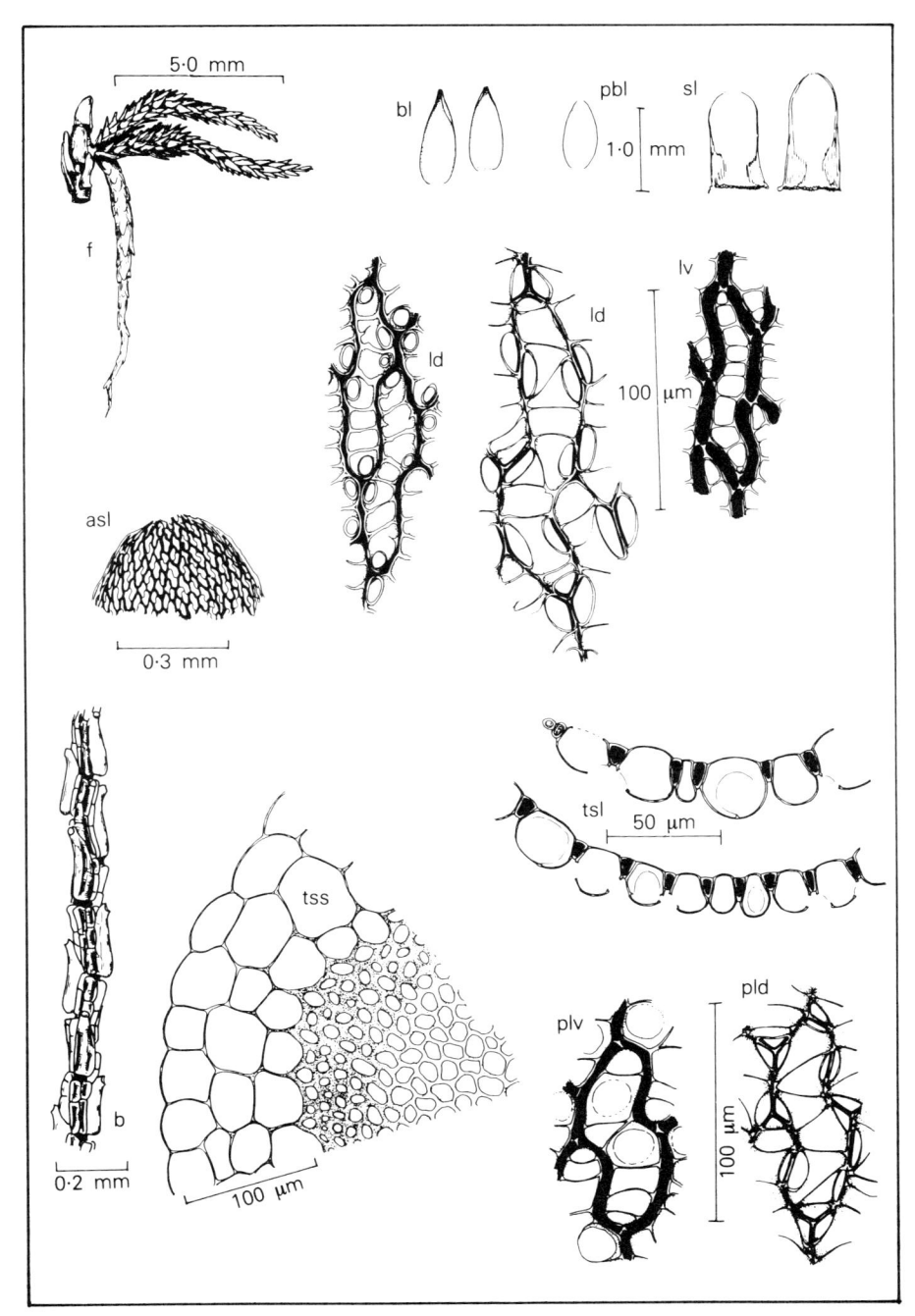

Figure 21. *Sphagnum fuscum*

10. SPHAGNUM QUINQUEFARIUM

Sphagnum quinquefarium (Braithw.) Warnst. (*Hedwigia,* **25,** 222. 1886)
S. acutifolium var. *quinquefarium* Braithw. (*Sphag. Eur. N. Am.,* 71. 1880)

PLANTS: Medium-sized to rather delicate, tall and stiff, rarely dense; pale green to yellow-green with zones or flecks of pink or pale red, rarely the whole plant red. **Fascicles:** More or less evenly and distantly spaced, not concealing stem; of 4–5 dimorphic branches; *spreading branches 3,* 15.0–25.0 mm long, tapering distally (except near capitulum); pendent branches 1–2, as long as, or longer than, the spreading, terete. **Stem:** Slender, 0.5–0.8 μm diameter; cortex of 3–4 layers of hyaline cells; outer cortical cells frequently with a single unringed, usually inconspicuous, pore or thinning (pores often very few, but sometimes present in more than 50% of cells: often undetectable on intact stems, but usually visible on stained strips of the easily detached cortex); internal cylinder well developed, yellowish to yellow-brown, with or without zones of red. **Branch anatomy:** Retort cells distinct; mostly solitary, but frequently with one or 2 cells above the retort cells swollen and perforated; internal cylinder pale green to pale brown or violet-red. **Stem leaves:** Erect; 1.1–1.3 mm long; triangular, triangular-ovate or triangular-lingulate, sharply tapered above; apex narrowly rounded, sometimes acute due to inrolled margins; border strong, expanded below. Hyaline cells intact on abaxial face, usually fibrillose in the upper parts of the leaf, often very weakly so, fibrils rarely absent; adaxial face resorbed. **Branch leaves:** *Strictly 5-ranked*, at least in the upper branches and the lower parts of the lower branches, closely set with an angle of attachment of about 40° so that branches have a prismatic appearance; 1.3–1.5 × ca 0.5 mm: lanceolate (proportionately narrower than, eg, *S. capillifolium*); apex narrow due to inrolled margins; border without resorption furrow. **Hyaline cells:** Up to about 160 × 25–35 μm below, smaller above. Adaxial pores more or less confined to 2–4 series of marginal cells, with one or a few large, round pores. Abaxial surface with several, up to 8, medium to large (11.0–19.0 μm diameter) ringed pores in the cell angles and along the commissures (pores more or less circular, but often appearing narrow due to convexity of cells). **Leaf TS:** Hyaline cells shallowly convex to almost plane on the adaxial surface, strongly convex on the abaxial. Photosynthetic cells triangular to slightly trapezoid, thin-walled with angular lumina; widely exposed on the adaxial leaf surface, narrowly to very narrowly exposed on the abaxial. **Fertile plants:** Monoecious, but plants often found in a unisexual condition. Antheridial bracts red or pink. Perichaetial bracts large (up to 4.8 × 2.1 mm) and sheathing; apex truncate or retuse; hyaline cells differentiated in upper third of bract, intact, without fibrils. Capsules locally frequent; spores slightly roughened, 21–24 μm diameter.

HABITAT: A species found on damp hillsides under dwarf shrubs or in open, usually *Betula*-dominated, woodland. Unlike most *Sphagnum* species, it is virtually absent from peatland, though it is common in areas of high humidity, and is most abundant in the oceanic or sub-oceanic areas of north-west Europe. In the more continental areas, it is confined to upland locations. Frequently, it is found mixed with *S. capillifolium* var. *capillifolium*, *S. palustre* and associated with *Calluna vulgaris*, *Betula* or *Vaccinium* species.

DISTRIBUTION: An oceanic species widely distributed in north and west Europe, eastern Asia and the Atlantic and Pacific coastal areas of North America. It is present throughout most of Europe and is most common in southern Scandinavia, though absent from the northern parts: in montane areas it reaches Italy, Yugoslavia and Romania, and the Pyrenees. It is absent from south-eastern Britain and local in northern England, Scotland, Wales and Ireland.

This species may be recognized in the field by its rather short and stiff branches which appear distinctly angular because of the arrangement of their leaves, closely set, but widely angled in 5 ranks. The fascicles themselves normally have 3 spreading branches, unlike other members of the section Acutifolia which have only 2, except sometimes near a stem bifurcation. Some green plants may be confused in the field with members of the section Cuspidata, eg *S. recurvum*, but the triangular, narrowly pointed stem leaves of *S. quinquefarium* are larger and held erect rather than reflexed (the more normal condition in members of the Cuspidata). Microscopically, the wider exposure of the photosynthetic cells on the adaxial surface will confirm the identity of *S. quinquefarium*.

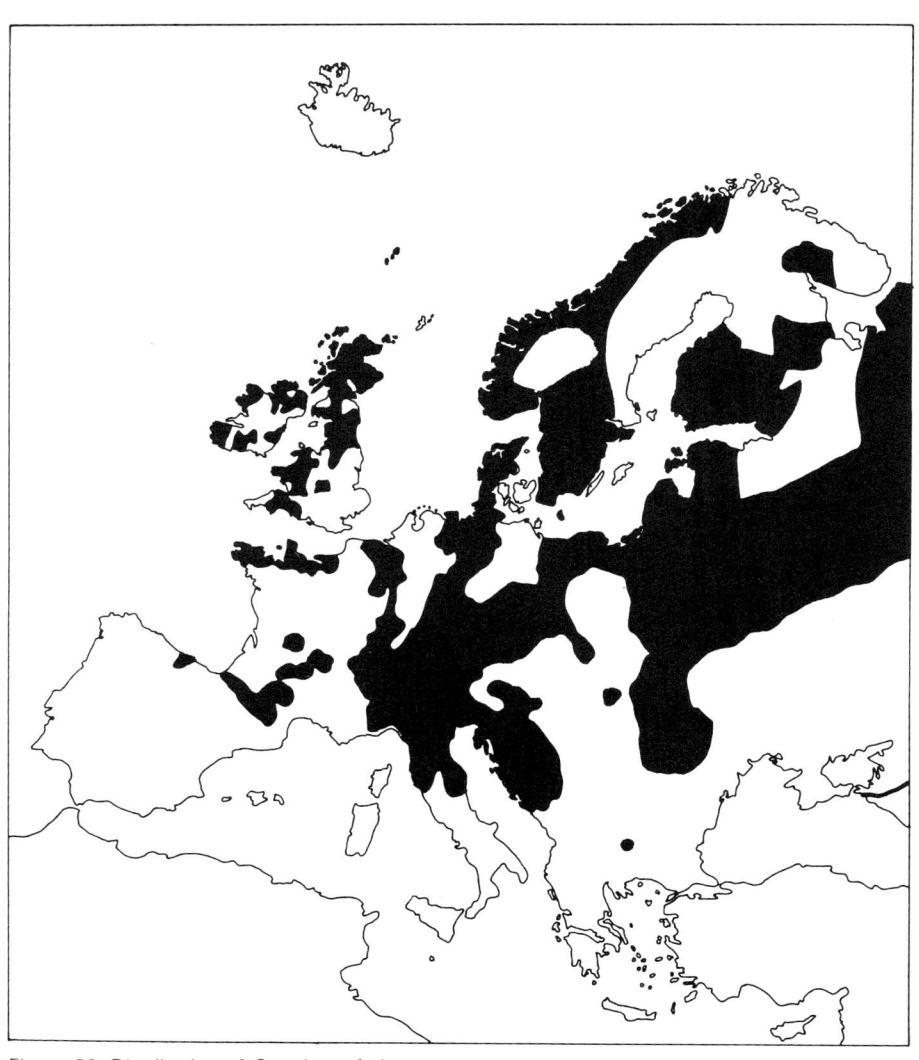

Figure 22. Distribution of S. quinquefarium

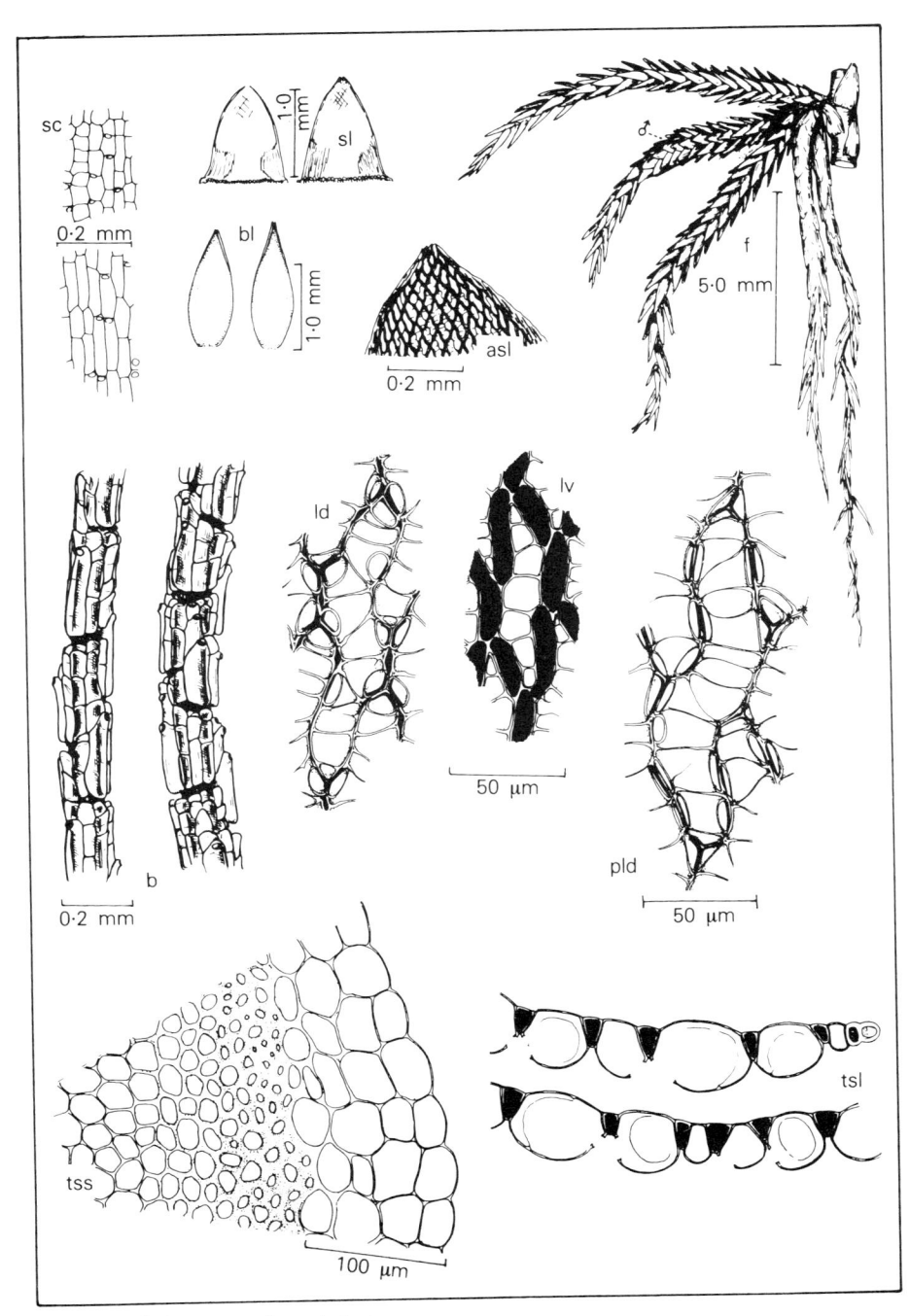

Figure 23. *Sphagnum quinquefarium*

11. SPHAGNUM CAPILLIFOLIUM

Sphagnum capillifolium (Ehrh.) Hedw. (*Fund. musc.*, 86. 1782)
?*S. nemoreum* Scop. (*Flora carniol.* 2nd ed., 305. 1772)
S. capillaceum (Weiss) Schrank (*Baierische flora*, 435. 1789)
S. acutifolium Ehrh. ex Schrad. (*Syst. Samml. krypt. Gew.*, **1,** 8. 1796)

PLANTS: Rather delicate, but variable in height and compactness; capitula well developed, pale or with small to extended patches of red in its outer layers. **Branch anatomy:** Retort cells distinct with well-developed rostra, confined to the internal tissues of branches. **Fascicles:** Of 3–4 quite strongly dimorphic branches; spreading branches 2, the upper ones rather short (5.0–9.0 mm) and ending abruptly, the lower ones much longer and distally attenuated. (The tapering distal portion of branches in this and many other species is etiolated and colourless, and is probably a response to deprivation of light following continuous apical stem growth and resultant immersion of the older branches). Pendent branches quite long, 15.0 mm or more, thin and pale, more or less appressed to the stem. **Stem:** Up to 0.7 mm diameter; cortex of 3–4 layers of hyaline cells without pores (diligent search may reveal an occasional pore in some specimens); internal cylinder well developed, pale or with small to extended patches of red in its outer layers. **Branch anatomy:** Retort cells distinct with well-developed rostra, solitary; other cortical cells about half as wide, variable in length, occasionally with a large pore, but never rostrate; internal cylinder pale green or reddish. **Stem leaves:** Erect and more or less appressed, sometimes spreading; lingulate to lingulate-triangular; apex broadly rounded-truncate with a rather narrow area of apical resorption, occasionally slightly inrolled; border strong, 2 or several cells wide above, widely expanded below to occupy one quarter to three-quarters of leaf base. Hyaline cells *weakly to strongly fibrillose in the upper half of leaf,* at least on the abaxial face. **Branch leaves:** 5-ranked or not; erect to erect-spreading, never squarrose; small, rarely exceeding 1.5 mm long, usually less than 1.4 mm; lanceolate to ovate-lanceolate; border 2 cells wide, without resorption furrow. Pendent branch leaves longer and narrower, more delicate, often with more frequent adaxial pores. **Hyaline cells:** Near apex 100 × 14–18 μm, much larger below, ca 160 × 25 μm. Abaxial surface with numerous, up to 8 per cell, medium to large (10–20 μm in mid-leaf) ringed pores in the cell angles and along the commissures, those of adjacent cells often occurring as opposed pairs or triplets. Adaxial surface, except near leaf margins, rarely with a large circular pore; 2–3 marginal series with one or a few large pores. **Leaf TS:** Hyaline cells shallowly convex to almost plane on the adaxial face, strongly inflated on the abaxial (making the more or less circular abaxial pores appear narrowly half-elliptical in surface view). Photosynthetic cells triangular to trapezoid, rather thin-walled with angular lumina; widely exposed on adaxial surface, narrowly to very narrowly exposed on the abaxial (never enclosed abaxially, but sometimes apparently so because of bulging adjacent hyaline cells). **Fertile plants:** Dioecious or, rarely, monoecious. Antheridial bracts crimson, rarely pale, resembling branch leaves. Perichaetial bracts large, up to 4.7 mm long, convolute; apex rounded or retuse, prosenchymatous; lower tissue of uniform, lax, thin-walled cells merging into thicker-walled, pitted cells; above differentiated into hyaline cells without pores or fibrils, rarely fibrillose near apex, and photosynthetic cells. Capsules occasional; spores yellow-brown, roughened, 24–28 μm diameter.

HABITAT: Forming small hummocks or more extensive mats on the drier parts of acid peatlands, on damp heathland, and on hillsides or in woods which are damp with rather acid, organic soils. It is primarily a plant of lightly shaded habitats, although it will grow in more open situations. It may form pure stands or be associated with other *Sphagnum* species, eg *S. papillosum* and *S. subnitens* in particular.

DISTRIBUTION: Circumboreal and widespread throughout Europe, Asia and North America. It is found from sea level to high altitude (eg 2700 m in the Alps) and in Europe extends from northern Scandinavia to Portugal. It is present throughout most of Britain.

S. capillifolium, not surprisingly in such a widespread and abundant taxon, shows considerable variation in colour, habit and minor morphological characters. There is no doubt that some of these variants are genetically determined and could be treated as distinct species, if sufficiently stable discontinuities between them could be recognized. In parts of Europe, such discontinuities are possibly more clearly marked than in Britain, and, consequently, several authors have recognized, for example, *S. rubellum* as a distinct species (eg Nyholm 1969; Isoviita 1966). In Britain, on the other hand, many find it difficult to accept *S. rubellum*, even as a variety. The position is complicated further by environmentally induced variation, as well as inconsistencies in the interpretation of the variants by different authors. A compromise solution is adopted here, in which *S. capillifolium* is treated as a single species made up of 2 variable varieties, var. *capillifolium* and var. *rubellum*.

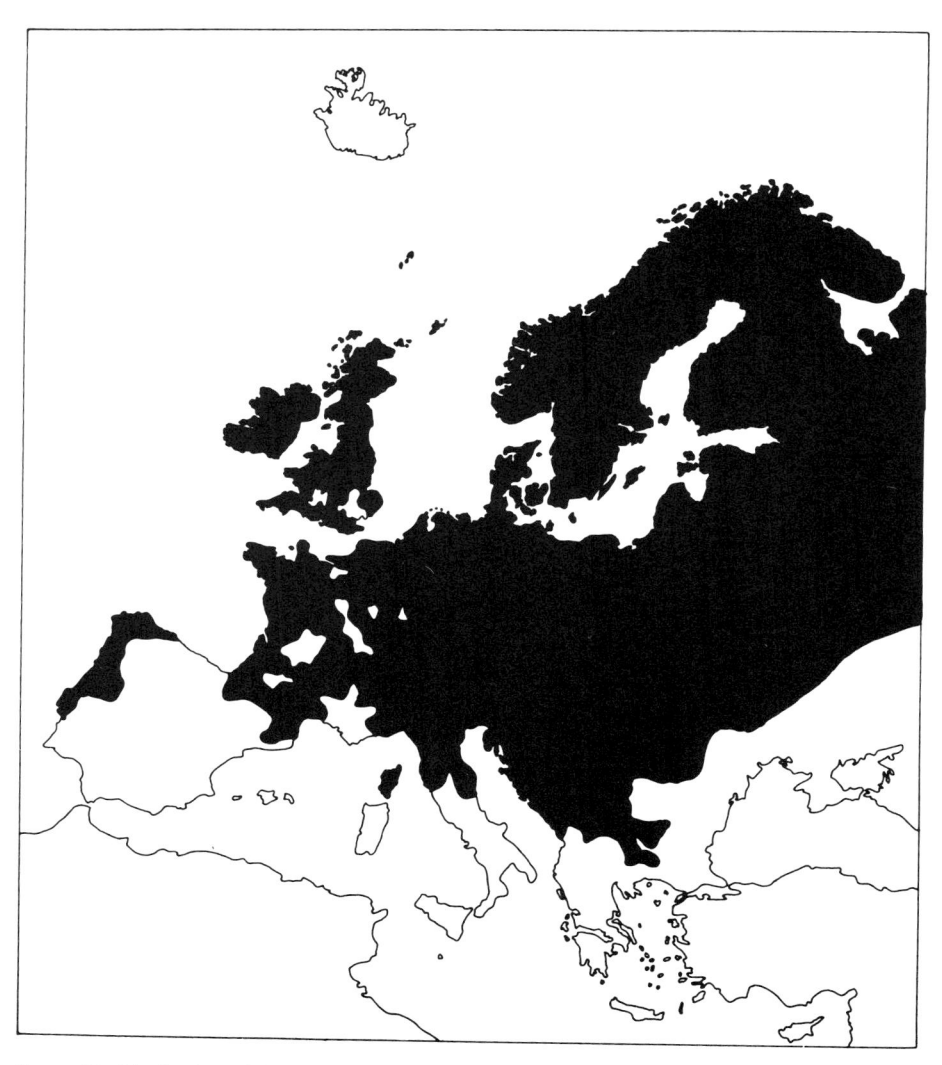

Figure 24. Distribution of S. capillifolium

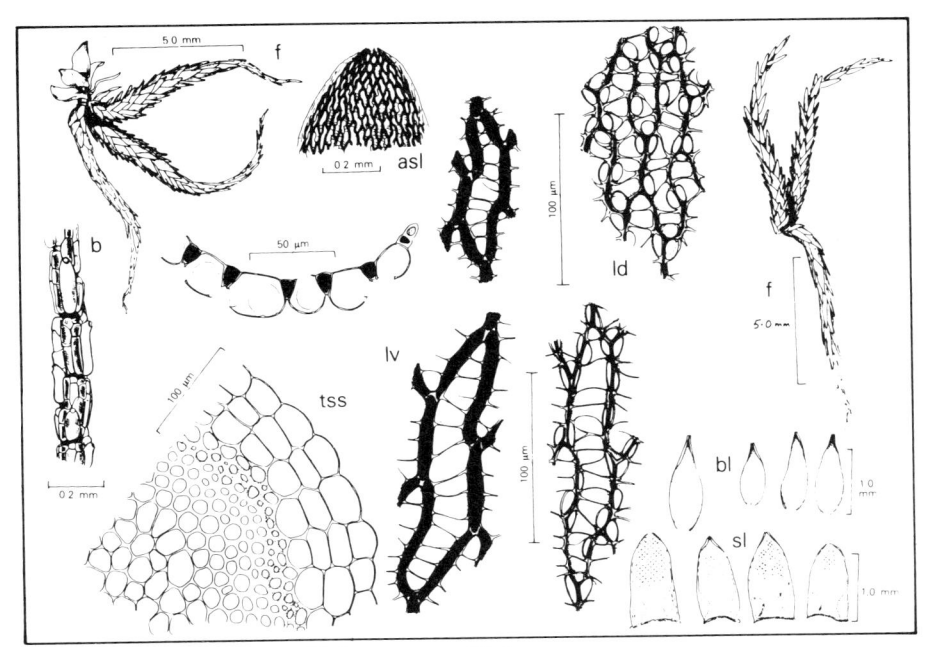

Figure 25i. *Sphagnum capillifolium* var. *capillifolium*

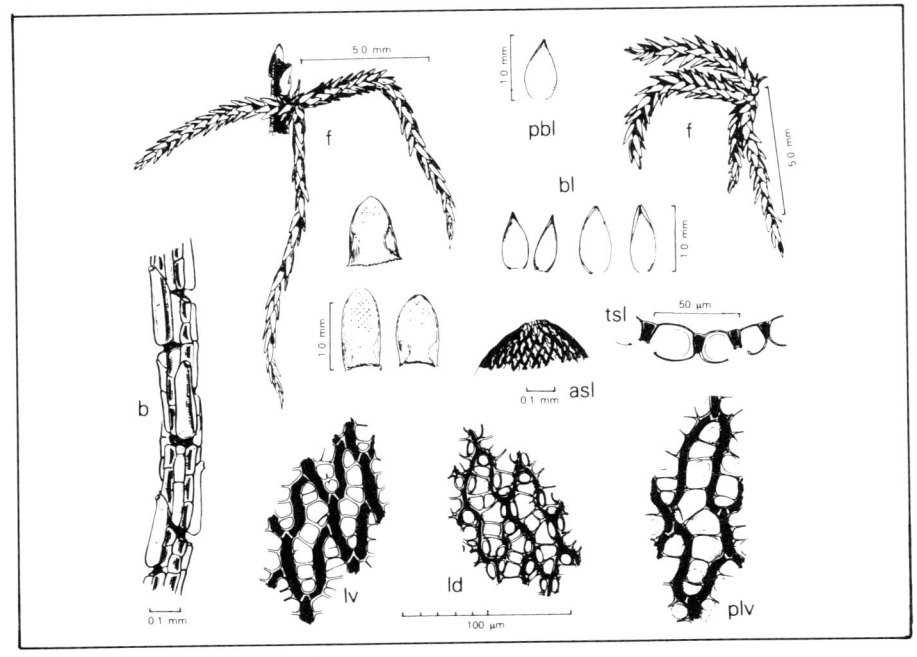

Figure 25ii. *Sphagnum capillifolium* var. *rubellum*

11a. *S. capillifolium* var. *capillifolium*

PLANTS: Lax or, often, *very compact: capitula usually more or less hemispherical;* nearly always with some, often pale, red flecks, or overall red to reddish brown, seldom deep crimson throughout. **Fascicles:** Closely packed and often hiding the stem completely. **Stem leaves:** Lingulate, but sometimes distinctly narrowed above; *fibrillose in the upper half; septa comparatively few.* **Branch leaves:** *Densely imbricated; rarely 5-ranked;* usually narrowed above to a more or less tubular apex (appearing distinctly acute). *Abaxial pores large,* mostly 10.0–15.0 μm in mid-leaf.

HABITAT: Forming wide, loose mats or smaller, more compact hummocks in a wide range of oligotrophic to very weakly mesotrophic habitats from upland blanket bog to open woodland, wet hillsides and the drier parts of lowland valley and basin mires.

DISTRIBUTION: Widespread throughout the northern hemisphere in both the lowlands and uplands, ascending to high altitudes. Common throughout Europe. Common in Britain, more particularly in the north.

11b. *S. capillifolium* var. *rubellum*

Sphagnum capillifolium var. *rubellum* (Wils.) A. Eddy, comb. nov.
S. rubellum Wils. (*Bryol. brit.*, 19. 1855)
?S. subtile (Russ.) Warnst. (*Krypt. Brandenburg,* 409. 1903)

PLANTS: *Usually rather lax; capitula flat;* with varying amounts of red, occasionally the whole plant deep crimson. **Fascicles:** Rarely closely packed, usually more or less evenly spaced and stem visible between fascicles. **Stem leaves:** Lingulate to almost rectangular; *weakly fibrillose to almost lacking fibrils,* rarely wholly without fibrils; *septa numerous,* often several per cell. **Branch leaves:** Laxly imbricate and *somewhat spreading; usually 5-ranked;* rather concave at apex with less inrolled margins, and appearing rather blunt. *Abaxial pores medium-sized to rather small,* (6.0–)8.0–12.0 μm in mid-leaf.

HABITAT: Common as hummocks in the drier parts of oligotrophic mires, on damp heath and in open, acid woodland. This variety seems to be more restricted to distinctly acid locations raised somewhat further above the water table than var. *capillifolium.*

DISTRIBUTION: Throughout the northern hemisphere, though more abundant in warmer, more sub-oceanic regions, and not reaching the same altitudes as var. *capillifolium*. In Europe, it is more common in the lowlands, and in the southern and western areas, though present throughout most of the area. It is present in much of Britain and is the more abundant variety in the south and west, where it largely replaces var. *capillifolium*.

In their extreme forms, var. *capillifolium* and var. *rubellum* offer no great problems of identification in the field or laboratory. However, none of the discriminatory characters, even those italicized above as being of special value, is sufficiently constant in all populations, or always sufficiently immune from environmental or seasonal influences, to permit consistent identification. Most cultured clones retain their more important features unchanged, but some gatherings, usually with intermediate characteristics in the field, can be induced to adopt either a '*capillifolium*' or a '*rubellum*' form, according to the combination of water level, exposure and mineral ion concentration in which they are grown. Even Wilson's herbarium, under the label '*S. rubellum*', contains a high proportion of specimens that would be interpreted here as *S. capillifolium* var. *capillifolium*. *Sphagnum subtile* Russ. and *S. tenerum* Sull. & Lesq. are 2 names that appear in treatments of European *Sphagnum*, but many herbarium specimens labelled *S. subtile* are attenuated forms of *S. capillifolium*, mostly of var. *capillifolium*, though a few are of var. *rubellum*. *S. tenerum* is an American species whose systematic position seems to lie between *S. molle* and *S. subnitens*. Putative European specimens (often labelled, for example, *S. nemoreum* Scop. var. *tenerum* (Sull. & Lesq.)) are mainly robust forms of *S. capillifolium* var. *capillifolium* or, in a few cases, *S. subnitens*. Whilst it is possible that the true *S. tenerum* may be found in western Europe, its presence there has yet to be established (it resembles *S. molle* in colour and habit, differing in its narrow, tapering, fibrillose and porose stem leaves, and in the absence of a resorption furrow along the branch leaf margin).

S. capillifolium var. *capillifolium* differs from small forms of *S. subnitens* in its (usually smaller) partially fibrillose stem leaves. In the field, *S. russowii* and *S. warnstorfii* may be indistinguishable from forms of *S. capillifolium* var. *rubellum* (see notes under these species for differences).

12. SPHAGNUM WARNSTORFII

Sphagnum warnstorfii Russ. (*Sber. naturfGes. Univ. Dorpat,* **8,** 315. 1887)
S. acutifolium var. *gracile* Russ. (*Beitr. Kennt. Torfm.,* 44. 1865)
S. warnstorfianum Du Rietz. in Sjörs (*Svensk bot. Tidskr.,* **38,** 405, 1944)

PLANTS: Rather small, lax; capitulum small, more or less flat (strongly resembling more robust forms of *S. capillifolium* var. *rubellum*); almost always with some *red or deep pink coloration,* often the whole plant deep crimson. **Fascicles:** Well spaced, with distinct 'internodes'; of (3–)4 dimorphic branches; spreading branches 2, long, up to ca 18 mm (except near the capitulum), tapering distally; pendent branches 1–2, variable in length, usually over 12 mm; (very rarely with 3 spreading branches and a single pendent one). **Stem:** Up to 0.7 mm diameter; cortex of 3–4 layers of hyaline cells without pores on the external surface: internal cylinder well developed, yellowish or, in the outer layers, violet-red. **Branch anatomy:** Retort cells distinct with moderately well-developed rostra, solitary; internal cylinder brownish to purple-red. **Stem leaves:** Erect and mostly appressed lightly to firmly to the stem; 1.1–1.4 mm long; lingulate to slightly triangular-lingulate; apex rounded to narrowly truncate, tip about 0.1 mm across; border strong, up to 5 cells wide above, strongly expanded below to occupy two-thirds to three-quarters of the leaf base. Hyaline cells shorter and somewhat rhomboid above, more elongated below; abaxial surface intact; fibrils usually present but faint, scattered and confined to cells just below apex, occasionally absent; septa numerous. **Branch leaves:** *Normally 5-ranked;* erect-spreading, inserted at 40°–45°; 1.1–1.4(–1.6) mm long; ovate to ovate-lanceolate; border 1–4 cells wide, without resorption furrow; apex narrowly rounded truncate with inrolled margins. Pendent branch leaves longer, narrower and more delicate. **Hyaline cells:** Rather sharply divided into 2 size classes, the upper rather small (70–80 × 15–18 µm), the lower much larger (160 × 25–30 µm). *Abaxial surface pores markedly dimorphic:* in the smaller hyaline cells, pores (except those in the apical cell angles) *very small* (aperture 2.0–3.0 µm diameter), *circular, heavily ringed,* rarely more than 5 µm diameter, including ring; in the large lower and lateral hyaline cells, pores large (10.0 µm or more), rather thinly ringed, often appearing half elliptical. Adaxial pores absent from upper mid-leaf or, if present, small, ca 40 µm diameter, few per cell. **Leaf TS:** Hyaline cells shallowly convex to almost plane on the adaxial face, strongly inflated on the abaxial; partial septa (fibrils) quite conspicuous. Photosynthetic cells triangular to trapezoid, rather thin-walled with angular lumina; widely exposed on the adaxial leaf surface, narrowly to very narrowly exposed on the abaxial (not enclosed on the abaxial though sometimes apparently so because of the bulging adjacent hyaline cells. **Fertile plants:** Dioecious. Male plants occasional. Antheridial branches deep crimson, often conspicuous, especially in paler plants. Female plants apparently rare; perichaetial bracts large, convolute with rounded or retuse prosenchymatous apices; lower tissue of uniform, lax, thin-walled cells, merging into thicker-walled ones; upper tissue differentiated into hyaline cells without pores or fibrils, except sometimes near apex, and photosynthetic cells. Capsules rare, unknown from the British Isles, occasionally produced in Scandinavia; spores yellow-brown, papillose, 24–26 μm diameter.

HABITAT: Forming tussocks and mats in eutrophic mires (or parts of mires) and flushes, or by stream-sides: never on acid mires or wet heaths. The only other *Sphagnum* species consistently associated with *S. warnstorfii* are *S. teres*, *S. squarrosum* and *S. contortum*, which are also able to tolerate base-rich conditions. More usually, this species is found with herbs (*Menyanthes trifoliata, Potentilla palustris, Parnassia palustris*), carices (*Carex rostrata, C. nigra*) and 'brown' mosses (*Scorpidium scorpioides, Drepanocladus revolvens, Campylium stellatum, Cratoneuron commutatum*). It is usually found where there is light shade, though it may occur in both open habitats and under quite dense scrub.

DISTRIBUTION: Circumpolar in the sub-arctic and northern boreal zones. More abundant in northern parts of Europe, but also present in the sub-alpine regions, where it extends to the Bulgarian mountains. In Britain, it is confined to northern districts.

S. warnstorfii shows very little variation between individuals, and entirely green, spiral-leaved plants are very rare. The normal red-flecked or crimson forms could only be confused with either *S. capillifolium* var. *rubellum* or *S. russowii*. Neither of these species are found in the eutrophic habitats occupied by *S. warnstorfii*. Microscopically, the presence of small pores on the abaxial surface of the branch leaf hyaline cells is diagnostic: the 2 species mentioned above may have small pores, but they are rarely less than 6.0 μm diameter. *S. fuscum* occasionally has small, ringed pores on its branch leaves, but the colour and habitat of that species are different: again, the pores are seldom less than 5.0 μm diameter.

Figure 26. Distribution of S. warnstorfii

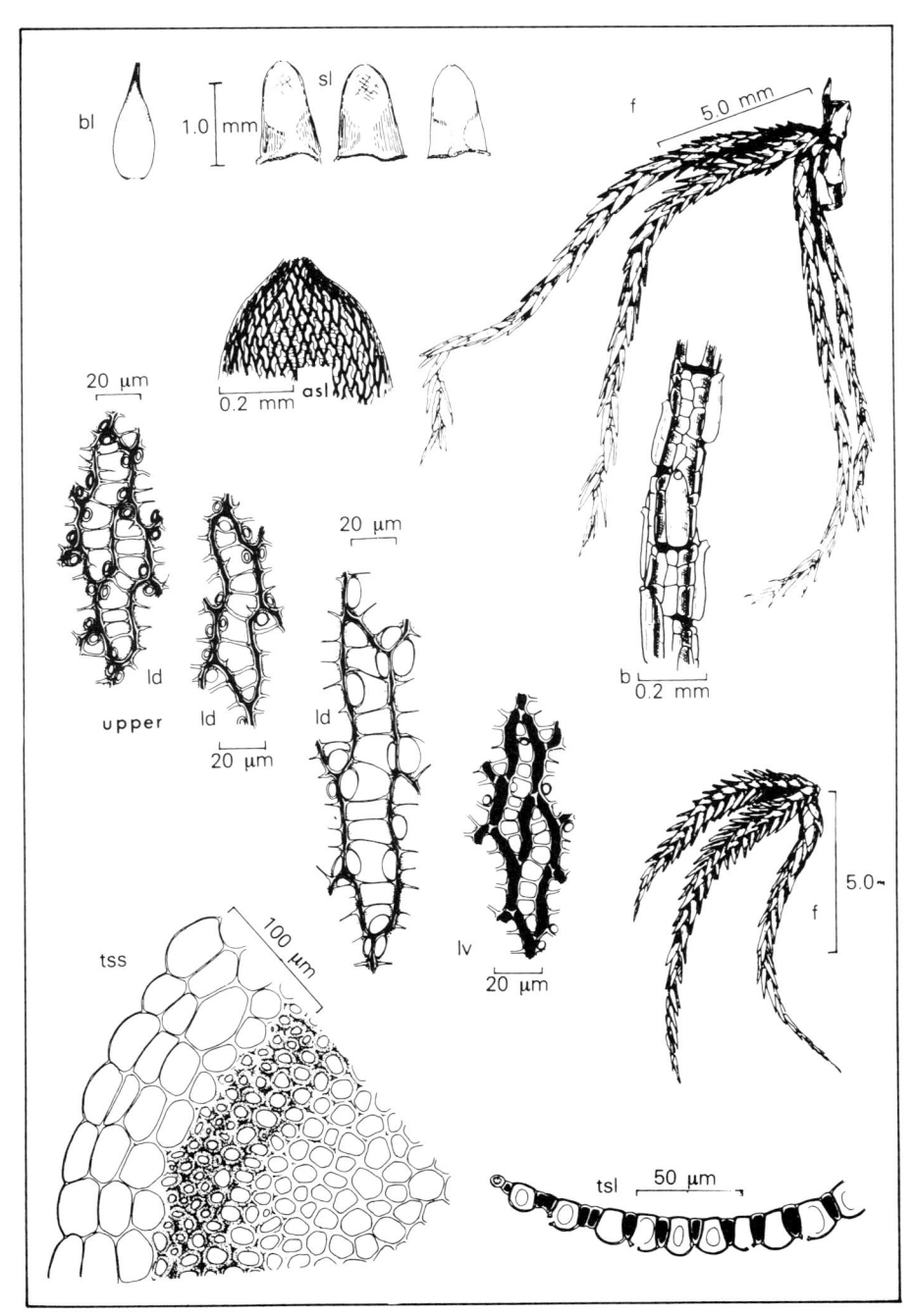

Figure 27. *Sphagnum warnstorfii*

13. SPHAGNUM RUSSOWII

Sphagnum russowii Warnst. (*Hedwigia,* **25,** 225. 1886)
S. robustum (Warnst.) Card. (*Bull. Soc. Hist. nat. Autun.,* **10,** 381. 1897)
S. girgensohnii var. *robustum* (Russ.) Dix. (*Handbk Br. Mosses.* 3rd ed., 17. 1924)

PLANTS: Medium-sized to delicate, often rather tall (resembling robust forms of *S. capillifolium,* especially var. *rubellum*); green with red or pink flecks, to deep red, sometimes the red inconspicuous and confined to bracts and the lower parts of stem leaves, very rarely absent. **Fascicles:** Rather distant, with distinct 'internodes'; of 3–4 dimorphic branches; spreading branches 2, long and slender, 20 mm or more (except in male plants), tapering distally; pendent branches as long as or longer than the spreading, thin and terete (very rarely one or more fascicles with 3 spreading branches and a single pendent one). **Stem:** To 0.8 mm diameter; cortex of 3–4 layers of hyaline cells, the outermost *occasionally to frequently with a single unringed, inconspicuous pore in usually fewer than 20%, rarely more than 30%, and probably never more than 60% of the cells;* internal cylinder pale, with or without some violet-red in its outer layers. **Branch anatomy:** Retort cells distinct, slightly rostrate, solitary; internal cylinder pale or, rarely, red or pale brown. **Stem leaves:** Erect, appressed to stem; 1.3–1.6 mm long; *lingulate, parallel-sided; apex broadly rounded, truncate and fringed across* the central 0.15–0.3 mm of *their tips;* border well developed, strongly expanded below. Hyaline cells wholly or partly resorbed on the adaxial surface; abaxial surface intact; fibrils thin, scattered, inconspicuous, occasionally absent; septa numerous, especially near the leaf margins. **Branch leaves:** Mostly 5-ranked, at least in the lower parts of some branches, occasionally spiral throughout; erect and somewhat spreading; 1.3–1.6 mm long; lanceolate to ovate-lanceolate, rather concave; apex rather broadly acute with margins not strongly inrolled; border 2 cells wide without resorption furrow. Pendent branch leaves narrower, more lanceolate, except at branch insertion where they are short and concave. **Hyaline cells:** Smaller near apex, ca 70 × 18–20 μm, gradually becoming longer towards leaf base (90–130 × 25–30 μm). Abaxial surface with several large (10–18 μm diameter) pores along the commissures (mainly appearing half-elliptical due to cell convexity); near leaf apex, pores smaller (6.0–11.0 μm diameter) and more or less circular. Adaxial surface without pores in lower mid-leaf; *near apex, with 1–5 large* (8–16 μm), *circular, unringed pores per cell,* occupying almost all of the interfibrillar spaces; 2–5 series of marginal hyaline cells with numerous large pores on both surfaces. Pendent branch leaves with numerous commissural pores throughout; many cells on the abaxial surface with 1–4 additional circular pores in the cell mid-line. **Leaf TS:** Hyaline cells more or less plane to shallowly convex on adaxial face, strongly convex on the abaxial. Photosynthetic cells triangular or trapezoid, with thin or moderately thickened cell walls; widely exposed on the adaxial leaf surface, narrowly to very narrowly on the abaxial. **Fertile plants:** Dioecious. Male plants frequent. Antheridial branches short, usually less than 10.0 mm long; bracts deep red. Functionally, female plants apparently very rare; inner perichaetial bracts very large, almost 5.0 mm long, convolute; apices truncate and minutely retuse; tissue relatively undifferentiated, except for a small zone below the apex which has distinct hyaline cells without pores or fibrils. Capsules rare; spores brown, coarsely papillose, 26–29 μm diameter.

HABITAT: Wet, mesotrophic, often wooded, mires, where it forms loose patches or small carpets. It grows with a variety of herbaceous species and *Sphagna* (either as isolated shoots or wider patches), including carpets of *S. lindbergii* and *S. riparium*, or hummocks of *S. fuscum* or *S. magellanicum*. In Fennoscandia, it is a lowland or low alpine plant, but further south it becomes more confined to upland areas, though it is still found in woodlands, where it tends to occur along the margins of soaks. It is shade-tolerant but may be found in more open situations on wet, rocky ground, as well as in open, upland mires.

DISTRIBUTION: Circumpolar in the boreal to sub-arctic regions of Eurasia and North America. Widespread in Europe, but with distinct northern tendencies and most common in the sub-arctic. Further south, it becomes increasingly rare in the lowlands but extends, in the mountains, as far as Macedonia and Bulgaria. It is mainly an upland species in Britain, though occurring occasionally in lowland areas, and is found from Wales northwards. It is also present in the northern part of Ireland.

Sphagnum russowii is a rather variable plant in Europe (though less so in Britain, British forms nearly always having some crimson coloration and at least some 5-ranking of the branch leaves, especially near the capitula). This species can be confused in the field with *S. capillifolium,* particularly with var. *rubellum,* but it is more robust than that plant, has longer branches, and stem leaves with broader apices. Under the microscope, the abundant adaxial pores in the branch leaves and the occurrence of pores in the stem cortex are the main features distinguishing it from *S. capillifolium, S. warnstorfii* and *S. subnitens*. Pale forms may be confused, at first sight, with *S. quinquefarium,* but the latter usually has 3 spreading branches in each fascicle and smaller, triangular, narrow-pointed stem leaves. Rarely, plants of *S. russowii* lack red pigment and have spirally arranged branch leaves throughout (var. *girgensoh-*

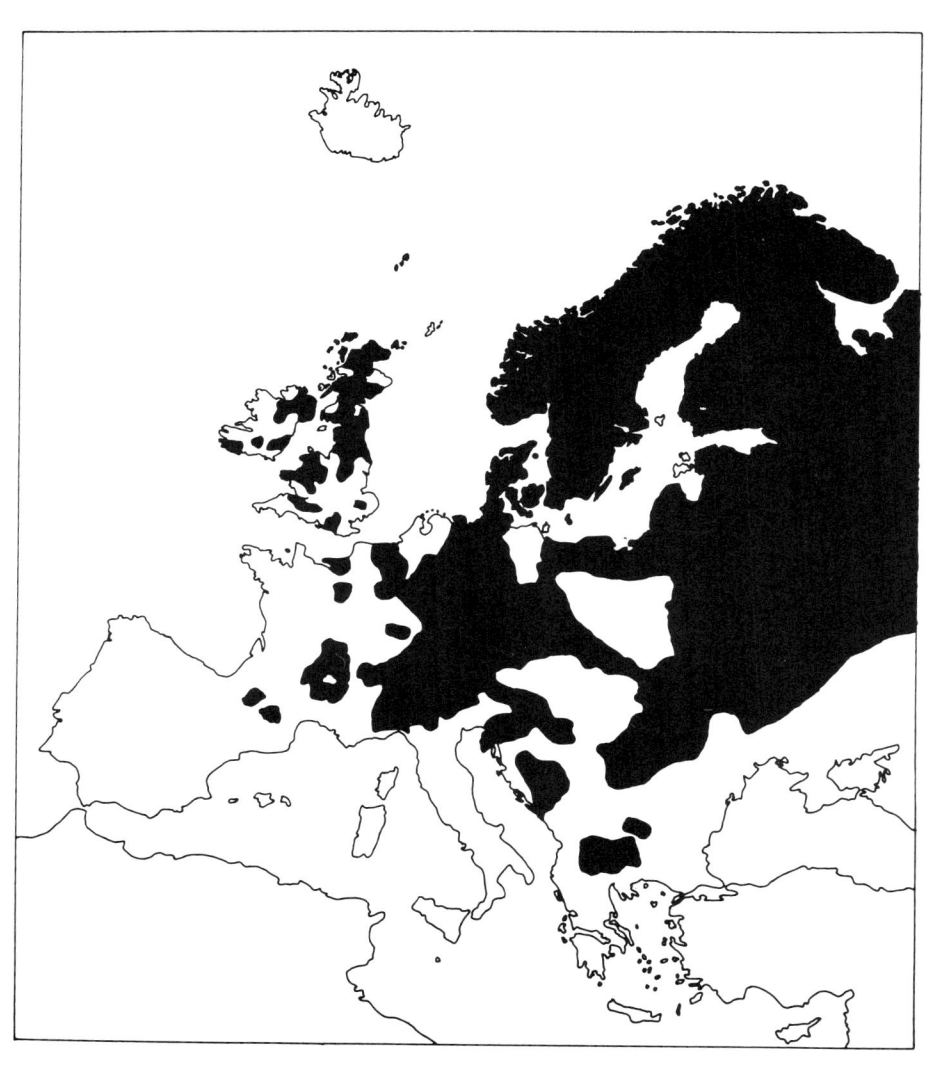

Figure 28. Distribution of *S. russowii*

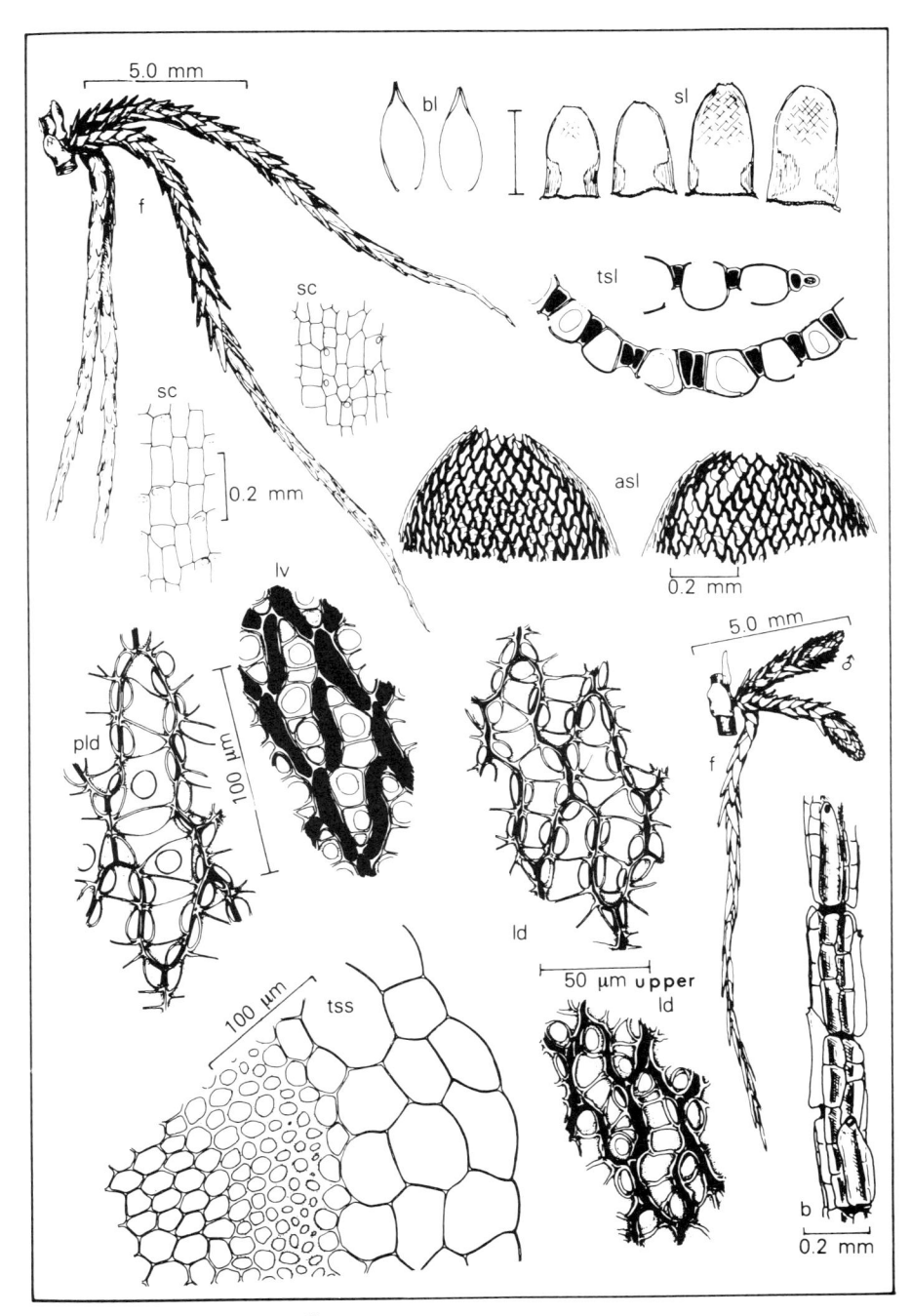

Figure 29. *Sphagnum russowii*

nioides Russ.): superficially, these are almost impossible to distinguish from *S. girgensohnii*. However, the latter species is usually more robust and has a larger stem bud, though its principal distinguishing characters are the more numerous and conspicuous stem pores and the expanded hyaline cells which occur in groups near the insertion of the stem leaves. As in many of the related species, male plants of *S. russowii* are often smaller than sterile or functionally female plants, and have much shorter (antheridial) branches.

14. *SPHAGNUM GIRGENSOHNII*

Sphagnum girgensohnii Russ. (*Beitr. Kennt. Torfm.*, 46. 1865)

PLANTS: Rather robust, occasionally smaller (then resembling green forms of *S. capillifolium*); terminal stem bud rather large but not very conspicuous; green to straw-coloured with pale green to pale brown stems, capitulum (and antheridial branches) often yellow-brown; never with any red. **Fascicles:** Well spaced or, occasionally, rather dense; of 3(–4) dimorphic branches; spreading branches 2, very long and tapering (often exceeding 25 mm), except in the capitulum where they are shorter and often distinctly clavate; pendent branches 1(–2), pale and cylindrical with closely imbricated leaves. **Stem:** Relatively stout, (0.6–) 0.8–1.0 mm diameter; cortex well developed with 2–3 layers of hyaline cells; outer cortical cells thin-walled, mostly not, or scarcely, elongated, *all or almost all with one (rarely 2) large, relatively conspicuous pores* (easily seen on intact stems even with light staining); internal cylinder thick, pale green to pale brown (darker in old, dried specimens). **Branch anatomy:** Retort cells moderately distinct, scarcely rostrate, solitary or in groups of 2–3 (the lowest always the largest); internal cylinder yellowish to pale brown. **Stem leaves:** Erect and appressed to the stem; 0.8–1.3 mm long; more or less rectangular to lingulate, usually distinctly wider at insertion and often also above the middle (ie sometimes *distinctly lyrate, or waisted*); border strongly expanded below, narrow above, vanishing in the apex; *apex wide, truncate and lacerate; parts of lower leaf tissue abruptly stretched, giving localized much expanded and often ruptured groups of hyaline cells.* Hyaline cells in upper mid-leaf *resorbed on both surfaces, rarely with any trace of fibrils;* in weaker or male plants, the adaxial surface sometimes intact and the abaxial with large resorption gaps; septa few and more or less confined to the peripheral parts of the leaf. **Branch leaves:** Never 5-ranked; rather densely imbricated, erect with slightly reflexed tips; rather large (for the section Acutifolia), 1.4–1.8 mm long; broadly lanceolate and appearing abruptly narrowed above due to strongly inrolled margins; apex somewhat cuspidate; border 1–2 cells wide without resorption furrow. **Hyaline cells:** Rather small near the apex (60 × 15–20 μm), larger towards mid-leaf (100 × 20–30 μm), longer near the insertion. Adaxial surface in upper third of leaf with rather large, ca 10 μm diameter, circular, faintly but distinctly ringed pores lying more or less along the cell mid-line; lower median cells mostly without pores. Abaxial surface, except near the leaf apex, with numerous large (9.0–18.0 μm) pores along the commissures; hyaline cells in peripheral parts of the leaf with numerous pores on both surfaces. Pendent branch leaves usually with additional free pores on the abaxial surface. **Leaf TS:** Hyaline cells shallowly convex on adaxial leaf surface, more strongly convex

Figure 30. Distribution of *S. girgensohnii*

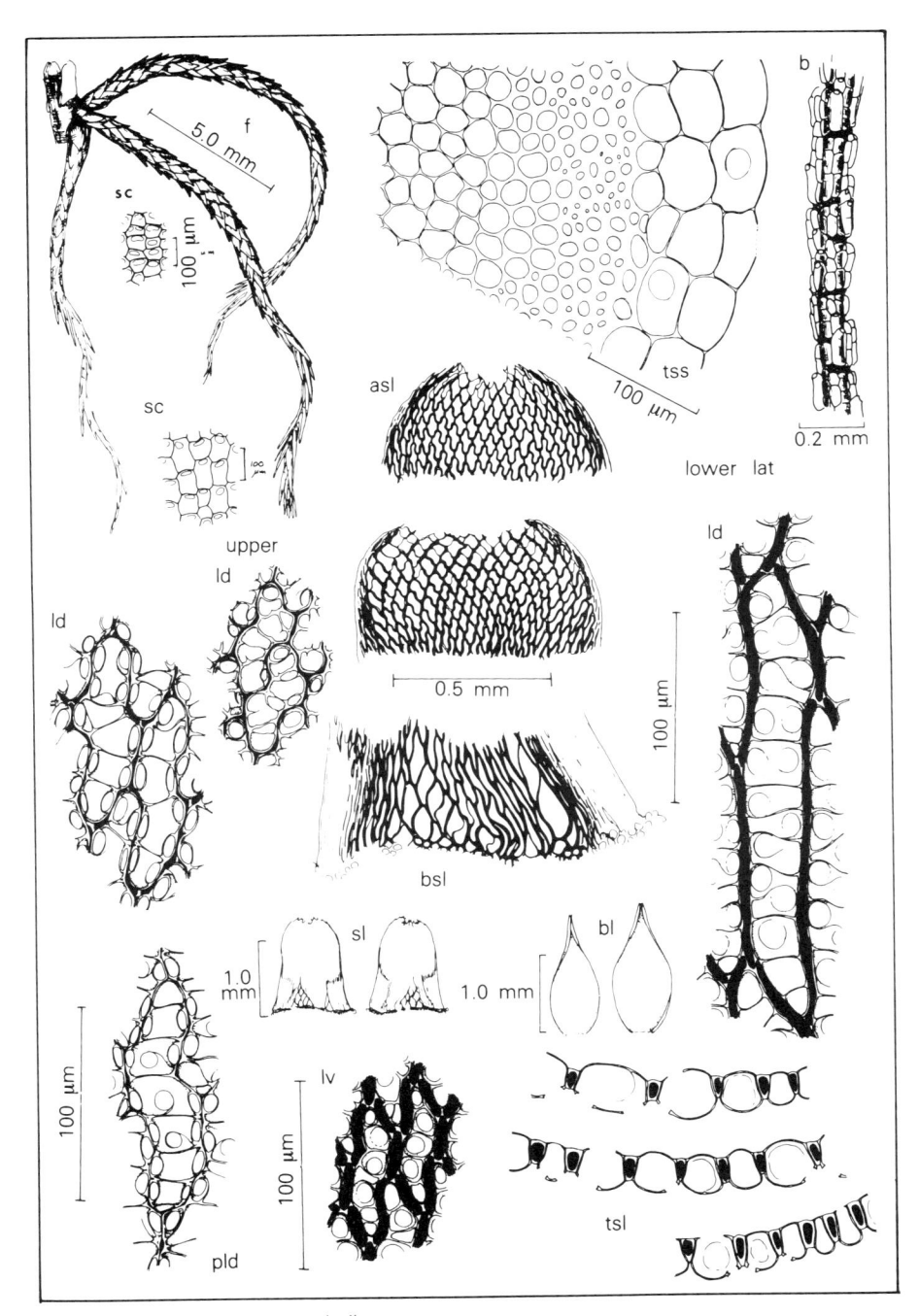

Figure 31. *Sphagnum girgensohnii*

on the abaxial. Photosynthetic cells triangular to trapezoid, sometimes with bulging sides, cell walls distinctly thickened, and lumina somewhat rounded; widely exposed on the adaxial leaf surface, narrowly to very narrrowly exposed on the abaxial. **Fertile plants:** Dioecious. Functionally male plants usually smaller, with short branches. Antheridial bracts shorter and wider than branch leaves, densely imbricate, brown. Perichaetial bracts large, up to 5.0 mm long and 2.0 mm wide; apices more or less retuse; tissue with clearly differentiated hyaline cells confined to upper third of bract or less, hyaline cells without fibrils or pores. Capsules rare; spores yellow-brown, slightly rough, 23–27 μm diameter.

HABITAT: This is a species of rather shaded sites, often with poorly developed peat deposits and significant mineral water influence. It is found, typically, in damp woodland, on grassy slopes, near ditches and in marginal parts of mires, especially if these margins carry fen woodland. It grows well above the water table and forms scattered, loose hummocks or mats, particularly under *Betula* or *Salix*. At the northern end of its range, it also occurs in more open mires and becomes abundant in sub-arctic *Eriophorum* peatlands where it is accompanied by *S. russowii*.

DISTRIBUTION: Widespread in the northern hemisphere. Scattered throughout lowland areas of Europe, but more abundant in sub-montane and sub-arctic areas: it is one of the most abundant *Sphagnum* species of northern Iceland and northern Scandinavia. Rare in the south of Britain, becoming more widespread and common in the northern parts of Wales, England, Scotland and Ireland.

This species can usually be recognized in the field by the shape of the stem leaves, which are somewhat waisted and tattered across the apex. In *S. fimbriatum*, with which it might be confused initially, the stem leaves are narrowest at the base and the fimbriation extends down the side of the leaf. In general appearance, *S. fimbriatum* is usually less stiff than *S. girgensohnii*, and has a more prominent stem bud. It may be distinguished from *S. teres*, which has a prominent stem bud, by the fascicles, which usually have 2 spreading and 1 pendent branch (3 + 2 in *S. teres*) and a paler stem: microscopically, the exposure of the photosynthetic cells is distinctly different. For differences between this species and *S. russowii*, see notes under that species.

Flatberg and Frisvoll (1984) have proposed a new species, *S. arcticum*. This appears to us to be no more than a form of *S. girgensohnii* typical of habitats at high latitudes or high altitude.

15. SPHAGNUM FIMBRIATUM

Sphagnum fimbriatum Wils. ex Wilson & Hooker in Hooker (*Flora antarct.*, 398. 1847)

PLANTS: Small and attenuated to rather robust; capitula small with *conspicuous, projecting stem buds;* bright green to pale yellow-green. **Fascicles:** Well spaced to moderately crowded; 3–5(–6) differentiated, but not strongly dimorphic, branches; spreading branches 2–3, thin and attenuated, (10–)20–30 mm long; pendent branches very long and thin, 25–30 mm or more, colourless. **Stem:** Thin to moderately thick, 0.4–0.8(–1.0) mm diameter; cortex well developed (but somewhat thinner than in most members of the Acutifolia), of 2–3 layers of hyaline cells; outer cortical cells with occasional to abundant large pores, 1(–2) per cell; internal cylinder pale, yellowish, sometimes *faintly* brownish with age. **Branch anatomy:** Variable; in smaller plants retort cells are usually distinct, much inflated but not rostrate, and solitary; in more robust forms, the principal retort cells are accompanied by 1–2 subsidiary cells which are also perforated; other cortical cells may also be perforate; internal cylinder pale green or yellowish. **Stem leaves:** Erect, closely appressed to stem and *forming a more or less complete foliar sheath* (virtually invisible when viewed against the stem and rather difficult to remove without damage, but easily seen in broken stems, eg if the capitulum is removed); 0.8–2.0 mm long; *shortly spatulate*, narrowest near insertion; border 1–3 cells wide, but largely lacking; apex very widely rounded and leaf *fimbriate around the whole of the upper part.* Lower lateral tissue prosenchymatous; the upper tissue lax, with hyaline cells more or less completely resorbed to leave an open mesh of photosynthetic cells and numerous septa; some median basal hyaline cells much enlarged (as in *S. girgensohnii*). **Branch leaves:** Never 5-ranked; closely imbricated in the lower parts of branches, lax above (then resembling pendent branch leaves in arrangement and shape); 1.1–2.2 mm long; lanceolate to somewhat ovate-lanceolate; lower half not markedly dilated; upper half gradually, occasionally somewhat abruptly, narrowed; apex more or less acute, with inrolled margins, not markedly reflexed; border 1–3 cells wide without resorption furrow. Pendent branch leaves laxly imbricated; lanceolate. **Hyaline cells:** Variable in size; smaller near apex, 60–90 × 15–20 µm; becoming larger towards insertion, up to 170 × 30–40 µm. Adaxial surface, at least in upper half of leaf, more or less plane; fibrils few; pores one or several, large (7.0–15.0 µm diameter), circular, unringed or thin-ringed. Abaxial surface convex; near leaf apex with few, medium-sized (8.0–10.0 µm diameter), ringed or unringed, but never thick-ringed, pores mainly in the cell angles; in lower part of leaf, pores numerous, regular, thin-ringed, in series along the commissures. Pendent branch leaf

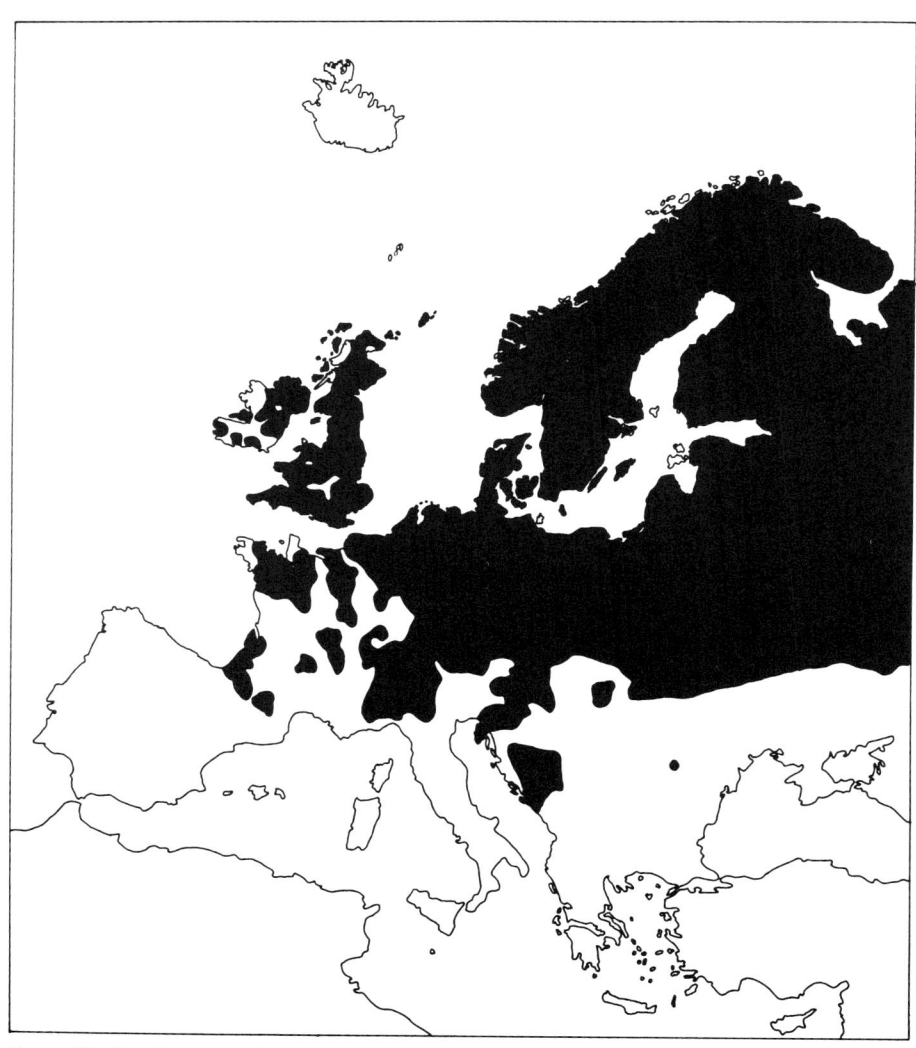

Figure 32. Distribution of *S. fimbriatum*

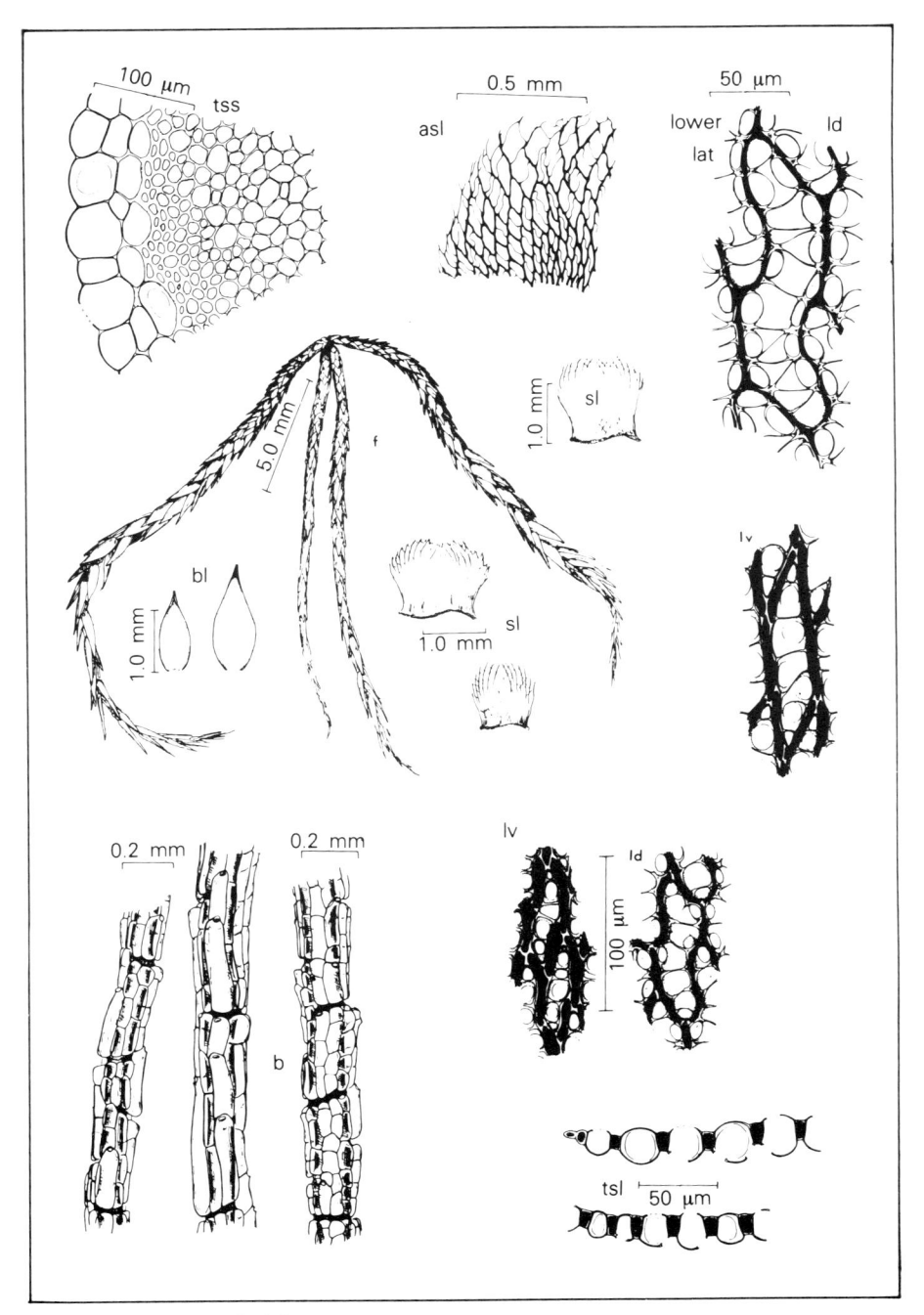

Figure 33. *Sphagnum fimbriatum*

anatomy similar to that of spreading branch leaves. **Leaf TS:** Hyaline cells more or less plane on adaxial leaf surface, convex on the abaxial. Photosynthetic cells trapezoid, rarely triangular, thin-walled; widely exposed on the adaxial leaf surface, narrowly exposed on the abaxial. **Fertile plants:** Monoecious. Antheridial branches often indistinguishable from sterile ones, occasionally yellowish. Perichaetial bracts large, up to 4.8 mm long; outer bracts broadly ovate, narrowed at apex, almost entirely prosenchymatous; inner 3–5 bracts distinctly spatulate, prosenchymatous below and sometimes at apex, *either side of apex tissue similar to that of stem leaves with at least partial resorption of the wide hyaline cells*, septa numerous. Capsules common, often abundant; spores almost smooth, 24–27 μm diameter.

HABITAT: Forming soft hummocks or loose carpets in damp, mesotrophic areas, usually with at least some shading. It is most frequent in wet *Salix* or *Betula* woodland and scrub, often with *Molinia*, but may also be found in more open situations on the grassy banks of streams or ditches, along lake margins, or in mesotrophic fen communities. It usually occurs as pure stands, but may be associated with *S. palustre, S. squarrosum* or *S. angustifolium* or, in the north, included in *S. lindbergii* or *S. riparium* carpets.

DISTRIBUTION: A very widespread species throughout the north temperate zone, extending into the Arctic. It also occurs along the Andes to the sub-Antarctic. It is common throughout most of Europe, but is more abundant in the lowlands. In Britain, it is widespread and mostly common, except for a few areas of central southern England, north-west Scotland and western Ireland.

Sphagnum fimbriatum, at a casual glance, appears to have no distinguishing features and may be mistaken for a weak, etiolated form of one of a number of other *Sphagnum* species, eg *S. recurvum*. The hard, conical stem bud is a useful field character, and is found, equally developed, only in *S. teres* and *S. squarrosum*. *S. teres* is a plant of rather more eutrophic habitats and normally has a dark brown stem and, like *S. squarrosum,* a more rectangular to lingulate stem leaf (*S. squarrosum* is normally also much more robust and has strongly squarrose branch leaves). The unique stem leaves of *S. fimbriatum* are best seen by pulling off a capitulum and examining the projecting fringe of their apices at the broken end of the stem. Only in the rare, northern, *S. lindbergii* are stem leaves of anything like similar form to be found, but that species is so different that confusion should not arise. In *S. girgensohnii*, the stem leaves are more oblong to lingulate and only fringed across the tip rather than all round the upper part of the leaf.

Although there appears to be little significant anatomical variation in this species, there is a considerable range of linear dimensions. The commonest form in the southern, lowland parts of its geographical range is slender, with branch leaves mostly less than 1.7 mm long. In northern and montane regions, this form is replaced by a more robust one resembling *S. girgensohnii*. Possibly 2 races exist, but insufficient work has been carried out to confirm this suggestion.

To conform with other treatments, *S. fimbriatum* has been included here in section Acutifolia, although it appears to be only distantly related, systematically, to other members of the group except *S. girgensohnii*. There do seem, however, to be a number of morphological characters linking this species with section Squarrosa, eg the stem bud, the distribution of pores, the structure of the stem leaves, as well as certain ecological similarities. Its geographical range is unique and only *S. magellanicum* shows a limited similarity.

SECTION SQUARROSA

Sphagnum sect. *Squarrosa* (Russ.) Schimp. (*Syn. Musc. eur.*, 835. 1876)

PLANTS: Small to robust, medium-sized to rather tall; with a *prominent, more or less conical, projecting stem bud;* green, yellowish or, infrequently, brownish. **Fascicles:** Of 4–6 (typically 5) moderately to strongly dimorphic branches. **Stem:** Slender to rather wide; cortex well developed, of 2–3 layers of inflated hyaline cells; outer layer rarely with pores, sometimes with 1(–2) pore-like thinnings (visible after staining); internal cylinder brown, sometimes green. **Branch anatomy:** Retort cells mostly distinct, occasionally rather poorly differentiated from the other cortical cells; usually in groups of 2–4; not, or scarcely, rostrate. **Stem leaves:** Erect, spreading or hanging; rather large, lingulate, broad above or slightly tapering, thin; apices with broad, very thin, multicellular, but ephemeral, borders (lacking photosynthetic cells) which are soon eroded so that older leaves are more or less fimbriate; *border thin throughout, not widened below.* Hyaline cells with resorption gaps on the abaxial surface (often almost completely resorbed); fibrils absent; septa confined mainly to the peripheral leaf areas. **Branch leaves:** Broadly ovate but abruptly contracted from half-way to tapering, narrow, sub-tubular apices which are weakly to strongly reflexed. **Hyaline cells:** Near apex rather short, elsewhere up to several times larger; internal commissural walls *frequently papillose.* Abaxial surface near apex often with a single pore in the apical angle; in mid-leaf, with few to fairly numerous (up to 15), rarely absent, large (12.0–40.0 μm) free pores which sometimes coalesce to form larger resorption gaps. Adaxial surface with pores absent or few to fairly numerous (up to 15), more numerous towards leaf margins, mostly along the commissures; near apex, often with a triple pore. **Leaf TS:** Hyaline cells plane to shallowly convex on abaxial face, more strongly convex on adaxial. Photosynthetic cells trapezoid, widely exposed on abaxial surface, more narrowly exposed on the adaxial. **Fertile plants:** Monoecious or dioecious. Antheridia on spreading branches. Inner perichaetial bracts large, narrow at insertion, *broad and retuse above, with wide, partially resorbed hyaline cells and eroded apices; hyaline cells differentiated throughout,* though narrower in lower part of bract; fibrils and pores absent; abaxial surface more or less resorbed throughout; adaxial surface more or less intact. Capsules typical of the genus, common or rare depending upon species.

Two circumboreal species confined to the northern hemisphere. Plants of mesotrophic to rather eutrophic habitats.

16. SPHAGNUM TERES

Sphagnum teres (Schimp.) Ångstr. in Hartm. (*Handbk skand. flora*, 417. 1861)

PLANTS: Small to medium-sized, lax to rather dense; capitula with *conspicuous, projecting, conical stem buds;* green to straw-coloured occasionally brown (particularly in exposed habitats). **Fascicles:** Of (3–)4–6 dimorphic branches; spreading branches 2–3, short to moderately long (10.0–20.0 mm); pendent branches 2–3, of variable length, but usually some as long as, or longer than, the spreading branches, thin and pale. **Stem:** Relatively strong, 0.6–1.0 mm diameter; cortex well developed, of 2–3 layers of hyaline cells, the outermost elongate and sometimes with 1(–2) thinnings ('shadow pores' visible only after heavy staining), rarely with true pores; internal cylinder *light to dark brown,* pale only in etiolated shade forms. **Branch anatomy:** Retort cells moderately distinct, not rostrate, sometimes solitary, but usually in groups of 2–3; internal cylinder pale brown. **Stem leaves:** More or less erect to variously spreading, but not closely appressed to stem; rather large, 1.2–1.5 mm long, lingulate to rectangular; *border thin, not expanded below,* vanishing below apex in older leaves (young stem leaves have a thin border which is soon lost); apex broadly truncate or rounded, eroded and more or less fimbriate. Hyaline cells without fibrils; *abaxial surfaces more or less completely resorbed;* sometimes near apex, both surfaces largely or wholly resorbed: *septa usually few, or absent,* except near the lower leaf margins, where they may be numerous. **Branch leaves:** On spreading branches, more or less *erect and closely overlapping with only the apices slightly divergent,* so that branches are markedly terete (only in shade forms are the leaf apices sometimes widely divergent, so that the plants may resemble very slender forms of *S. squarrosum*); rather large (1.8–2.1 × 1.3–1.4 mm); broadly ovate, concave, abruptly narrowed from half-way to more or less narrow apices. Pendent branch leaves varying from short and ovate near branch base to elongate and lanceolate at its distal end. **Hyaline cells:** Relatively short and wide, 70–100 × 20–30 µm in upper mid-leaf, longer above insertion. Adaxial surface with partly ringed or unringed pores, most of which, in the upper part of the leaf, have been *much enlarged by resorption* (12–40 µm diameter), and are somewhat irregular in outline (especially in comparison with the neat, circular pores of, for example, section Acutifolia); resorption gaps present in apical angles. Abaxial surface with 1–6 pores similarly enlarged by resorption. Hyaline cells of pendent branch leaves similar to those of spreading branches, except that pores are less modified, more regularly circular, and distinctly, often thinly, ringed. **Leaf TS:** Hyaline cells more convex on adaxial face; internal commissural walls smooth or, frequently, *lightly to rather strongly*

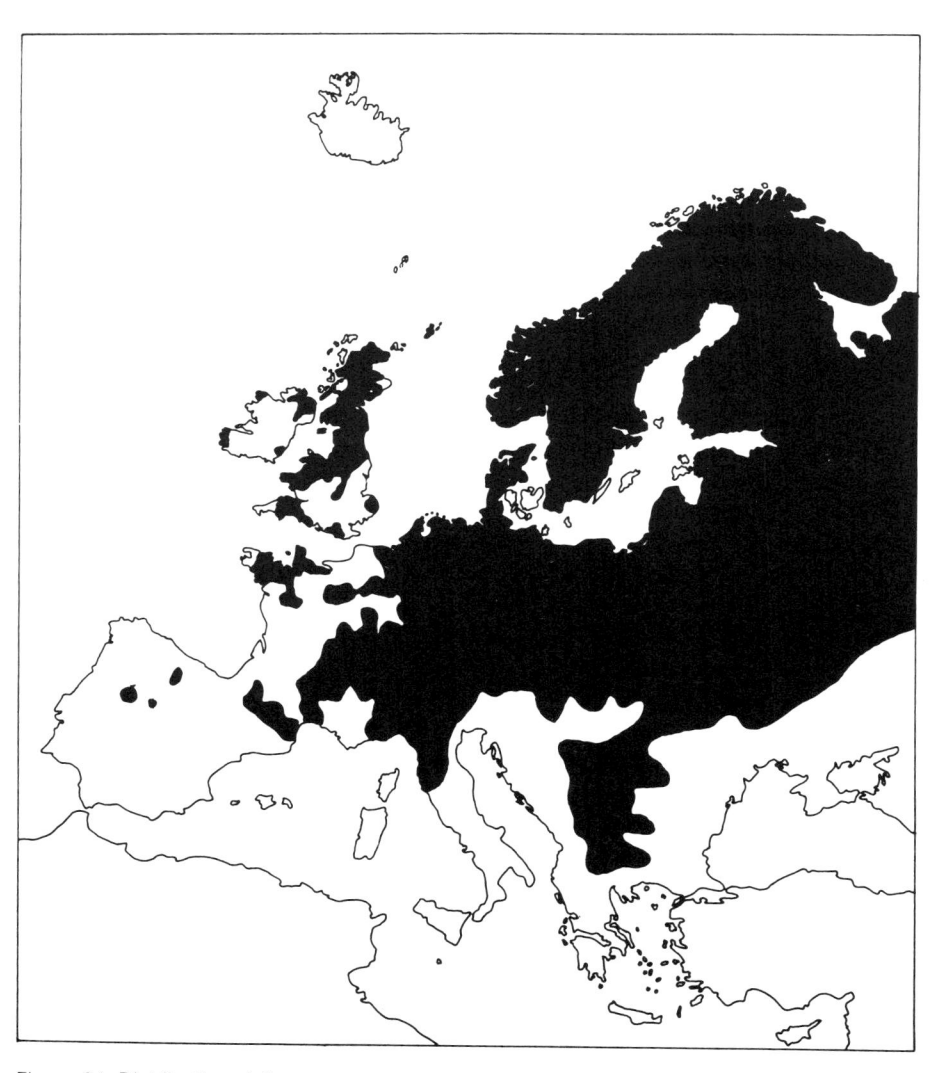

Figure 34. Distribution of *S. teres*

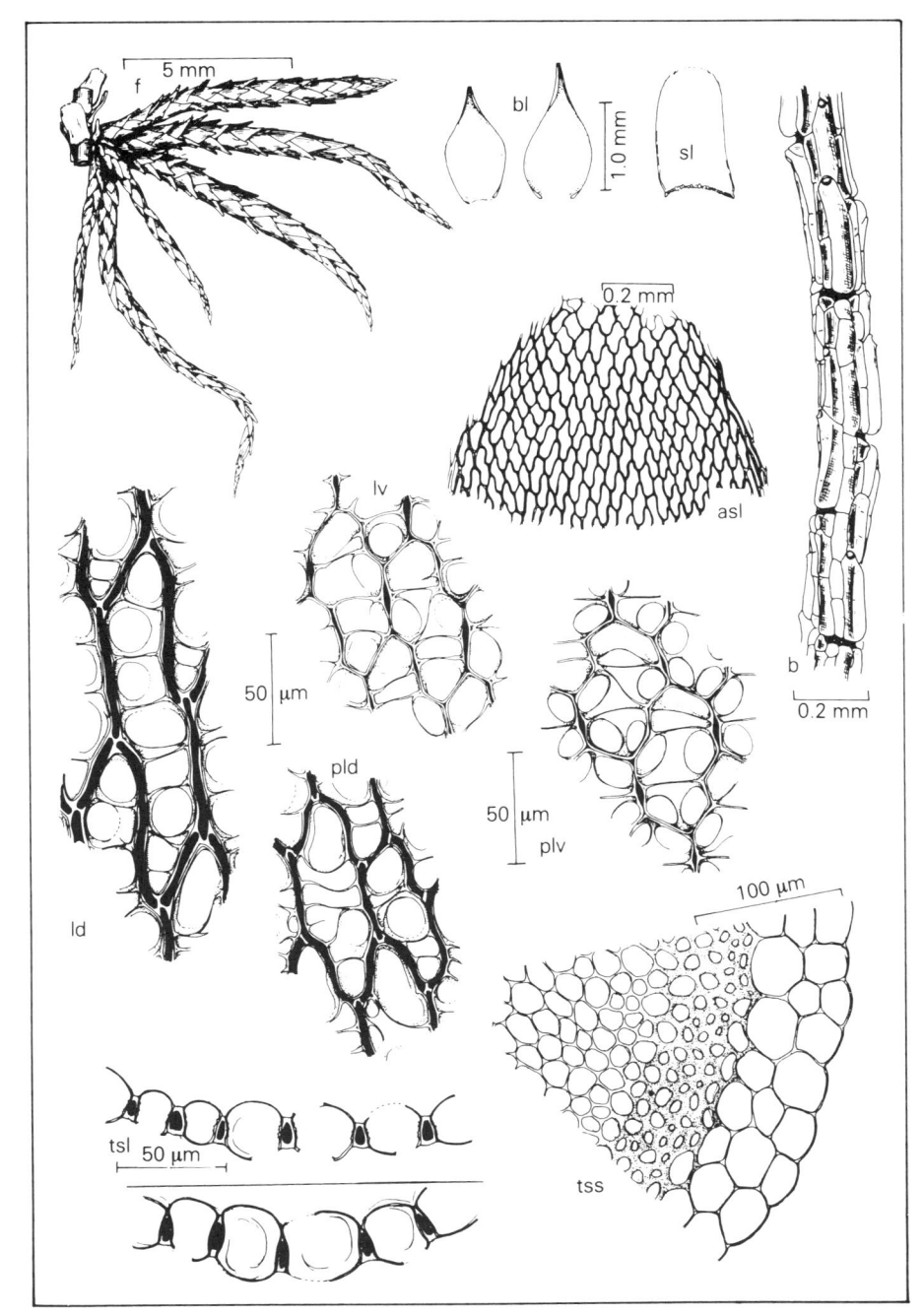

Figure 35. *Sphagnum teres*

papillose. Photosynthetic cells narrowly oval-triangular to trapezoid, more widely exposed on abaxial leaf surface; walls, especially the abaxial, moderately to strongly thickened. **Fertile plants:** Dioecious. Male plants often smaller than sterile ones. Antheridial branches short, the bracts brownish and densely imbricated, otherwise resembling branch leaves. Perichaetial bracts large, broadly spatulate; apices broadly retuse; hyaline cells differentiated throughout, narrow below, broad and short near apex, and with considerable resorption of cell walls, at least on the abaxial surface. Capsules rare (very rare in Britain); spores light brown, roughened, 23–27 μm diameter.

HABITAT: A widespread species because of its tolerance of a wide range of variation in trophic status, water level and shade. It is found in a variety of damp or wet mesotrophic to somewhat eutrophic habitats, including open fen, scrub, fen woodland, flushes and stream-sides. At the more eutrophic end of its range, it may be associated with *S. warnstorfii* or *S. contortum* and *Phragmites australis,* whilst at the other extreme where mineral water influence is less pronounced it may be found with *S. angustifolium* or *S. russowii.* In wet carpets, it can be associated with *S. riparium* or it may grow on more solid peat accompanied by *S. squarrosum.* Although tolerant of quite dense shade and found in scrub or mature fen woodland (with *S. fimbriatum*), it is usually most abundant in open situations where it forms low hummocks or mats, often between tussocks of *Carex* or *Juncus* species.

DISTRIBUTION: Very widespread, but often localized, in the northern hemisphere, extending as far south as the Himalayas and the mountains of Colorado, but common only in the sub-arctic or sub-alpine regions. It occurs throughout Europe from the lowlands to the montane zone and is confined to the latter at the southern end of its range (Pyrenees, northern Italy and Yugoslavia). In Britain it is widespread but, except in some localized northern upland areas, is not abundant; in the lowlands, it is comparatively rare.

In the field, *S. teres* is not a very colourful or distinctive plant—"... recognised ... more by elimination than by special characteristics" (Crum 1976), although it can be recognized, with experience, by the form and colour of its branches, dark stems coupled with a prominent stem bud in the capitulum, and by its habitat. Under the microscope, confusion is only possible with the closely related *S. squarrosum,* but the latter, except under rare and exceptional circumstances, is a plant with a very different appearance. *S. teres* has broader and shorter hyaline cells than any members of the section Subsecunda, and is readily separated from section Acutifolia by its abaxially exposed photosynthetic cells and lack of a widened border in its stem leaves.

The most likely member of the section Acutifolia to be confused with *S. teres* in the field is *S. girgensohnii,* but the two may be distinguished macroscopically by the dark brown stem of *S. teres.* Initially, there may be some field confusion between this species and *S. angustifolium,* but the prominent stem bud of *S. teres* and its sharply contracted branch leaves should readily distinguish the two taxa.

In shade, plants tend to be green, slender, and often show a tendency to squarrose branch leaves. Such rare forms have been given the varietal epithet *'squarrosulum'* (Schimp.) Warnst. At the other extreme, robust plants of open habitats, in which resorption of hyaline cells is especially extensive, have been called var. *reticulata* C. Jens. These both seem to be habitat-induced forms, with no real taxonomic significance.

17. *SPHAGNUM SQUARROSUM*

Sphagnum squarrosum Crome (*Samml. deutsch. Laubm.*, 24. 1803)

PLANTS: Robust, with spreading leaves, large capitula and conspicuous stem buds; pale green to yellow-green, rarely (in exposed alpine or arctic habitats) pale brown. **Fascicles:** Distant to rather closely set; strong with 4–6 dimorphic branches; spreading branches 2–3, 20.0–30.0 mm or more, tapering distally; pendent branches 2–3, 8.0–30.0 mm or more, attenuated distally (often one of the branches of a fascicle intermediate between normal spreading and pendent types). **Stem:** Strong, 0.7–1.3 mm diameter; cortex well developed, but relatively thin in relation to stem thickness, of 2–3 layers of hyaline cells; superficial layer often with indistinct thinnings ('shadow pores' visible only after intense staining); internal cylinder dark brown, at least peripherally, pale only in shade forms. **Branch anatomy:** Retort cells *often relatively indistinct from other cortical cells*, in groups of (1–)2–4, internal cylinder pale brown or yellowish. **Stem leaves:** Erect, spreading or hanging; lingulate; 1.2–1.8 × ca 0.8 mm; apices broadly rounded-truncate with ephemeral borders 2–3 cells wide, mostly lost in older leaves (but more persistent than in *S. teres*) which are then eroded or more or less fimbriate; border thin and eroded above, not widened below. Hyaline cells without fibrils, largely, to almost completely, resorbed on the abaxial surface; septa few or absent in upper mid-leaf, often numerous near the leaf margin. **Branch leaves:** *Large,* 2.3–3.3 mm long; *lower half erect, broadly ovate and concave* (leaf bases forming a sheath round the branch), then *abruptly narrowed to a sharply reflexed, acuminate limb* with pronounced 'shoulders' (ie 'squarrose'), giving the branches a prickly appearance; very rarely, all leaves are incumbent, so that the plant resembles a robust form of *S. teres*). Pendent branch leaves all incumbent, ovate and concave near the branch base, becoming lanceolate above; not abruptly contracted. **Hyaline cells:** In the squarrose limb, relatively small (70–100 × 15–22 μm) but longer and proportionately narrower in mid-leaf (forming an indistinct 'vitta'), and towards the lower margins very large and wide (eg 200 × 30–35 μm). Adaxial surface with a variable number of pores: in limb, 2–6 large, mainly distinctly ringed, pores per cell; lower lateral cells with numerous medium-sized pores along the commissures; median 'vitta' usually with few pores; most cells with a resorption gap in the apical angle, but pores otherwise *regular and not enlarged due to extended resorption*. Abaxial surface similar to the adaxial or somewhat less porose in the limb. Hyaline cells of pendent branch leaves more regular in size and porosity, but otherwise similar to those of the spreading branches. **Leaf TS:** Hyaline cells very shallowly convex on both faces near the apex, elsewhere biconvex, but more strongly inflated on the adaxial surface; internal commissural walls smooth or indistinctly (very rarely strongly) papillose. Photosynthetic cells narrowly oval-triangular to trapezoid, reaching both surfaces, but more widely exposed on the abaxial leaf surface; walls slightly to strongly thickened. **Fertile plants:** Monoecious (although apparently sometimes functionally unisexual). Antheridial bracts densely imbricated, often yellowish or pale brown, resembling branch leaves but rather smaller and with less divergent apices. Inner perichaetial bracts large, with relatively narrow insertion; broad and retuse above, with abaxial resorption gaps; apex eroded; hyaline cells differentiated and distinct almost or quite to insertion. Capsules common, often abundant; spores yellow-brown, papillose, variable in size from different populations, 22.0–30.0 μm diameter.

HABITAT: A widespread and common species of damp mesotrophic to slightly eutrophic areas. Like *S. teres*, it is tolerant of a wide range of shade conditions, but is usually more abundant in wooded areas than in open communities. Unlike *S. teres*, it is usually restricted to rather drier, less eutrophic conditions, although it is occasionally found as floating mats, or with *S. warnstorfii*. It forms loose mats between *Carex* and *Juncus* species or *Molinia* tussocks, in fens and flushes, or along shaded stream and pond banks, or wider carpets in *Betula*, *Salix* or *Alnus* carr.

DISTRIBUTION: Widespread throughout the northern hemisphere, from the sub-arctic to the warm temperate zones of Europe (and to the Azores), Asia and North America. Present in the uplands and lowlands throughout Europe, but more abundant at lower altitudes than the closely related *S. teres*, though, like that species, it is confined to mountains in the southern part of its range. Present in suitable localities throughout Britain.

Typical plants are usually recognized readily by their robust habit, bright green colour and strongly squarrose branch leaves, which give them a spiky, 'bottle brush' appearance. Very weak plants may resemble some forms of *S. teres*, in which case microscopic examination is necessary to make a positive identification (though even here there may be some difficulty because of gradation of characters in extreme forms of both species). Some shade forms of *S. palustre* may also resemble *S. squarrosum*, but the former species may be recognized by the roughened, cucullate apices of its branch leaves and the spiral fibrils in the branch cortex. Squarrose-leaved forms of *S. compactum* may be distinguished by their minute stem leaves.

Figure 36. Distribution of *S. squarrosum*

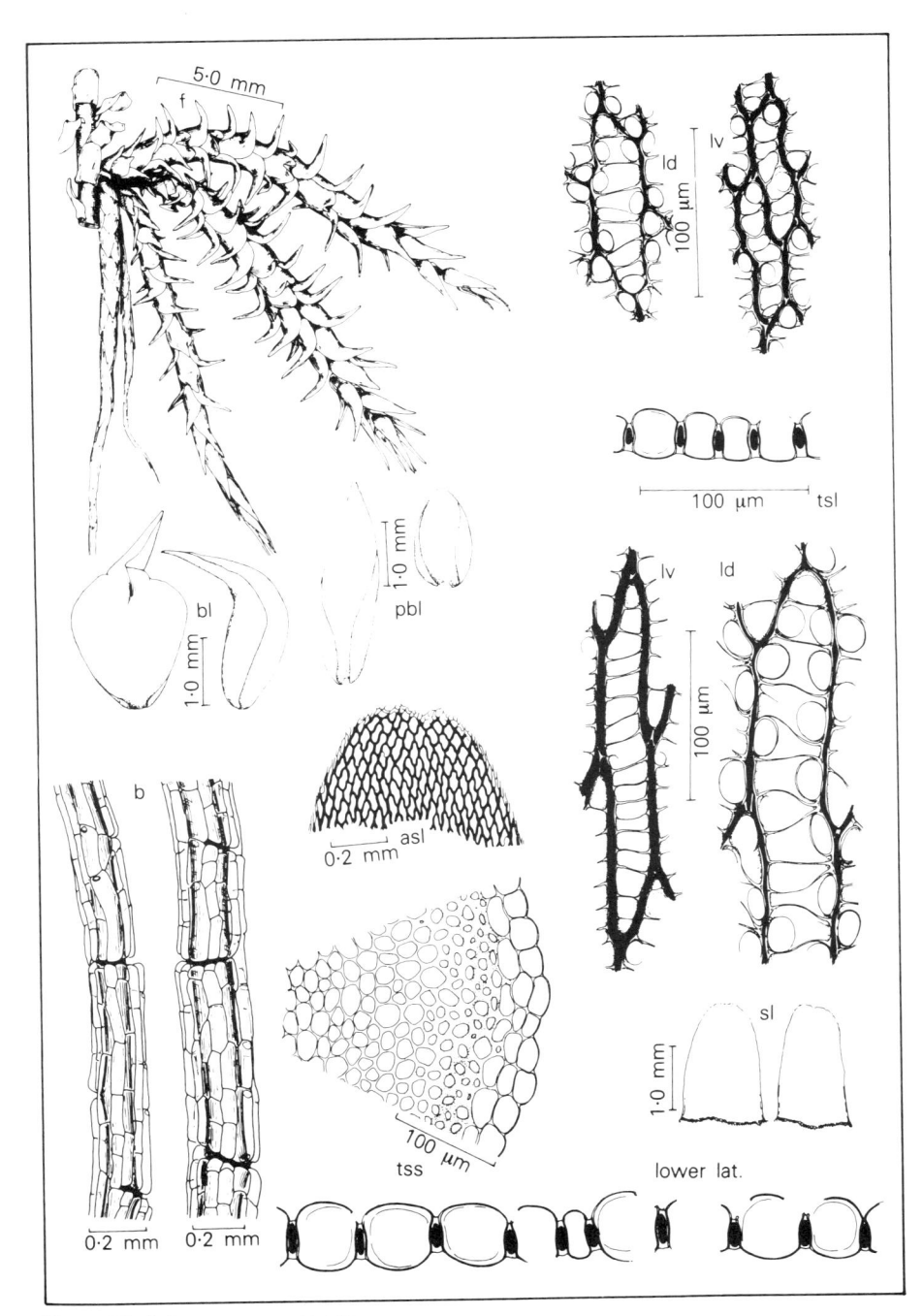

Figure 37. *Sphagnum squarrosum*

SECTION INSULOSA

Sphagnum sect. *Insulosa* Isov. (*Ann. bot. fenn.*, **3,** 231. 1966)
Sphagnum sect. *Truncata* Horrell (*J. Bot., Lond.*, **38,** 119. 1900) Nom. illeg.

This section contains a single species, *S. aongstroemii*, so that the characters given for this species may be regarded as those of the section as a whole.

18. *SPHAGNUM AONGSTROEMII*

Sphagnum aongstroemii Hartm. (*Handbk skand. flora*, 399. 1858)

PLANTS: Medium-sized or, occasionally, small and dense, turgid (like a delicate *S. papillosum*); yellow-green or straw, occasionally with some ochre coloration; capitula without visible stem buds. **Fascicles:** Of 4–5(–6) branches; 2(–3) spreading branches fat, abruptly to moderately tapered; 2–3 pendent branches much weaker and often shorter than spreading branches, or tapering and slightly longer; commonly one of the branches intermediate in character between spreading and pendent types. **Stem:** Rather thin, 0.8–1.1 mm diameter; cortex well developed, of 3–4 layers of much inflated hyaline cells which are more or less square in surface view and without pores; internal cylinder, in section with moderately thick walls and rather irregular lumina, yellowish. **Branch anatomy:** Spreading branches 6.0–15.0 (–20.0) mm long; cortical cells large; retort cells distinct, in groups of 2–3, moderately protuberant, but scarcely rostrate; internal cylinder yellow. **Stem leaves:** Erect or, below, hanging; rather small (1.1–1.2 × 0.8 mm); lingulate to rectangular; resorbed across the wide, truncate or rounded-truncate apex; border evanescent above, moderately to strongly expanded above insertion; adaxial surface intact, without fibrils; hyaline cells resorbed on abaxial surface; septa usually numerous. **Branch leaves:** Large; very concave; erect and incumbent, or the lower somewhat spreading, occasionally almost squarrose; *broadly ovate-triangular;* lower leaves usually with strongly inrolled upper margins, the upper more shallowly concave; apex (at least in the lower leaves) *widely truncate, ca 0.2 mm across the tip, with 8 or more distinct teeth,* in upper leaves often eroded and less conspicuously truncate; border narrow, 2–3 cells wide, without resorption furrow. Pendent branch leaves ovate to ovate-lanceolate, concave, not markedly convolute. **Hyaline cells:** In mid-leaf 17–21 × 90–110 μm, wider towards lower margins; abaxial faces with rather small to medium-sized (8.0–10.0 μm), strongly ringed pores scattered along the commissures and in most or all of the cell angles; pores in lower lateral cells usually more numerous and often larger; adaxial surface without pores but with 1–3 small pseudopores in some lateral angles.

Pendent branch leaf hyaline cells with numerous pores in series along the commissures on the abaxial surface (round, but appearing elliptical due to convexity of cell walls); on the adaxial surface, with numerous circular pores near the commissures. **Leaf TS:** Hyaline cells shallowly convex; internal commissural walls smooth; photosynthetic cells elliptical, with oval lumina; narrowly exposed on both leaf surfaces via strongly thickened walls (often with slightly wider exposure on the adaxial surface). **Fertile plants:** Dioecious. Antheridial bracts smaller than normal leaves, borne near the ends of fertile branches; inner female bracts large, up to 5.0 mm long, convolute, ovate, narrowed above to obtuse, usually minutely retuse apices composed of intact, thick-walled prosenchymatous cells; lower halves of bracts of uniform, pitted, prosenchymatous cells; upper halves differentiated into hyaline cells and thin-walled photosynthetic cells; small area just below apex usually weakly fibrillose. Female plants frequent but capsules uncommon; spores yellowish, rather strongly papillose; 22.0–23.0 μm diameter.

HABITAT: Forming low hummocks or tufts in weakly minerotrophic locations in open *Salix* scrub or dwarf shrub communities, often near the margins of oligotrophic and weakly mesotrophic mires, beside streams or lakes or, occasionally, on wet rocks. *S. aongstroemii* is a species of rather open habitats and, although it may form pure hummocks, it often grows in mixed stands with other species of *Sphagnum*, eg *S. lindbergii*, *S. russowii*, *S. girgensohnii* or *S. squarrosum*.

DISTRIBUTION: A circumpolar species occurring only rarely south of the Arctic Circle. Locally frequent in arctic Scandinavia and Russia, but rare further south; it reaches central Sweden and Lithuania. Absent from the British Isles.

This arctic species most closely resembles a small member of the section *Sphagnum* or, in its laxer states, a robust *S. tenellum*. Its most characteristic features in the field are the broad, very concave, widely truncated branch leaves (especially noticeable in leaves from lower branches, which, under a lens, have a distinctive 'sheared off' appearance). *S. teres* may bear a superficial resemblance to small forms of this species, but has narrower, more finely pointed branch leaves. *S. auriculatum* normally has darker stems and lacks the thick, hyaline cortex of *S. aongstroemii* (this is readily seen with the aid of a lens).

The monotypic section Insulosa is maintained, principally because *S. aongstroemii*, like *S. wulfianum*, combines anatomical features of some section Acutifolia (especially *S. girgensohnii*) species with others found in members of section Squarrosa. Merging it with either of these sections would

Figure 38. Distribution of *S. aongstroemii*

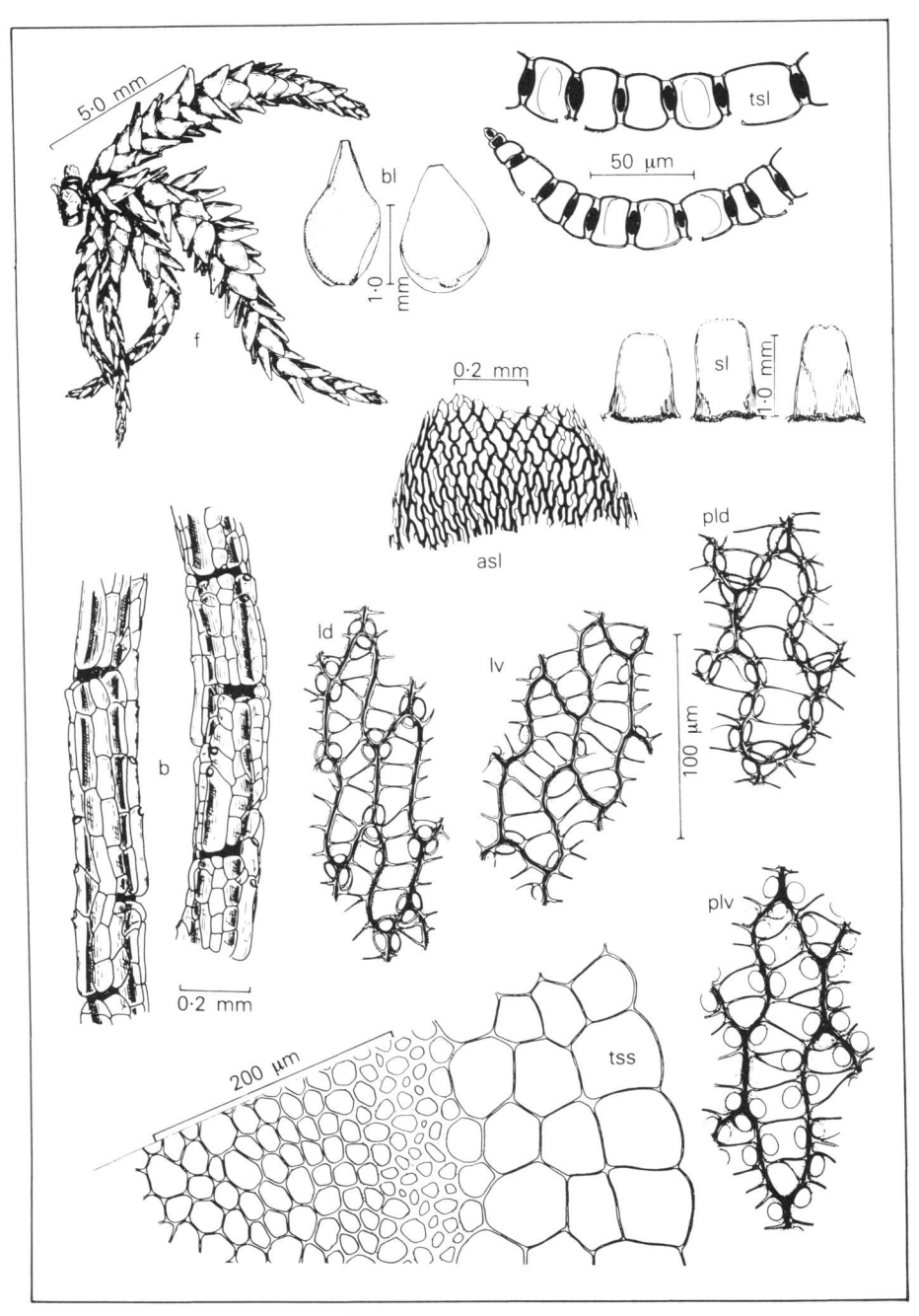

Figure 39. *Sphagnum aongstroemii*

create difficulties in maintaining the Squarrosa as an entity distinct from the Acutifolia. Section Insulosa does not appear to possess any unique characters, but has a combination of features which prevent its easy inclusion in either section Acutifolia or section Squarrosa. The stem leaves closely resemble those of *S. girgensohnii* and the perichaetial bracts are similar to some in members of section Acutifolia. The stem cortex is also very similar to that of *S. girgensohnii*, although lacking pores. Branch leaf pores, although rather small, are similar in form and arrangement to those of members of the Acutifolia (eg the strong tendency for the development of 'triplets'). Other features of branch leaf anatomy suggest closer affinities with section Squarrosa. On balance, anatomical features of *S. aongstroemii*, superficial appearance apart, suggest the closest phylogenetic relationship to be between section Insulosa and section Acutifolia.

SECTION POLYCLADA

Sphagnum sect. *Polyclada* (C. Jens.) Horrell (*J. Bot., Lond.*, **38,** 119. 1900)

This section contains only one species, *S. wulfianum,* so that the characters italicized below under the species are also those which are diagnostic of the section.

The systematic position of the section Polyclada is somewhat obscure. Morphological features suggest close relationships with section Acutifolia and section Squarrosa. Similarities with the former include the solitary retort cells, perichaetial bract morphology and, perhaps, the *S. warnstorfii*-like pores. In common with section Squarrosa are the form and attitude of the spreading branch leaves, the pore structure in the pendent branch leaves, and the form and apparent papillosity of the photosythetic cells. In *S. teres,* there is also a distinct tendency for occlusion of the photosynthetic cells on the adaxial leaf surfaces. Although reduced in size, the anatomy of the stem leaves is very similar to that found in section Squarrosa. The view taken here is that this section is more closely related to the section Squarrosa (an approach first adopted by W P Schimper in 1857), but that it has some relict features of the section Acutifolia. It is considered to add further evidence of an ancient derivation of section Squarrosa from section Acutifolia, and it may be appropriate here to draw attention to the 'Squarrosa' features found in the more divergent species of section Acutifolia, such as *S. fimbriatum.*

19. *SPHAGNUM WULFIANUM*

Sphagnum wulfianum Girgens. (*Arch. Naturk. Liv.-, Est.-u Kurlands,* ser. 2, 173. 1860)

PLANTS: Medium-sized to slender but relatively tall; stiff; *capitula dense, often almost spherical, with closely packed branches,* with visible terminal stem bud; green to olive-brown. **Fascicles:** Regular, rather distant to occasionally dense; *of numerous, 8 or more, branches;* spreading branches usually 4–5, 8.0–15.0(–20.0) mm long, stiff, narrow but not tapering, densely foliated; pendent branches 4–6, variable in length and vigour, but at least some much longer than the spreading branches, up to 25 mm long or more, *thin but not conspicuously terete or pale.* **Stem:** Rigid; relatively strong, 0.7–1.2 mm diameter; in section rounded-pentagonal; cortex well developed, 3-layered, the outermost layer thin-walled and ephemeral, soon collapsing and often completely lost on older parts of stems (so that old stems appear 2-layered with firm outer walls), the inner layers thick-walled and usually

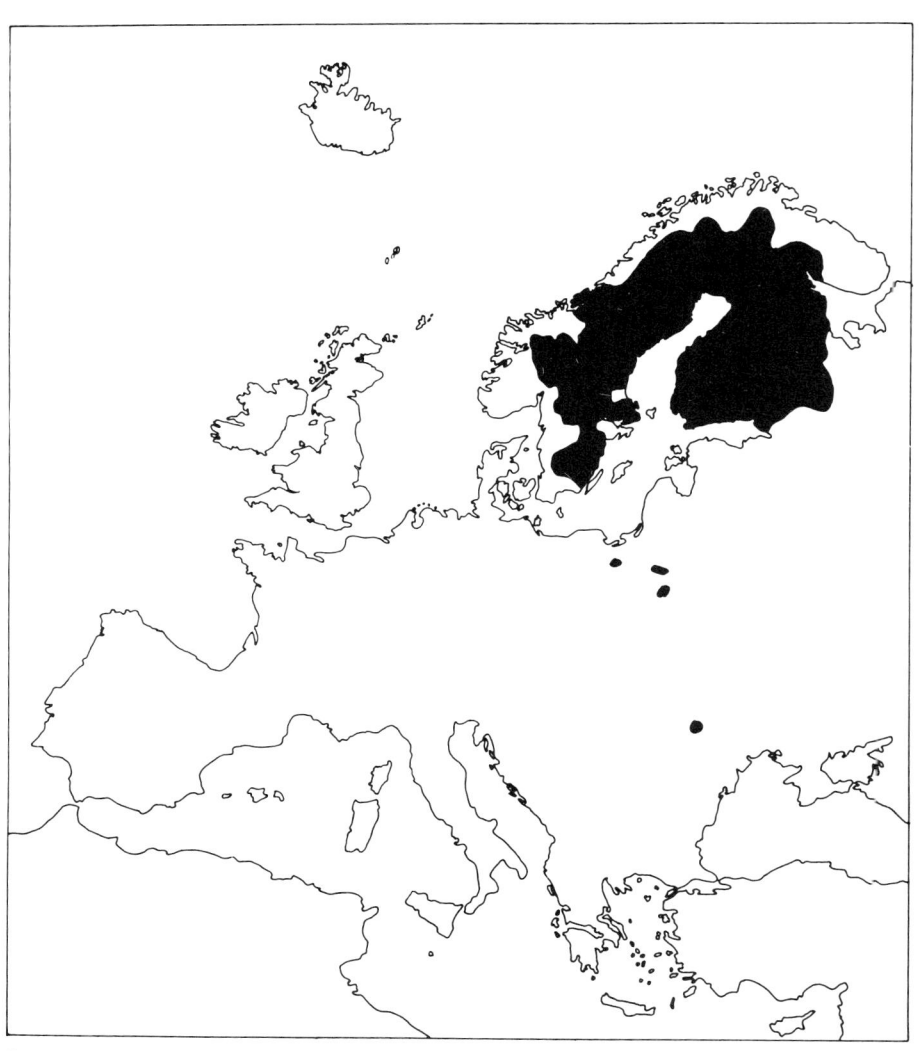

Figure 40. Distribution of *S. wulfianum*

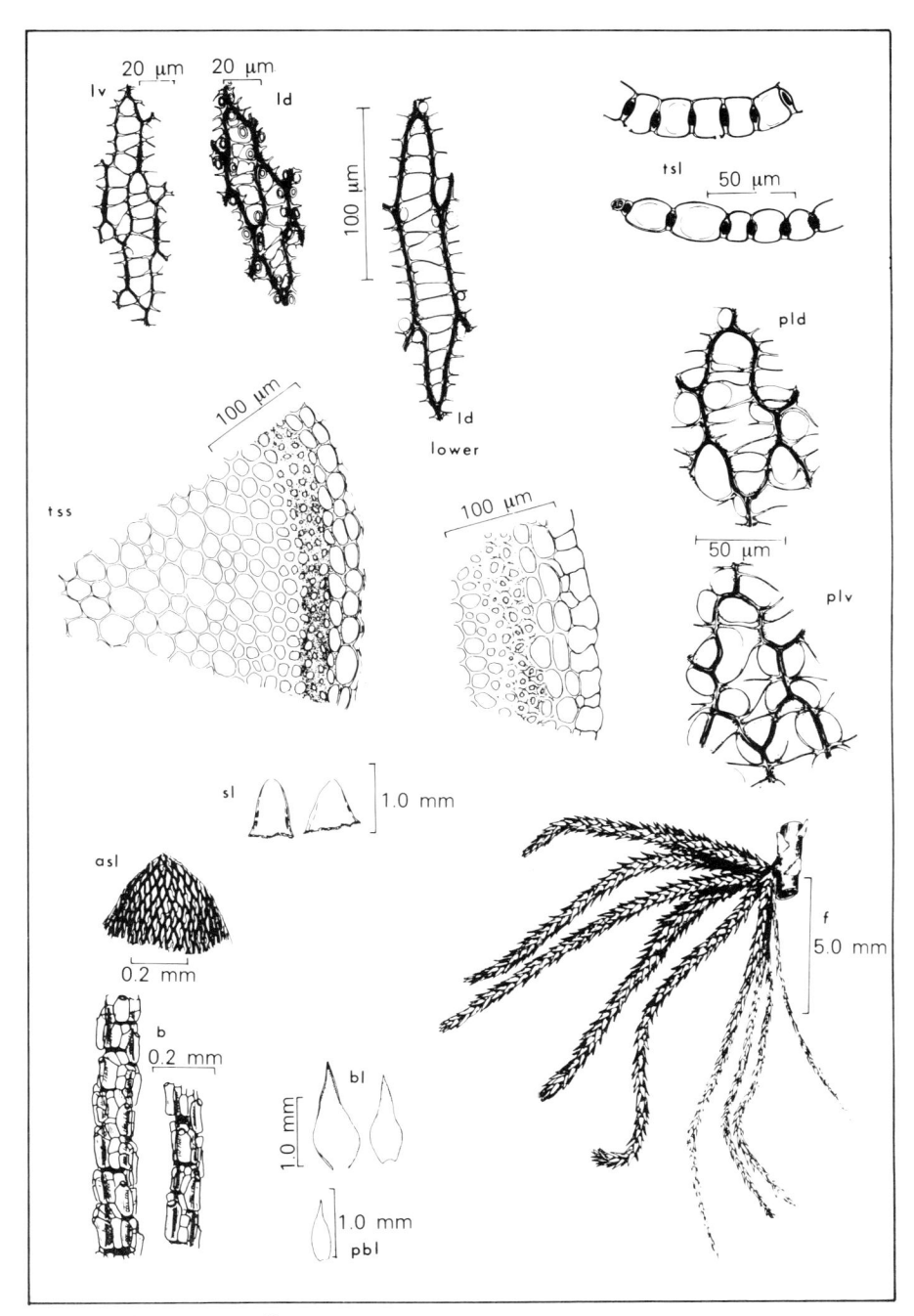

Figure 41. *Sphagnum wulfianum*

concolorous with the internal cylinder: internal cylinder well developed, thick, the outer layers often with almost occluded lumina, dark brown (to almost black in surface view). **Branch anatomy:** Retort cells distinct (sometimes indistinct towards branch base), solitary, not or slightly rostrate; internal cylinder brown. **Stem leaves:** Erect, or more or less spreading or hanging; minute, 0.6–0.9 mm long; rounded-triangular; apices rounded, resorbed; border narrow, slightly widened (to about 6 cells) below, vanishing below apices. Hyaline cells without fibrils; resorbed on abaxial surface; adaxial surface intact except near leaf apices; septa numerous; tissue above insertion thin-walled and weak (leaves often folded back along this line of weakness to appear hanging). **Branch leaves:** Rather small, 1.1–1.6 mm long; *densely imbricated* on the branches, mostly spirally arranged but occasionally 5-ranked (never consistently so); slightly to strongly recurved, the apices spreading at an angle of 40°–80°, sometimes sharply bent out from the middle (subsquarrose). Leaves ovate, sharply narrowed from about half-way to finely tapering, almost acute, apices (resembling a slender version of *S. teres*); coronate at insertion due to backwardly projecting hyaline cells. Pendent branches less densely foliate; leaves erect, lanceolate (smaller than spreading branch leaves but not differing markedly in outline). **Hyaline cells:** Near apex, small and narrow, ca 70 × 15–18 μm; longer and wider below, ca 170 × 20 μm; a few at 'shoulders' up to 30 μm or more wide. Abaxial face of upper cells with few to several *very small* (2 μm or less), *thick-ringed pores* near to, but clear of, the commissures (similar to pores of *S. warnstorfii*); cells in lower part of leaf with few, often unringed and poorly defined medium-sized (6.0–9.0 μm diameter) pores; about 1–3 marginal cell series in mid-leaf usually with somewhat larger pores. Adaxial surfaces without, or with 1(–2) pores in some lateral angles. Pendent branch leaf hyaline cells short and *wide, with large pores and resorption gaps* on both leaf surfaces. **Leaf TS:** Hyaline cells almost plane to shallowly convex; internal commissural walls usually minutely papillose. Photosynthetic cells narrowly oval, with oval lumina, mostly immersed on the adaxial leaf surface, rarely, here and there, also immersed on the abaxial face, mostly reaching, and narrowly exposed (via thick walls), on the abaxial surface. **Fertile plants:** Monoecious or dioecious. Antheridial branches concolorous with, or distinctly darker than, sterile branches; bracts scarcely differing from branch leaves. Inner perichaetial bracts large, ovate, tapering to *more or less acute apices; apices of more or less uniformly prosenchymatous tissue;* hyaline cells intact, without fibrils, sometimes distinct to near insertion, sometimes only in upper half of bract. Capsules occasional, usually remaining rather short-stalked; spores yellow-brown, slightly roughened to coarsely papillose, about 22 μm diameter.

HABITAT: A species of moist lowland coniferous forest, in which it grows as loose carpets on damp mineral soil or peat. Typically, it is associated with *S. girgensohnii* or *S. capillifolium* var. *capillifolium* and *Dicranum undulatum* within *Pinus* or *Picea* forest, often with an under-flora of dwarf shrubs. Rarely, it occurs in more open conditions in dwarf shrub communities of the sub-arctic zone.

DISTRIBUTION: An incompletely circumboreal species with distinctly continental tendencies, but absent from eastern Asia and the central interior of North America. It is more or less confined within the coniferous forest belt between 50° N latitude and the Arctic Circle. In Europe, it is distinctly northern-continental, found mainly in Fennoscandia but extending through parts of western Russia to eastern Poland and northern Romania. Absent from the British Isles.

The densely crowded branches (8 or more per fascicle, of which 4–5 are spreading) and the almost spherical capitulum are features which distinguish this species from other members of the genus. *S. compactum* also has crowded branches, but these are fewer per fascicle (3–5, of which 1–2 are spreading) and are usually directed upwards, so hiding the capitulum which, in the present species, is conspicuous. In *S. compactum,* the branch leaves are large; in *S. wulfianum,* they are small.

SECTION HEMITHECA

Sphagnum sect. *Hemitheca* Lindb. ex Braithw. (*Sphag. Eur. N. Am.,* **30,** 85. 1880)

This is a monospecific section and the description of *S. pylaesii* (especially the italicized characters) may be taken as that of the section as a whole. In superficial appearance, there are strong resemblances to some weak, aquatic forms of some species of the section Subsecunda, especially where those plants have been stained by humic acids in mire water. However, the very markedly dimorphic cells of the branch hyaloderm of *S. pylaesii,* in addition to the lack of pores and other, more minor, structural differences, set the species and the section apart from Subsecunda. Sporophyte variation in *Sphagnum* is minimal for such a diverse genus, so that the characteristically small capsules of *S. pylaesii* lend additional support to the separation of Hemitheca from Subsecunda.

20. *SPHAGNUM PYLAESII*

Sphagnum pylaesii Brid. (*Bryol. univers.,* 749. 1827)
S. pylaiei Braithw. (*Mon. Microsc. J. Trans.,* **13,** 231. 1875)

PLANTS: Small, but often much elongated; dull, olive-green to *purple-brown to almost black; often rather short and lacking branches* (but with stems dividing remotely and irregularly) and having the appearance of a pleurocarpous moss; capitula scarcely developed. **Fascicles:** *Lacking or, when present, of only 1–2 small branches.* **Stem:** Thin, up to 0.5 mm diameter; cortex well developed, of (1–)2(–3) layers of hyaline cells mostly with intact walls lacking fibrils; internal cylinder brown or red-brown. **Branch anatomy:** Branches (where present) up to 10.0 mm; cortical cells distinctly dimorphic; retort cells large, much inflated, slightly protuberant at distal ends, mainly in linear groups of 2–3; other cortical cells *short, small.* **Stem leaves:** *Closely imbricated;* large, 1.5–2.0(–2.5) mm long, 1.0–1.3 mm wide, ovate, concave; apices rounded and eroded, of ephemeral, partly resorbed cells (a thin hyaline border often detectable under a lens); border narrow or, usually, *border of narrow cells absent and the leaf margin consisting of reduced, fibrillose hyaline cells;* hyaline cells fibrillose, but *pores absent* (rarely with a minute, inconspicuous pore in the basal angle of the adaxial surface), occasionally with interfibrillar resorption gaps on the abaxial surface (not seen in European material); photosynthetic cells almost, or quite, as wide as hyaline cells, in TS more or less rectangular or trapezoid with oval lumina and thick walls, slightly to distinctly more widely exposed on the adaxial surface. **Branch leaves:**

Small, *much smaller than the stem leaves*, up to 1.2 × 0.6 mm, ovate; apices rounded-obtuse; border of a single cell series with a *resorption furrow*. **Hyaline cells:** Narrow, 10.0–12.0 × 80–110 μm; adaxial face flat, intact or with a minute, inconspicuous pore in the apical and/or basal angle; abaxial face flat to slightly convex, without pores. **Leaf TS** (stem leaf or branch leaf): hyaline cells scarcely larger than photosynthetic cells; partial septa (fibrils) conspicuous; internal commissural walls smooth; photosynthetic cells trapezoid, thick-walled with oval to rounded trapezoid lumina; more or less equally exposed on both surfaces via the thick upper and lower walls; cell walls brown or purple. **Fertile plants:** Dioecious. Antheridial branches resembling sterile ones, but more tumid; female bracts resembling stem leaves but larger. Fertile plants unknown in Europe but found occasionally in North America; capsule very small, hardly emerging from upper bracts.

HABITAT: In Europe, a plant of upland areas under oceanic influence, although it has been reported from more inland montane areas in eastern North America. Found in hollows subject to periodic flooding by weakly minerotrophic water or on damp marginal areas receiving runoff from moderately oligotrophic peatlands. In Finistère, it is a characteristic species of an oceanic facies of Rhynchosporetum, in which the cover of vascular plants is low, and where it may be accompanied by other members of the section Subsecunda, particularly *S. subsecundum* subsp. *inundatum*.

DISTRIBUTION: Widespread in eastern North America. Extremely local in Europe, known only from Finistère and a single locality in north-west Spain.

In the field, this species is more likely to be overlooked as a pleurocarpous moss rather than be mistaken for any other species of *Sphagnum*. Although other *Sphagna* do produce unbranched 'stolons', these have pale, almost hyaline leaves and stems. *S. pylaesii* is very rarely pale and, especially in its unbranched form, tends to develop a deep red-brown colour. Short forms lacking branches have been named var. *sedoides* (Brid.) Lindb. whilst longer, usually dull olive-brown to blackish plants are typical of the species. A series of intermediate forms occurs in North American material and this throws some doubt on the value of var. *sedoides* as a taxonomic entity. Finistère plants are of the *'sedoides'* type, but the Spanish material does have some branches.

Figure 42. Distribution of *S. pylaesii*

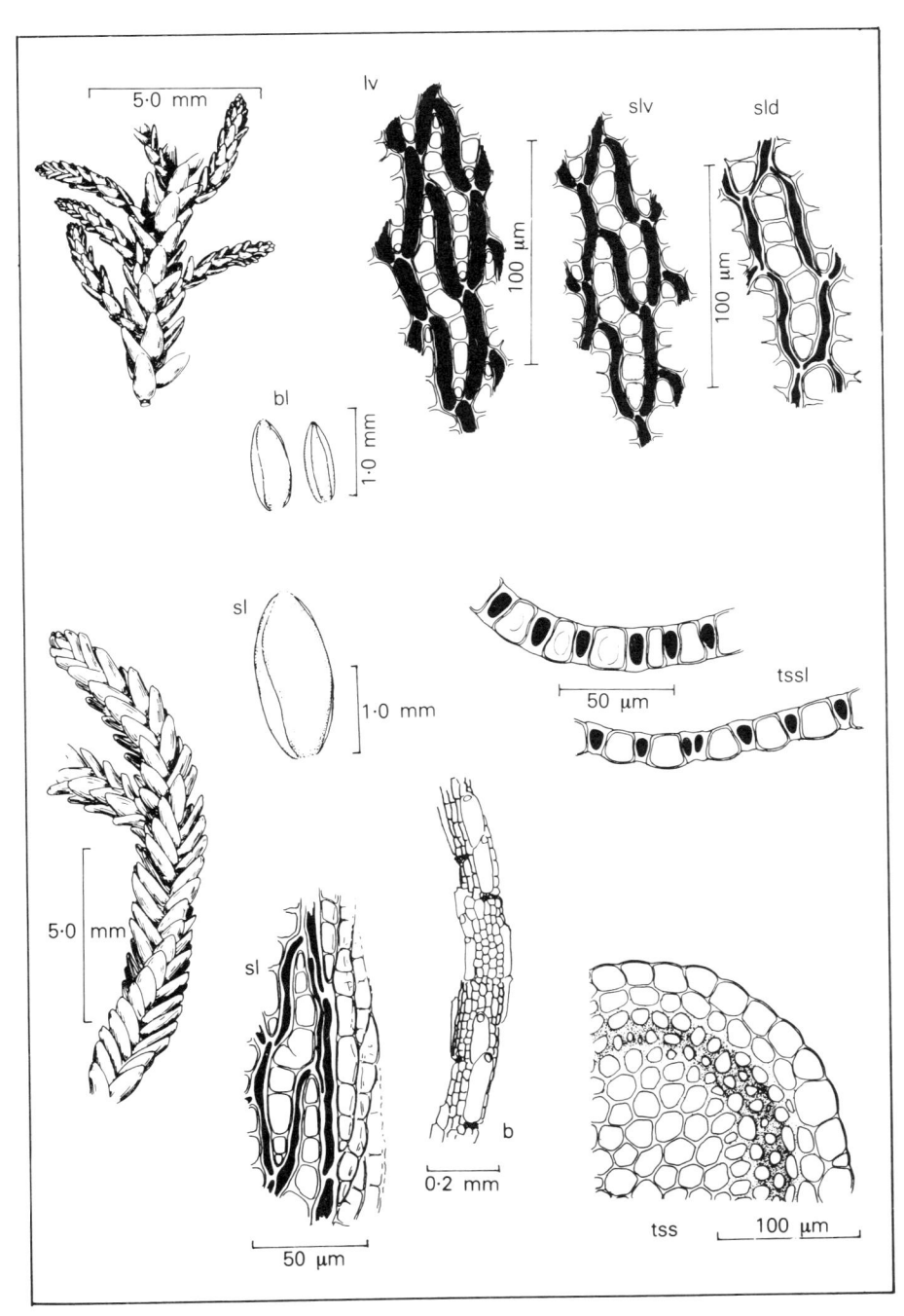

Figure 43. *Sphagnum pylaesii*

SECTION SUBSECUNDA

Sphagnum sect. *Subsecunda* (Lindb.) Schimp. (*Syn. musc. eur.* 2nd ed., 843. 1876)

PLANTS: Extremely variable in size, colour and ecology. **Fascicles:** Sometimes absent, often few-branched, but in some species with up to 7, rarely 8, branches which are *not, or weakly, dimorphic*. **Stem:** Cortex of 1–3 layers of hyaline cells; fibrils absent from cell walls; outer walls without pores, or with a single, indistinct one, at the distal end of the cell; internal cylinder distinct, pale or dark. **Branch anatomy:** Retort cells usually distinct (except in basal regions of some of the more robust species), in groups of 2–4. **Stem leaves:** Except in a few species, usually large to very large in comparison with the branch leaves, usually strongly fibrillose in at least the upper third (exceptions: *S contortum, S. subsecundum* subsp. *subsecundum*); border 3–6 cells wide, *not, or only slightly, widened near leaf insertion*. **Branch leaves:** Except in the distal parts of branches of some aquatic forms, *proportionately broad*, ovate to ovate-lanceolate, sometimes not much longer than broad, usually markedly concave, with obtuse to truncate, dentate or eroded, sometimes hooded (but not scabrid) apices. **Hyaline cells:** (in the European species) *Proportionately narrow*, 6–10 times as long as wide in mid-leaf (resembling those of section Cuspidata); pores, when present, usually distinctly, and often heavily, ringed, *small*, rarely more than 6.0 μm diameter, scattered or in series along the commissures. **Leaf TS:** Internal commissural walls smooth. Photosynthetic cells mostly *barrel-shaped with median, oval lumina* and thick upper and lower walls; in European species, rarely asymmetrically exposed or fully enclosed (but normally with slightly wider exposure on one or other surface). **Fertile plants:** Dioecious (as far as known). Inner perichaetial bracts large and obtuse, nearly always fibrillose near the apex.

A more or less cosmopolitan section containing about 40 species.

Section Subsecunda is probably the largest, and is certainly the most heterogeneous subgeneric group of the genus *Sphagnum*. Moreover, many of its species possess a combination of characters which suggest that it may be the antecedent of the other European sections. There are also indications of phylogenetic links between this and the sections Acutifolia, Cuspidata and Rigida in other parts of the world. In Europe, however, the Subsecunda are clearly defined and only rarely should there be any difficulty in assigning members correctly to the section, even if the species cannot be determined. The branch leaf hyaline cells resemble those of section Cuspidata in their narrow outlines, but normally differ in the frequent to abundant small, ringed pores along the commissures on the abaxial leaf surface and the absence of large, unringed pores from the adaxial. The branch leaves, at least when flattened, are broader than those in the Cuspidata and the photosynthetic cells differ in shape and position. In the field, there is an occasional resemblance of some species to *S. compactum* or *S. palustre*, but there is never the combination of large branch leaves and minute stem leaves of the former, or the thick, fibrillose stem cortex of the latter.

21. *SPHAGNUM SUBSECUNDUM*

Sphagnum subsecundum Nees. in Sturm. (*Deutschl. Flora,* **2** (17), 3. 1819)

PLANTS: Rather small (sometimes somewhat resembling *S. tenellum*), low growing to rather tall; various shades of yellow-brown variegated with orange (never deep red), green only in dense shade. **Fascicles:** Rarely dense; 5–6 branches; 2–3 spreading, rarely more than 15 mm long; pendent 2–4, weaker than, but otherwise similar to, spreading. **Stem:** 0.4–0.7 mm diameter; cortex a single, well-defined layer of inflated hyaline cells, mostly with one or occasionally 2 pores or thinnings; internal cylinder brown to almost black, at least in part, rarely entirely pale, green (mostly in shade forms). **Branch anatomy:** Retort cells distinct, protuberant, in linear pairs; internal cylinder light brown. **Stem leaves:** Mostly hanging, a few sometimes spreading; *very small,* 0.7–1.1 mm long; triangular-lingulate; apices broadly rounded, obtuse and eroded, usually concave and somewhat cucullate; border narrow, not widened above insertion. Hyaline cells on abaxial surface without, or with very few, pores; on adaxial surface near apex, with 2 or several circular, more or less unringed pores, 7.0–12.0 µm diameter, these becoming fewer and being replaced by 1–2 resorption gaps away from apex; fibrillose only near apex, rarely lacking fibrils, although fibrils often incomplete. **Branch leaves:** *Small,* 0.9–1.2(–1.4) mm long; weakly 5-ranked; ovate, the *lower usually asymmetric, curved and secund with inrolled margins,* the upper symmetric and concave. **Hyaline cells:** Small and narrow, 11.0–15.0(–18.0) × 80–130 µm; abaxial surface with numerous (up to 40), minute (2.0–4.0 µm), usually ringed pores in series along the commissures; adaxial surface without pores, or pores few and inconspicuous in a few cell angles. Pendent branch leaf hyaline cells similar to spreading branch ones, or slightly wider with larger pores. **Leaf TS:** Hyaline cells shallowly convex; internal commissural walls smooth. Photosynthetic cells barrel-shaped or elliptical with median, oval lumina and narrow, but strongly thickened, abaxial and adaxial walls; narrowly exposed on both leaf surfaces but slightly more so on abaxial. **Fertile plants:** Dioecious. Antheridial branches hardly distinct, but more densely foliated in fertile region. Inner perichaetial bracts about 3.5 mm long, more or less ovate-spatulate with obtuse apices; near apex on adaxial surface, fibrillose and with frequent pores (resembling those of the stem leaves); hyaline cells differentiated to insertion. Capsules uncommon; spores papillose, 30–35 µm diameter.

HABITAT: Not as hydrophilous as the related species *S. auriculatum* or *S. subsecundum* subsp. *inundatum,* and not found in habitats subjected to prolonged inundation. It is a plant of mesotrophic mires and, although sometimes found with *S. contortum,* it avoids the base-rich habitats usually favoured by that species. It is also absent from the more oligotrophic areas in which *S. auriculatum* commonly grows. It is most usually found in oligotrophic or mesotrophic flushes, along stream-sides or on wet, peaty slopes, though usually avoiding areas of dense shade.

DISTRIBUTION: Circumboreal, extending southwards to the Himalayas and New Guinea, but absent from Africa and South America. Widespread in western Europe and locally common in Scandinavia. Uncommon to rare in the British Isles, particularly in the south and east where it is absent from large areas. Locally frequent in hilly districts in the west.

The ovate, rather than lanceolate, distal branch leaves, together with the dark stems and very small stem leaves, should prevent confusion of this plant with those from any of the other sections. It most closely resembles *S. contortum,* except for its usually darker stem: in doubtful cases, the single-layered stem cortex is diagnostic. Small forms of the closely related dihaploid *S. subsecundum* subsp. *inundatum* are sometimes difficult to distinguish from subsp. *subsecundum*. Normally subsp. *inundatum* is larger in all its parts, rarely has more than 5 branches per fascicle, and has more extensively fibrillose and porose stem leaves. *S. tenellum* may bear a superficial resemblance to *S. subsecundum,* but a lens is sufficient to discern the large stem leaves and pale stems, with more or less symmetrical branch leaves throughout, of that species.

There have been many, varied, treatments of *S. subsecundum* and its allies in Europe. Among the more recent discussions of their status and relationships are those by Rahman (1972), Hill (1975) and Eddy (1977).

21a. *S. subsecundum* subsp. *inundatum*

Sphagnum subsecundum subsp. *inundatum* (Russ.) A. Eddy (*J. Bryol.,* **9,** 313. 1977)

S. inundatum Russ. (*Arch. Naturk. Liv.-, Est.- u. Kurlands,* ser. 2, **10,** 405. 1894) Excl. parte

S. auriculatum var. *inundatum* (Russ.) M.O. Hill (*J. Bryol.,* **8,** 439. 1975)

PLANTS: Medium-sized, rarely robust, with dark, elongated or, occasionally congested, stems; green to brownish, variegated with yellow and orange (never deep red), wholly green only in shade. **Fascicles:** Well spaced and not concealing stem completely, sometimes closely set; of *5 branches,* rarely 4 or 6; spreading branches 2–3, 10.0–18.00 mm long, arched but rarely curved and

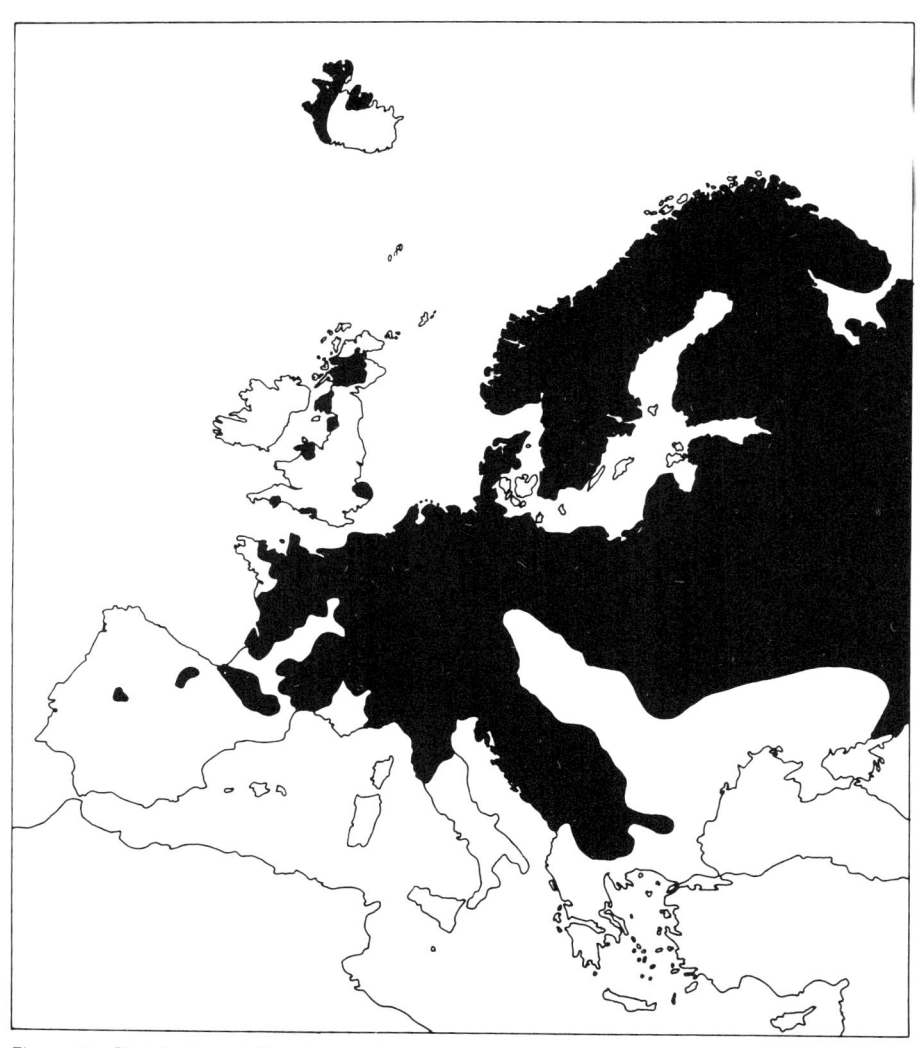

Figure 44. Distribution of S. *subsecundum*

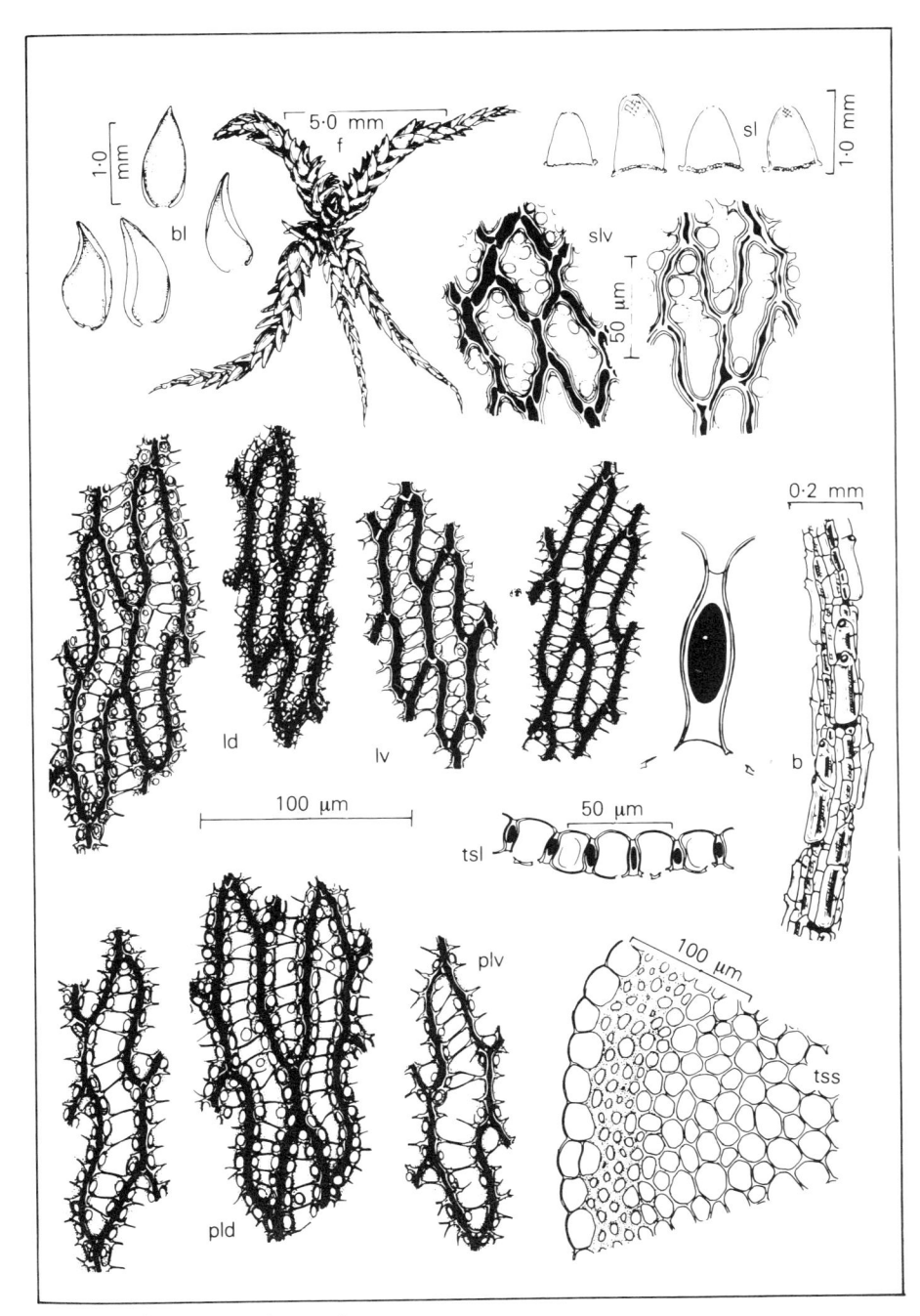

Figure 45. *Sphagnum subsecundum*

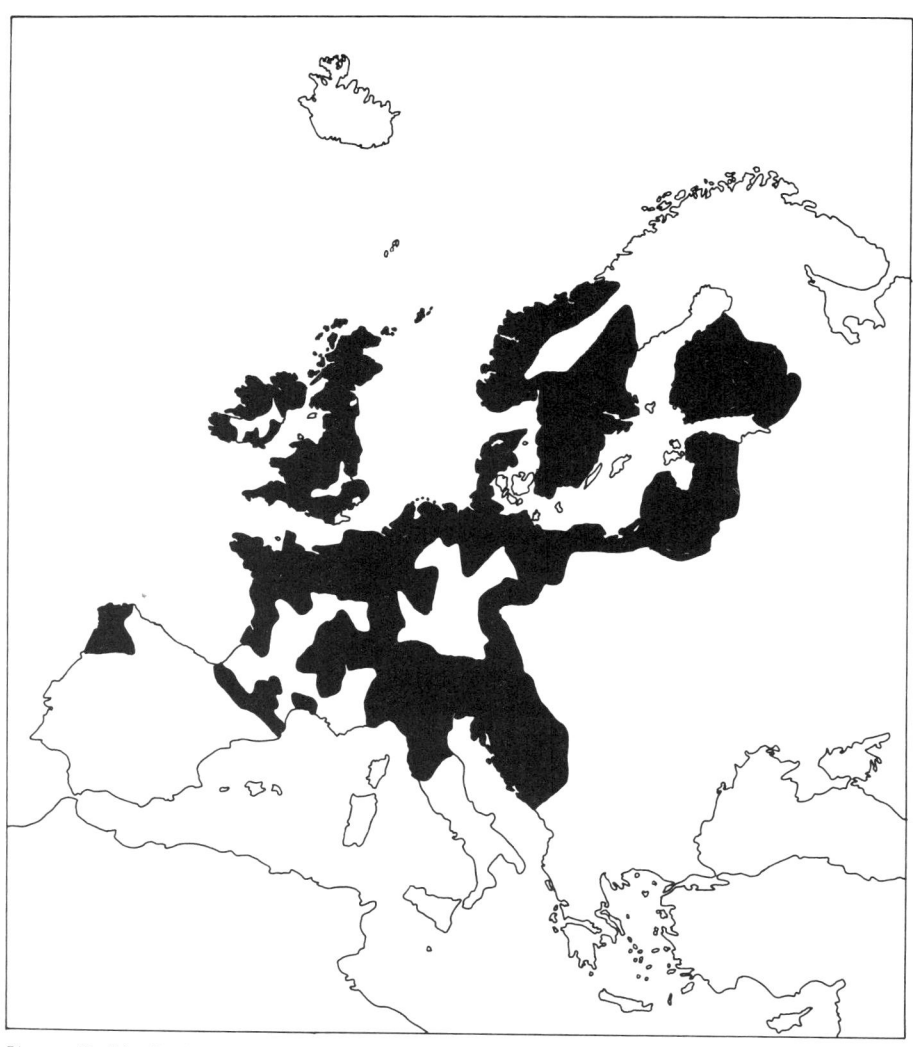

Figure 46. Distribution of S. *subsecundum* subsp. *inundatum*

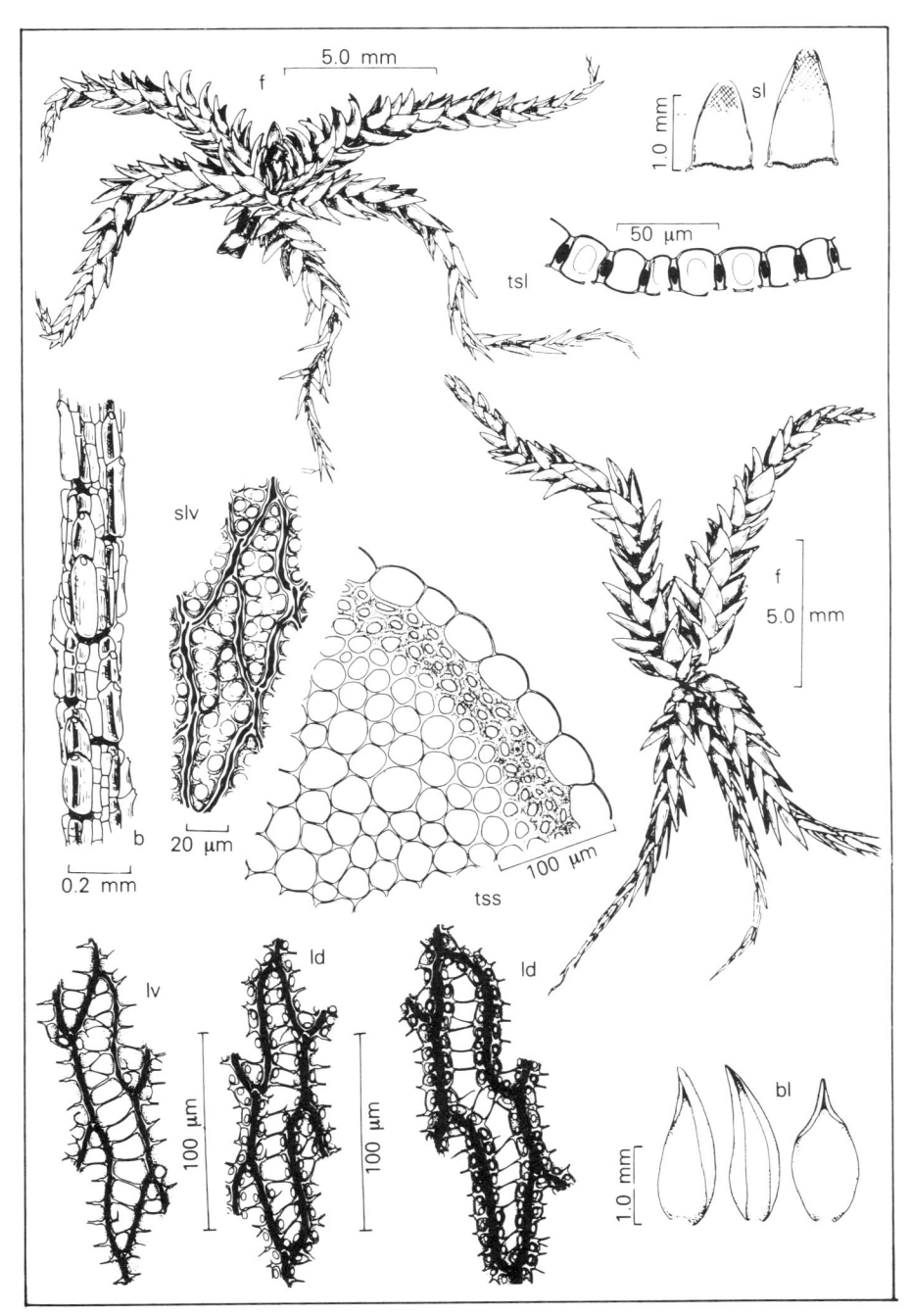

Figure 47. *Sphagnum subsecundum* subsp. *inundatum*

contorted, each bearing *35–45 fully developed leaves* (excluding the 3–5 reduced leaves at the branch base and the smaller leaves at the distal end of the branch. As the leaves are basically in 5 ranks, they may be counted along in rows); pendent branches 2–3, paler, weaker, but not markedly different from spreading ones; more or less as long as spreading branches, but variable. **Stem:** Dark and rather rigid; cortex a single layer of inflated hyaline cells, some of which have a large circular pore or thinning (revealed by staining); internal cylinder dark brown to almost black. **Branch anatomy:** Retort cells distinct, mainly in linear pairs; internal cylinder light brown. **Stem leaves:** Spreading or hanging; 1.2–1.4(–1.6) mm long; triangular-ovate to triangular-lingulate, widest at insertion; apex rounded; *fibrillose above but rarely to beyond one third from apex;* abaxial surface with few or no pores, except near apex; adaxial surface, in fibrillose area, with many small to medium-sized (6.0–8.0 μm) pores near the commissures. **Branch leaves:** (1.2–)1.4–1.8 (–2.2) mm long; weakly 5-ranked; ovate, the lower usually asymmetric, curved and secund, the upper symmetric and concave (similar to subsp. *subsecundum* but larger). **Hyaline cells:** In upper mid-leaf 15.0–20.0 × 130–170 μm; adaxial surface with few, or no, pores; abaxial surface with very small, 2–6 μm diameter, ringed pores along the commissures. **Leaf TS:** Hyaline cells shallowly convex; internal commissural walls smooth. Photosynthetic cells barrel-shaped or elliptical with median, oval lumina and narrow, but strongly thickened, abaxial and adaxial walls; narrowly exposed on both leaf surfaces but slightly more so on abaxial. **Fertile plants:** Dioecious. Antheridial branches hardly distinct. Perichaetial bracts very large, the inner up to 5.5 mm or more, long; ovate-lanceolate from narrow insertions; apices obtuse to subacute; fibrillose to about half-way; hyaline cells distinct almost, or quite, to insertion; adaxial surface, near apex, with numerous pores, the abaxial more or less lacking pores. Capsules occasional to frequent; spores dark brown; coarsely papillose, *large,* 35–42 μm diameter.

HABITAT: Like *S. subsecundum* subsp. *subsecundum,* this subspecies is found along stream-sides and in flushes, though it is more tolerant of both waterlogging and shade. It also occurs in mesotrophic mires, in communities containing *Schoenus nigricans, Juncus acutiflorus* or *Carex* spp. (eg *C. rostrata*), and there it is commonly associated with bryophytes such as *Drepanocladus revolvens, Calliergon giganteum* and *Scorpidium scorpioides.*

DISTRIBUTION: Circumboreal. Its range is much the same as that of *S. subsecundum* subsp. *subsecundum,* though it extends less far into southeast Asia. It is more common in Europe than subsp. *subsecundum* but does not extend as far to the north and east. Although more common in Britain than subsp. *subsecundum,* it is, like that plant, more scattered in the south and east.

In both Britain and Europe, it may be less common than published records indicate because, in the past, there has been much confusion between this taxon and *S. auriculatum*.

This is essentially an enlarged and vigorous version of *S. subsecundum* subsp. *subsecundum* and may be little more than a dihaploid variant of that taxon. Weak plants may be difficult to distinguish from etiolated forms of subsp. *subsecundum*, but size alone, as well as the more extensively fibrillose stem leaves, will usually characterize subsp. *inundatum*. *S. auriculatum*, a much more common and polymorphic species, includes forms which may be more difficult to distinguish from subsp. *inundatum*. Consideration of all the characters italicized above should yield positive identification in most cases: doubtful plants will nearly always prove to be *S. auriculatum*.

22. *SPHAGNUM AURICULATUM*

Sphagnum auriculatum *Schimp.* (Mém. prés. div. Sav. Acad. Sci. Inst. Fr., **15**, 80. 1857)

PLANTS: Extremely polymorphic; medium to very large (rarely as small as *S. subsecundum*); lax or dense; forming pure or mixed carpets, floating, or submerged; colour very variable, green, yellow, red-brown or, sometimes, with some distinct red coloration (especially in aquatic forms); capitula well developed in terrestrial forms and often with the upper branches tapered and curved, so appearing horn-like, and arranged so as to resemble a corkscrew, when viewed from above; in aquatic forms, capitula relatively small and branches straighter. **Fascicles:** Distant to very dense, with 3–4, rarely 5 well-developed branches which vary in size but are not, otherwise, markedly dimorphic (weakest branch usually rather short, pale and deflexed); branches blunt or tapering, usually tumid, short or, occasionally in aquatic forms, long. **Stem:** Sometimes rather firm in terrestrial forms, but usually soft and thin, 0.4–0.9 mm diameter; cortex a single layer of thin-walled or distinctly (rarely strongly) thickened hyaline cells rarely with pores; internal cylinder well developed, yellow-green (especially in shade), brown or almost black. **Branch anatomy:** Retort cells often indistinct near bases of stronger branches (then a number of ordinary cortical cells may have pores), otherwise distinct and in linear groups of (2–)3–4; internal cylinder brown. **Stem leaves:** Erect, spreading or hanging; large, but smaller than branch leaves, (1.1–)1.3–1.8 mm long; lingulate to spatulate, or somewhat rectangular and broadest some distance above insertion; apices widely rounded-truncate, eroded, concave; border thin, not expanded below; hyaline cells fibrillose at least in upper third of leaf, often almost, or quite, to insertion; pores in fibrillose hyaline cells various, usually more numerous on abaxial surface but, sometimes, more numerous on adaxial surface. **Branch leaves:** Large to very large, (1.3–)1.5–2.5 mm or more, smaller in plants from exposed arctic or alpine locations; usually *35 or fewer fully developed leaves per branch* (cf *S. subsecundum* subsp. *inundatum*); more or less symmetrical, ovate, concave, rarely the lower somewhat curved and secund; erect to suberect, so that branches are tumid; apices truncate, dentate or eroded, often hooded; border 2–3 cells wide. **Hyaline cells:** Narrow, 15–20 × 100–200 μm, pores small, 2.0–6.0 μm, ringed, variable in number and distribution; on abaxial surface few to, typically, numerous (up to 40 or, exceptionally, 50 per cell lying along commissures); on adaxial surface, few or absent, sometimes numerous, rarely more so; near leaf apex, on abaxial surface, sometimes with additional free pores in the cell mid-line. Pendent branch leaves similar to those of spreading branches, but hyaline cells sometimes wider with larger, more well-defined pores. **Leaf TS:** Hyaline cells slightly biconvex; internal commissural walls smooth. Photosynthetic cells (except in immersed, shaded forms) barrel-shaped, thick-walled, with oval lumina; exposed on both surfaces but slightly more widely so on abaxial. **Fertile plants:** Dioecious. Antheridial branches scarcely distinct; bracts more concave and more densely imbricated than branch leaves. Inner perichaetial bracts large to very large, 3.5–5.5 mm long, broad above and narrowed to blunt, obtuse apices; hyaline cells differentiated to leaf insertion or almost so, in the upper half of bract fibrillose and usually with scattered pores. Capsules occasional; spores obscurely papillose, 29.0–34.0 μm diameter.

HABITAT: In a wide range of wet, oligotrophic to mesotrophic habitats from sea level to the alpine zone, though most abundant at lower altitudes. Commonly it is submerged in bog pools or ditches, though it also occurs as scattered stems or mats along the margins of pools or oligotrophic streams, in flushes, or in wet, periodically inundated hollows in oligotrophic parts of mires.

DISTRIBUTION: Circumboreal, but with distinct oceanic tendencies, in Europe, north-eastern Asia and North America, and extending from North Africa to the Arctic. Throughout most of Europe but more common in the western, oceanic parts. Common throughout Britain, except for parts of the south-east.

Sphagnum auriculatum is the most variable European species of the genus. Although generally absent above the average water table of undisturbed bogs, and seldom occurring in extensive carpets, it occupies a wide range of oligotrophic, aquatic and semi-aquatic habitats. Although not a basiphile, this species is relatively resistant to eutrophication and may persist in polluted areas near conurbations where few other *Sphagna* survive. Most features of the plant are variable and have been used as the basis for a range of treatments by different authors. The most extreme 'splitting' of the taxon was carried out by Warnstorf (eg 1911), whose treatment was accepted uncritically by Sherrin (1927), with the result that Sherrin's handbook lists 9 species and a great number of named varieties and forms. Whilst much of the plasticity of the species relies on genetic variability, detection of this is obscured by responses to environmental influences. Hence, any logical division of the species into realistic infraspecific taxa is impracticable without detailed genecological studies.

Despite the protean nature of the species, which includes forms superficially similar to other members of the section Subsecunda, it can usually be identified in the field and only rarely provides major difficulties under the

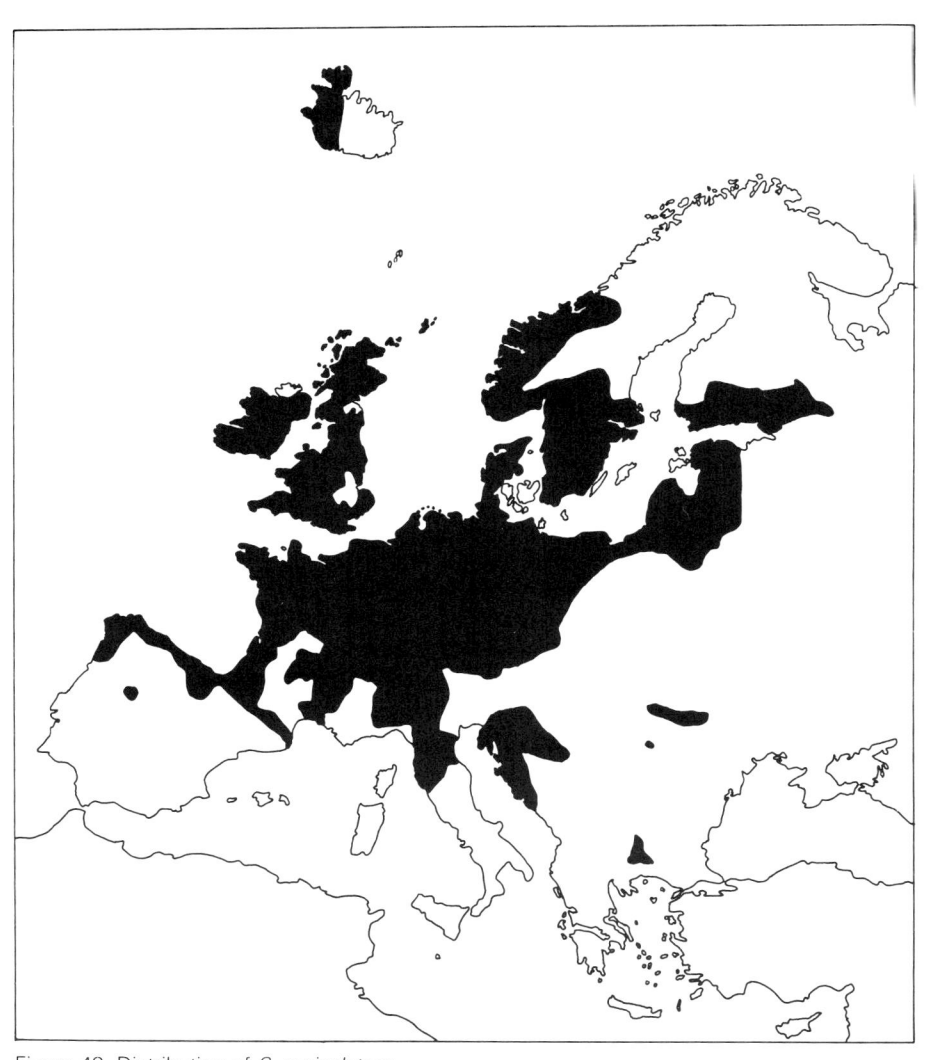

Figure 48. Distribution of *S. auriculatum*

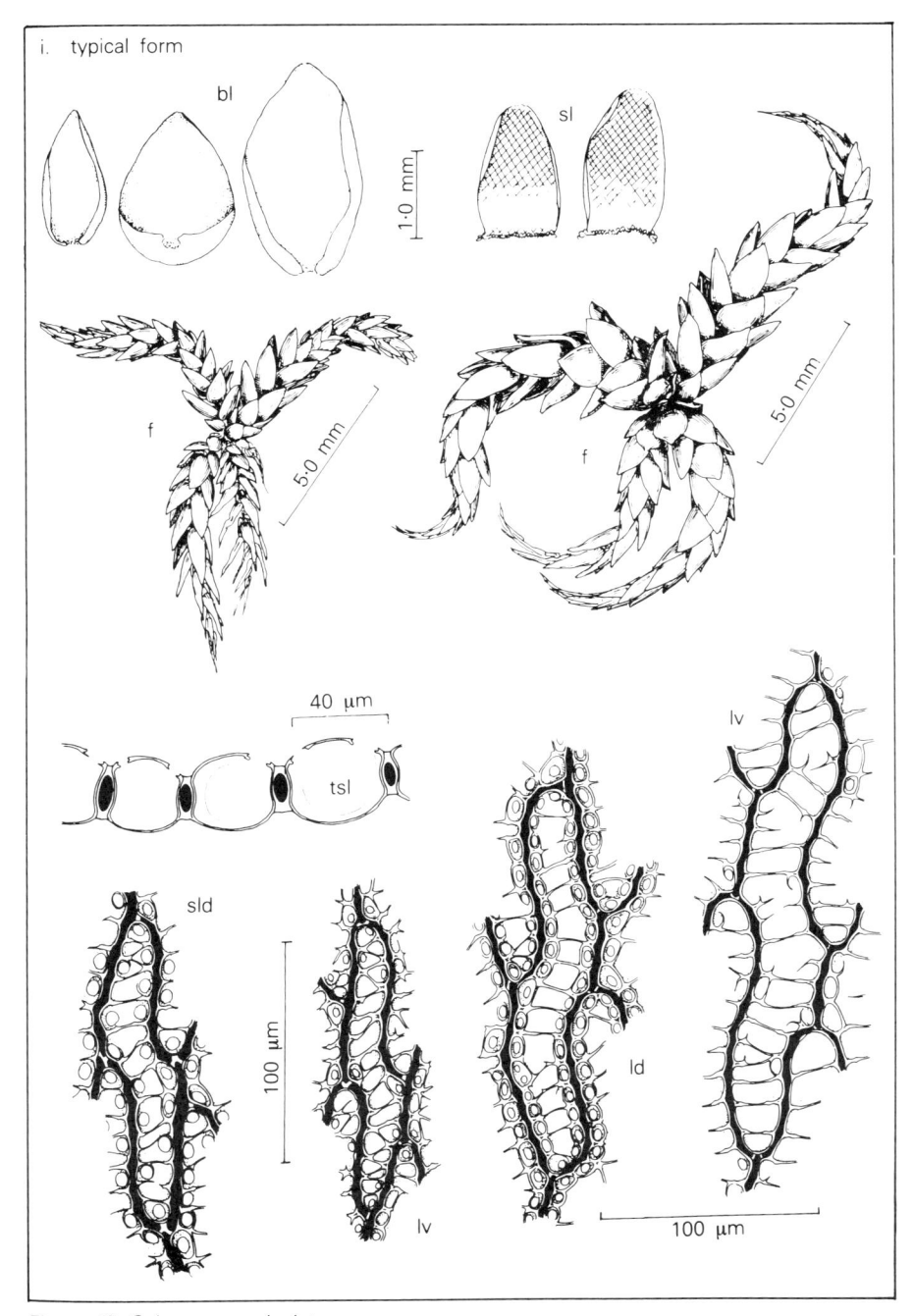

Figure 49. *Sphagnum auriculatum*

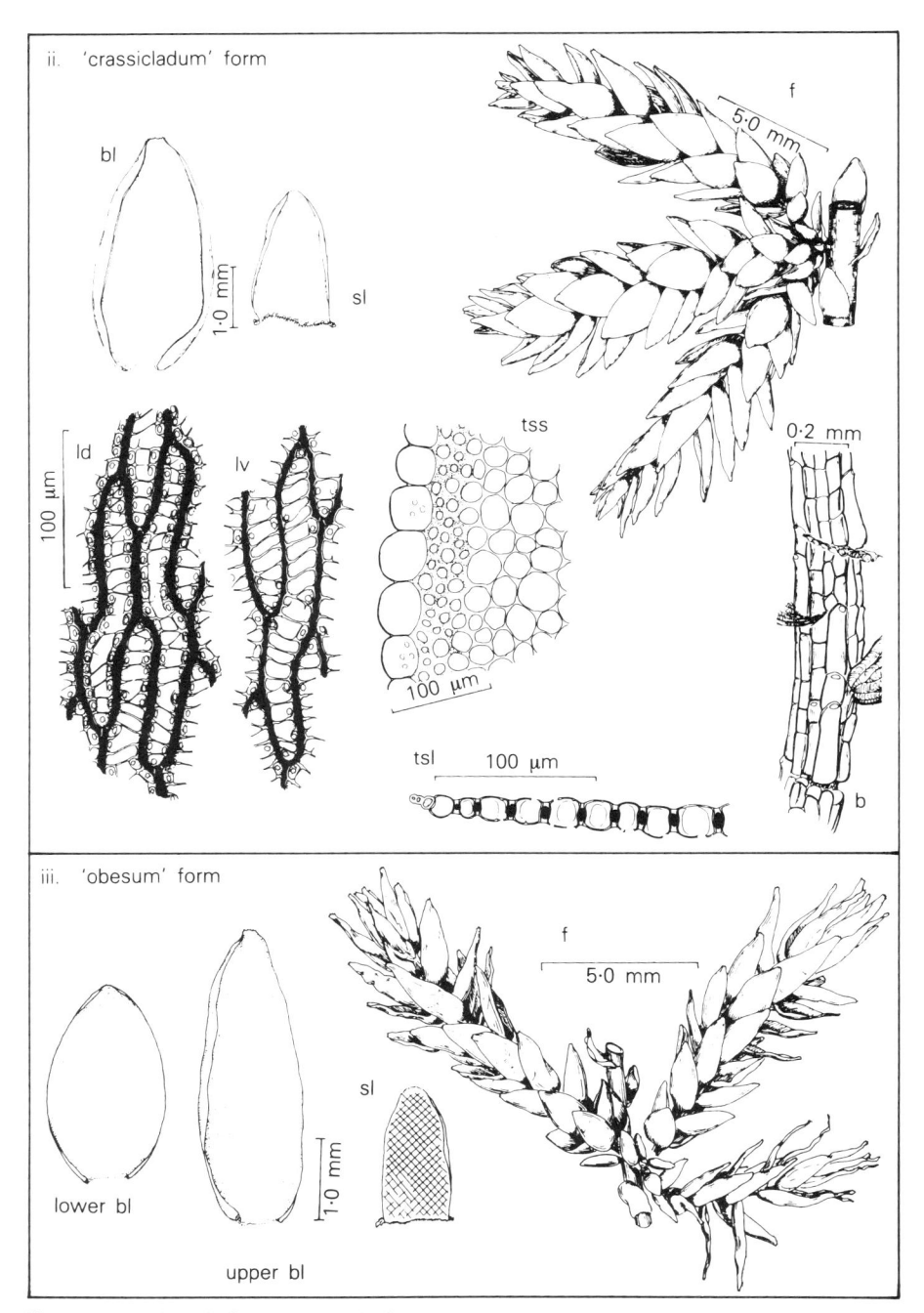

Figure 49 continued. *Sphagnum auriculatum*

microscope. The lax, aquatic forms are very distinctive and are often, at least partially, a rich, dull red. They can also be dull green with some resemblance of *S. platyphyllum,* but that species is smaller than most submerged forms of *S. auriculatum,* with shorter branches and more markedly concave leaves: it is further distinguished by its larger stem leaves and 2–3-layered cortex. There has been much confusion between *S. auriculatum* and *S. subsecundum* subsp. *inundatum,* in most cases because of the emphasis placed on the distribution of stem leaf pores. When typical, the plants are quite dissimilar, but robust shade forms of *S. subsecundum* can superficially resemble *S. auriculatum.* In such cases, identification should be based upon examination of a number of characters and not a reliance on a single feature. Doubtful specimens are usually forms of the more common and more variable *S. auriculatum.* Some very small plants from arctic or alpine localities may belong to a distinct subspecies or variety, or may simply be plants dwarfed by extreme climatic conditions; they have the dimensions of *S. subsecundum* but are, otherwise, typical *S. auriculatum.*

23. *SPHAGNUM PLATYPHYLLUM*

Sphagnum platyphyllum (Braithw.) Warnst. (*Sphag. eur.* no. 187. 1884;
S. laricinum var. *platyphyllum* Braithw. (*Mon. Microsc. J. Trans.*, **13**, 230.
1875)

PLANTS: Usually lax; dull, pale green, olive-green to brownish, sometimes partly purple, rarely yellow-brown; capitula poorly developed, with large, oval, projecting stem buds. **Fascicles:** Rather distant (very rarely absent); rarely with more than 3 branches; one branch usually weak and deflexed, but not otherwise markedly different from the spreading branches; branches *relatively short and blunt-ended.* **Stem:** Relatively weak, up to 0.9 mm diameter; *cortex 2-layered,* but rather irregular and in parts one or 3-layered; outer cortical cells often with a single large pore or area with thin cell wall (revealed only after staining); internal cylinder pale to light brown, *never dark or blackish brown.* **Branch anatomy:** Branches seldom exceeding 12.0 mm; retort cells moderately distinct, in linear series of 3–4, not rostrate; internal cylinder pale or light brown. **Stem leaves:** Large, *almost as large as, or larger than, branch leaves, ovate-spatulate, strongly concave;* apices rounded-obtuse, somewhat eroded; border narrow, without resorption furrow; stem leaf tissue identical to that of branch leaves, fibrillose almost, or quite, to insertion. **Branch leaves:** Fairly large, 1.4–2.2 × 1.1–1.3 mm; concave and suberect so that branches appear tumid, widely ovate; apices narrow, rounded-obtuse (under a hand lens often appearing mucronate because of inflexed upper margins), eroded rather than dentate; border strong, 2–3 cells wide, without resorption furrow. **Hyaline cells:** Uniform, rather small in relation to leaf size, 15.0–20.0(–25.0) × 110–150 μm in mid-leaf; abaxial surface variously, but very rarely serially, porose; pores minute, 2.0–3.0 μm, ringed or unringed, few to many, scattered along the commissures, intermixed with pseudopores; adaxial surface usually lacking pores, but with imperfectly formed pseudopores. **Leaf TS:** Hyaline cells biconvex; internal commissural walls smooth. Photosynthetic cells barrel-shaped with more or less central oval lumina and strongly thickened abaxial and adaxial walls; more or less equally exposed on both surfaces. **Fertile plants:** Dioecious. Male branches scarcely distinguishable from sterile ones; perichaetial bracts large, 4.5 mm long, ovate-spatulate; fibrillose and porose in upper half or third; more or less identical to the stem leaves; lower tissue with differentiated hyaline cells. Male plants apparently rare in Europe; capsules very rare, unknown in Britain; spores about 33 μm diameter.

HABITAT: Usually found as loose mats in wet, mesotrophic to mildly eutrophic locations, where it grows submerged or close to the water surface.

It is especially typical of *Carex*-dominated soakways and seasonally flooded sites, eg pool, river and lake margins or fens, where it is associated with *Phragmites australis, Juncus* spp., *Typha* spp. and *Drepanocladus revolvens*. It is rarely found with other *Sphagnum* species. Occasionally, it may be found in pools or runnels over mineral ground.

DISTRIBUTION: Circumpolar in the arctic and sub-arctic regions of Europe and Asia, but uncommon in western and central North America, though in the east it extends to Alabama and Louisiana. Scattered throughout Europe, but with distinctly north-eastern tendencies, being most common in Scandinavia, but reaching Portugal and Bulgaria. A plant of the lowland and sub-alpine zones, but becoming increasingly montane towards the southern part of its range. Peculiarly rare in the British Isles, where it is recorded from only a few localities in Wales, north-west England and west Scotland, and a single locality in Ireland.

Sphagnum platyphyllum could easily be overlooked in the field as a dull variant of the common and polymorphic *S. auriculatum*, the only other European species to which it bears any real resemblance. *S. platyphyllum*, however, has blunt branches with comparatively short and more concave branch leaves, and stem leaves which, unlike those in *S. auriculatum*, are not significantly shorter than the largest branch leaves. The 2-layered stem cortex is the best microscopic confirmation.

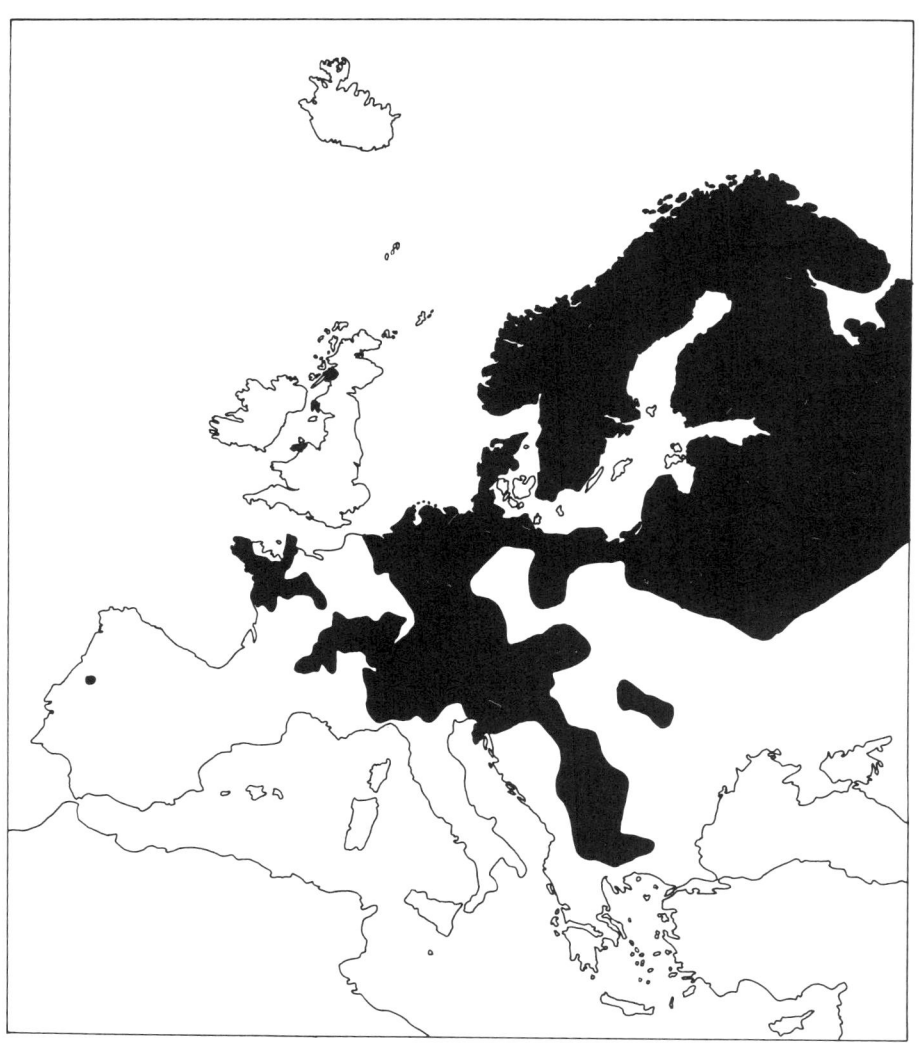

Figure 50. Distribution of *S. platyphyllum*

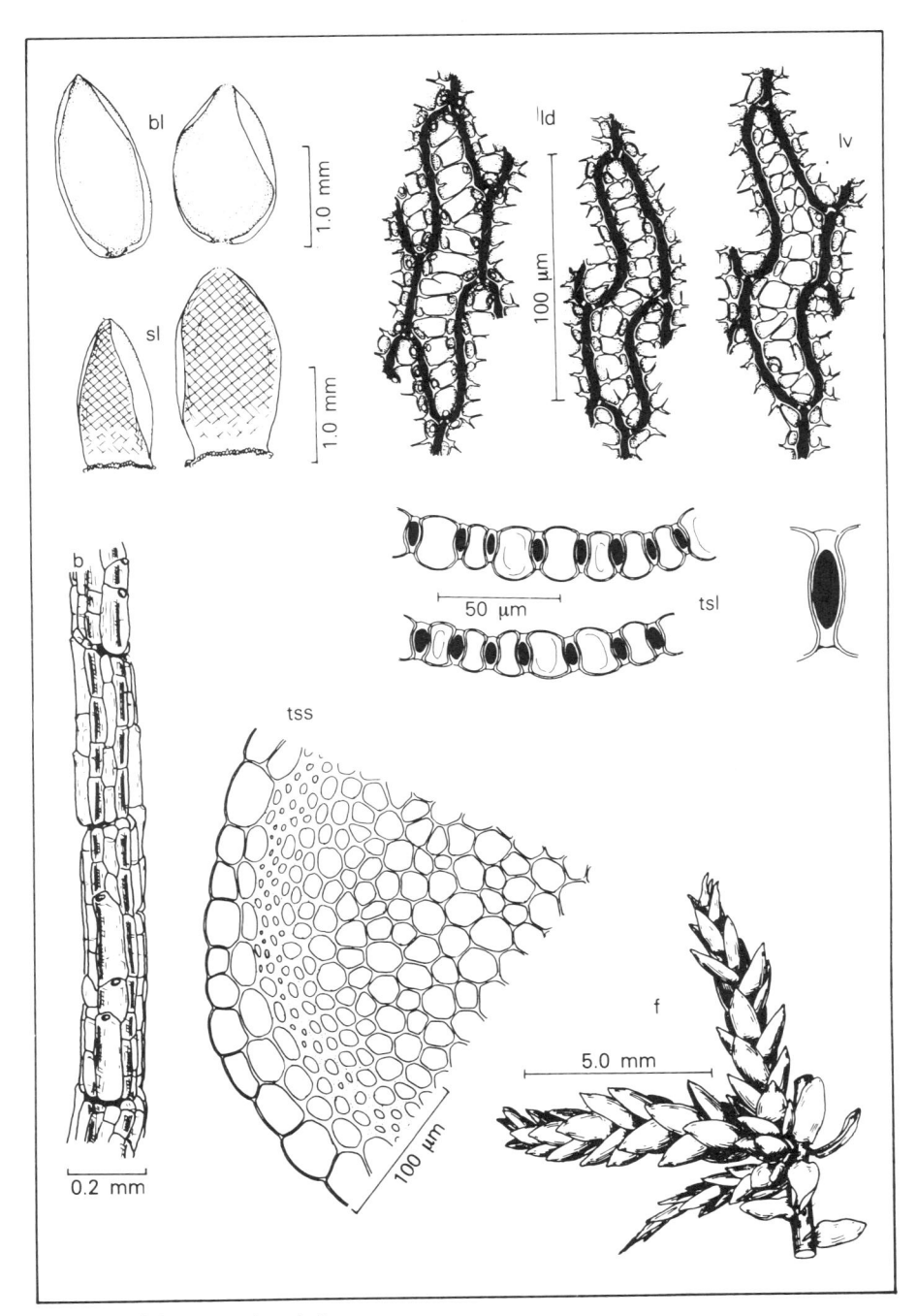

Figure 51. *Sphagnum platyphyllum*

24. SPHAGNUM CONTORTUM

Sphagnum contortum Schultz. (*Prodr. fl. stargard.*, suppl. 1, 64. 1819)
S. laricinum (Wils.) Spruce ex Ångström (*Öfvers. K. VetenskAkad. Förh.*, **21**, 197. 1864)

PLANTS: Rather slender (similar to *S. subsecundum* subsp. *subsecundum*); green, variegated with yellow or brown to ochre. **Fascicles:** Rather closely set, occasionally distant, with at least 5, commonly 6–7, somewhat dimorphic branches; spreading branches 3–4, 8.0–15.0 mm long, occasionally somewhat curved; pendent branches 3–4, up to 20.0 mm long; weaker but not markedly different anatomically. **Stem:** Rather stiff; 0.5–0.8 mm diameter; cortex well developed, of *2–3 layers* of distinct hyaline cells; outer faces of cells mostly with a single large pore or thinning at the distal end; internal cylinder pale, *yellow-green to light brown (never dark brown to blackish)*. **Branch anatomy:** Retort cells distinct, mainly in linear pairs with the lower of each pair larger and shortly rostrate; internal cylinder yellow-green to yellow-brown. **Stem leaves:** Spreading or hanging; small, 0.9–1.1 × 0.6–0.7 mm; lingulate to lingulate-triangular, more or less concave above due to inrolled margins; apices broadly rounded, eroded; border not, or slightly, widened below; hyaline cells near apex fibrillose, rarely lacking fibrils; on adaxial surface with small (3.0–6.0 μm) circular pores in the upper and lateral angles (similar to those of *S. subsecundum*, but usually smaller); *tissue in the lower part of the leaf with hyaline and photosynthetic cells of more or less equal width, the transition between this and the upper third of the leaf being quite abrupt.* **Branch leaves:** Ovate to ovate-lanceolate, the lower often narrow and secund, curved and asymmetric (as in *S. subsecundum*); 1.1–1.8 mm long; more or less 5-ranked, but not conspicuously so; pendent branch leaves mostly symmetrical, incumbent, but anatomically similar to spreading branch leaves. **Hyaline cells:** Small and narrow, 11.0–15.0 × 70–150 μm; uniform throughout leaf; adaxial surface usually lacking true pores, but sometimes with few to several pseudopores; abaxial surface variable, usually with numerous, minute, often ringed, pores along the commissures, but these *rarely in uninterrupted series,* pores sometimes few, but rarely absent; pores of pendent branch leaf hyaline cells as in spreading branches. **Leaf TS:** Hyaline cells almost plane on both surfaces; internal commissural walls smooth. Photosynthetic cells oval-triangular with thick abaxial walls and oval lumina; exposed on abaxial surface but *shallowly enclosed on, or just reaching, the adaxial surface* (with stronger abaxial bias than *S. subsecundum*). **Fertile plants:** Dioecious. Antheridial bracts brown-ochre; inner female bracts 2.5–4.0 × 2.2–2.8 mm, widely ovate with rounded apices, concave; apex fibrillose and with adaxial pores (as in stem leaves);

hyaline cells differentiated almost, or quite, to insertion. Capsules rare (not recorded from Britain); spores brown, papillose, ca 25 μm diameter.

HABITAT: One of the few *Sphagnum* species able to tolerate high base status in peat and peat waters. It grows in a range of eutrophic habitats from open mixed fen to more dense *Phragmites/Carex* swamp and open fen woodland (carr), as either pure mats or as scattered stems among other species, eg *S. teres, S. squarrosum* or *S. warnstorfii*. Occasionally, it may be found floating in open water.

DISTRIBUTION: A circumboreal species of Europe, northern and eastern Asia, and North America, particularly the maritime provinces of Canada and New England. Widespread, but scattered, throughout most of central and north-west Europe, except the extreme north of Fennoscandia. More abundant in the lowlands, but reaching 2000 m in the Alps. Its distribution in Britain is patchy and it is absent from most of southern England and Wales, the central valley of Scotland and central Ireland.

This species bears a strong resemblance to *S. subsecundum* subsp. *subsecundum* and may be difficult to separate from that species in the field, although stem colour is usually helpful. Unfortunately, pale-stemmed forms of *S. subsecundum* are not rare, especially in unfavourable habitats (and seem to be locally plentiful in North America). Other field characters, small colour differences and slightly narrower leaves in *S. contortum*, together with different habitat preferences, are not always reliable either. The best microscopic character is the 2–3-layered stem cortex. Rarely, forms are found which, because of their narrow leaves and ochre to brown colour, bear a strong resemblance to *S. recurvum*. Even under the microscope, the branch leaf shape and the markedly abaxial bias of the photosynthetic cells may appear more typical of section Cuspidata. However, the ventriporose stem leaves have no counterpart in section Cuspidata, and only *S. jensenii* has branch leaf pores that bear any similarity to those of *S. contortum*.

S. platyphyllum has, at times, been treated as a variety of *S. contortum*, but the 2 taxa differ in several fundamental characters and seem to be only distantly related.

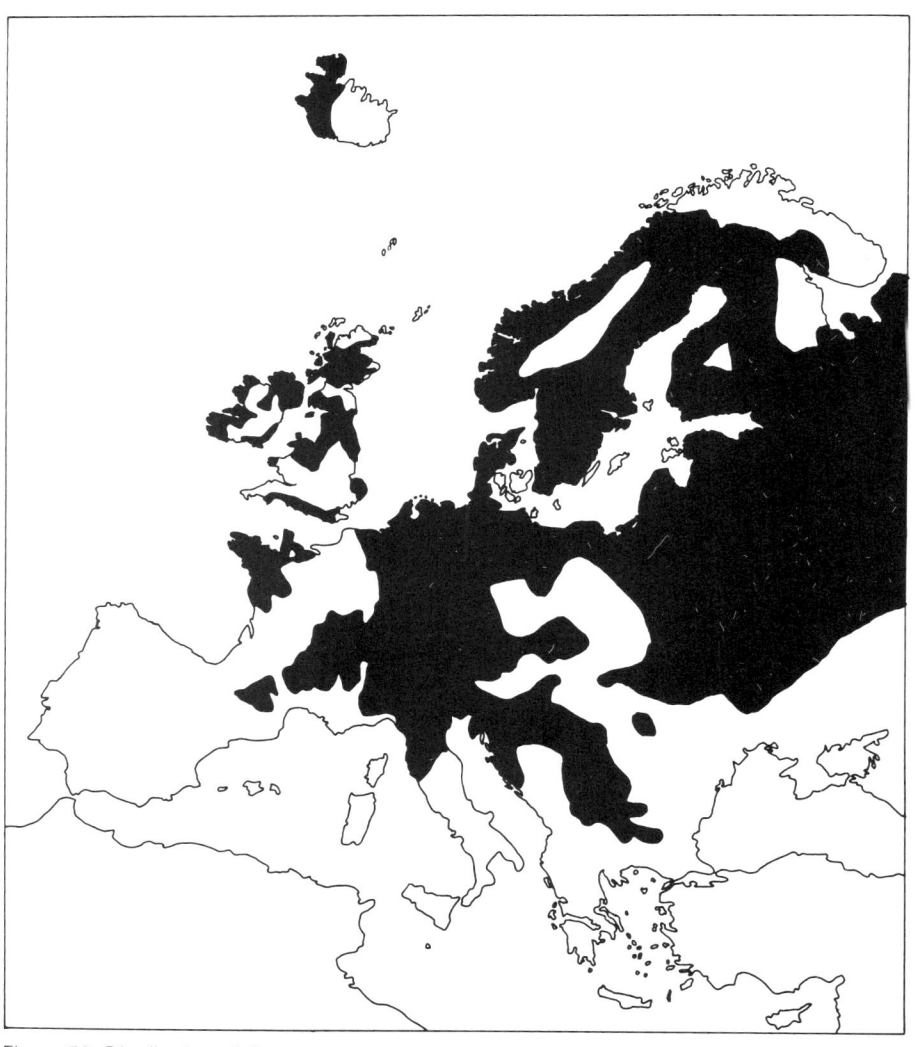

Figure 52. Distribution of *S. contortum*

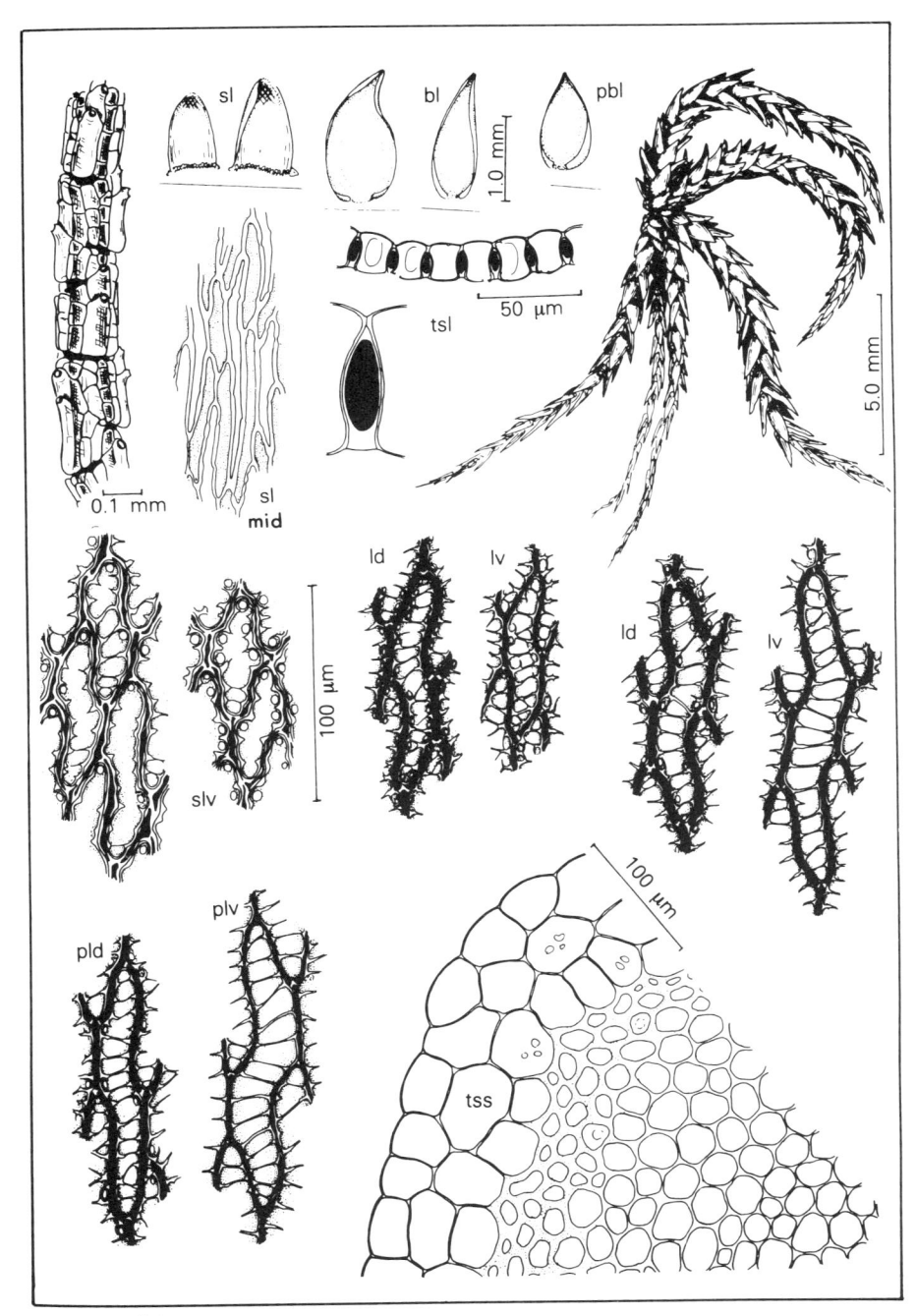

Figure 53. *Sphagnum contortum*

SECTION CUSPIDATA

Sphagnum sect. *Cuspidata* (Lindb.) Schimp. (*Syn. musc. eur.* 2nd ed., 829. 1876)

Plants: Mostly medium-sized, occasionally small (cf *S. capillifolium*) or robust (cf *S. squarrosum*); mostly green, yellow-green or brown, sometimes orange, but never wine-red. **Fascicles:** Of 4–5, rarely 3, (*S. balticum* and forms of *S. cuspidatum*) branches which are monomorphic (*S. cuspidatum*) or distinctly dimorphic. **Stem:** Cortex of 2–3 layers of thin-walled or slightly thickened cells, or poorly differentiated; outer surfaces never with pores. **Branch anatomy:** Retort cells distinct, usually in linear groups of 2, but up to 4 in some species, only in *S. lindbergii* in groups of 4–6. **Stem leaves:** Hanging or spreading; small or large; fibrillose or not; acute or fimbriate; border strong, always expanded below into patches of prosenchymatous tissue which may be almost confluent across the leaf base. **Branch leaves:** Narrow, lanceolate to linear on spreading branches, sometimes ovate on pendent branches; border strong without resorption furrows; apices truncate, dentate, but mostly acute due to inrolling of leaf margins; distinctly 5-ranked in many species, but sometimes (eg *S. flexuosum*) not, or only irregularly so. **Hyaline cells:** Narrow; strongly fibrillose; abaxial face with few (1–3) small pores, or a single resorption gap, confined to the apical angle, rarely with more numerous small to medium-sized unringed pores not confined to the commissures (*S. majus, S. jensenii, S. obtusum*); adaxial face rarely without pores; mostly with 2–6 or more medium-sized, unringed or thinly ringed circular pores, mainly adjacent to cell angles (these pores rarely absent, at least in the upper half of leaf, or replaced by pores of a different type, eg *S. jensenii, S. obtusum,* some forms of *S. riparium*). **Leaf TS:** Hyaline cells plane to shallowly convex, highly inflated only in forms of *S. riparium* and *S. lindbergii;* internal commissural walls smooth. Photosynthetic cells triangular to trapezoid, sometimes with bulging walls (*S. lindbergii*); *always widely exposed on abaxial surface, narrowly exposed or deeply immersed on the adaxial surface;* lumina mostly distinctly wider abaxially, but sometimes oval in some arctic species. **Fertile plants:** Dioecious. Antheridia borne on spreading branches; bracts brown to orange. Inner perichaetial bracts *without fibrils or fibrillose only at apex;* more or less uniformly prosenchymatous in lower half of bract.

The majority of species in this cosmopolitan section, which shows its greatest diversity in the sub-arctic and cool temperate zones of the northern hemisphere, occur in consistently wet areas, mainly in oligotrophic to slightly mesotrophic mires. The section Cuspidata has 30–35 species.

Although, superficially, the species of the Cuspidata may resemble some of the Acutifolia, there is no direct relationship between these 2 sections.

However, there are species that clearly demonstrate the affinity of section Cuspidata and section Subsecunda (eg the North American species *S. mendocinum*). Although *S. contortum* (section Subsecunda) and *S. jensenii* (section Cuspidata) are clearly referable to their phyletic groups, each has features which, taken individually, might leave some doubt as to their true systematic positions. On arguments similar to these, Le Roy Andrews (1915) proposed merging the 2 sections, but there would be little advantage in following this procedure. The European species, at least, are fairly easily recognized as members of the section Cuspidata. The more numerous, more closely imbricated, linear to lanceolate branch leaves, the large adaxial branch leaf pores, the leaf section, and the strongly expanded stem-leaf border separate all the European species of section Cuspidata from section Subsecunda.

S. tenellum is included in section Cuspidata by most present-day authors, and there is probably a relationship between it and this section. However, such a relationship would seem to be a remote one, and the grounds for separating *S. tenellum* off into a section of its own are at least as strong as those used to separate the Squarrosa and Polyclada from section Acutifolia.

25. SPHAGNUM CUSPIDATUM

Sphagnum cuspidatum Hoffm. (*Deutschl. Flora*, **2,** 22. 1796)

PLANTS: Varying from rather compact terrestrial forms to very lax, often pale and plumose, free-floating, aquatic forms; green, yellow-green or whitish green, seldom (except in male branches) with pronounced secondary pigmentation. **Fascicles:** Closely set to very distant, with 3–4, occasionally 5, branches, 8.0–12.0 mm long (occasionally much longer in some floating forms); branches varying in vigour, but not dimorphic, rarely with the weaker appressed to the stem. **Stem:** Rather thin, 0.4–0.8 mm diameter, brittle or flexuose; pale green to yellow, rarely with areas of translucent, faintly pink or brown coloration; cortex distinct, of 2–3 layers of moderately inflated hyaline cells; internal cylinder green or yellow, rarely faintly brown or dull pink. **Branch anatomy:** Retort cells moderately distinct, mainly in linear pairs, not rostrate; internal cylinder green or translucent pink. **Stem leaves:** Hanging or spreading, rarely appressed to stem; triangular to triangular-ovate, *always longer than wide;* mostly more than 1.2 mm long; border strong, usually widely expanded below; apex obtuse to narrowly truncate, dentate or eroded, concave or appearing subacute due to inrolling of upper leaf margins; *strongly fibrillose in upper third, and often beyond,* rarely (except in etiolated or submerged plants) fibrillose throughout; fibrillose tissue resembling that of branch leaves. **Branch leaves:** Narrowly lanceolate to linear (except a few at branch base), those from the middle of a branch at least 3 times, and often more than 6 times, as long as wide; straight or curved and secund, suberect to widely patent; 1.7–5.0 mm long, or more; basically, but not distinctly, 5-ranked, frequently spirally arranged. **Hyaline cells:** Long and narrow, 12.0–18.0 × 100–120 μm in upper mid-leaf in terrestrial forms, 16.0–20.0 × 160–230 μm in aquatic forms. Abaxial surface without, or with only a few, pores, mostly with only a small resorption gap in, or near, the apical angle; with or without 1–3 pseudopores in the lateral angles. Adaxial surface with 3–7 rather small (4–5 μm) unringed, circular pores mainly adjacent to the cell angles, sometimes lacking pores. **Leaf TS:** Hyaline cells plane on abaxial face, shallowly to strongly convex on the adaxial. Photosynthetic cells often almost as wide as the hyaline, *trapezoid and exposed on both leaf surfaces,* rarely triangular and, very rarely, enclosed on the adaxial surface. **Fertile plants:** Dioecious. Antheridial branches yellow-brown to orange (sharply contrasting with sterile branches), occasionally concolorous with the rest of the plant; archegonial bracts sometimes shorter than branch leaves, but not differing anatomically. Inner perichaetial bracts very large (3.5–5.5 mm or longer), ovate, convolute (female branches in aquatic plants often considerably elongated); apices rounded-truncate to slightly retuse, mostly of prosenchymatous tissue; lower apical tissue fibrillose, at least in part; lower tissue of bract lacking differentiated hyaline cells. Capsules frequent; spores yellow-brown, strongly papillose, 28.0–35.0 μm diameter.

HABITAT: A common species of wet, oligotrophic sites, where it is often floating or submerged in pools, although it also occurs along pool margins or in wet hollows and soaks. It is found mostly in ombrotrophic bogs or acid fens, but may also grow in ditches or oligotrophic flushes. It forms pure stands or may be associated with other species, eg *S. auriculatum* or *S. majus* in pools or partly submerged lawns, or with *S. pulchrum*, *S. recurvum* forma *fallax* or *S. papillosum* around pool margins, in wet hollows or more extensive carpets.

DISTRIBUTION: Widespread, but with oceanic tendencies, in Europe, eastern Asia (Japan), eastern North America and Greenland. Absent from northern Scandinavia, but present throughout most of the rest of Europe and extending to Portugal. Widespread and common in suitable habitats throughout most of Britain, except parts of the south-east.

Aquatic plants of this species, with their very narrow leaves and floating habit, are among the first *Sphagna* to be recognized in the field. The leaves are commonly so finely drawn out that submerged plants have been compared to saturated animal fur and described as resembling drowned cats. In bog pools, they often form an association with aquatic forms of *S. auriculatum*, where the colours of the 2 species contrast strongly. Terrestrial plants are not so distinctive and bear a superficial resemblance to other species of this section and, to a lesser extent, of other sections. Certain forms of *S. flexuosum* and *S. recurvum* forma *fallax* may resemble *S. cuspidatum*, although only rarely are the middle branch leaves of either of these species as long and narrow as in shorter-leaved forms of *S. cuspidatum*. *S. flexuosum* is usually more robust, with thicker stems and blunt, non-fibrillose stem leaves. *S. recurvum* forma *fallax* has larger pores on the adaxial branch leaf surfaces, strongly reflexed stem leaves, which lie appressed to the stem, and photosynthetic cells which are triangular in cross-section and mostly enclosed on the adaxial leaf surface. The rarer, northern species, *S. majus*, is normally a brown plant, but pale shade forms may closely resemble *S. cuspidatum*. However, the numerous abaxial pores in the branch leaves of *S. majus* will distinguish that species from *S. cuspidatum*.

Sphagnum cuspidatum has a number of growth forms which appear to be responses to a reduction in light intensity and increased nutrient levels (particularly increased available nitrogen). The most common effects of this combination of habitat factors are the production of branch leaves with

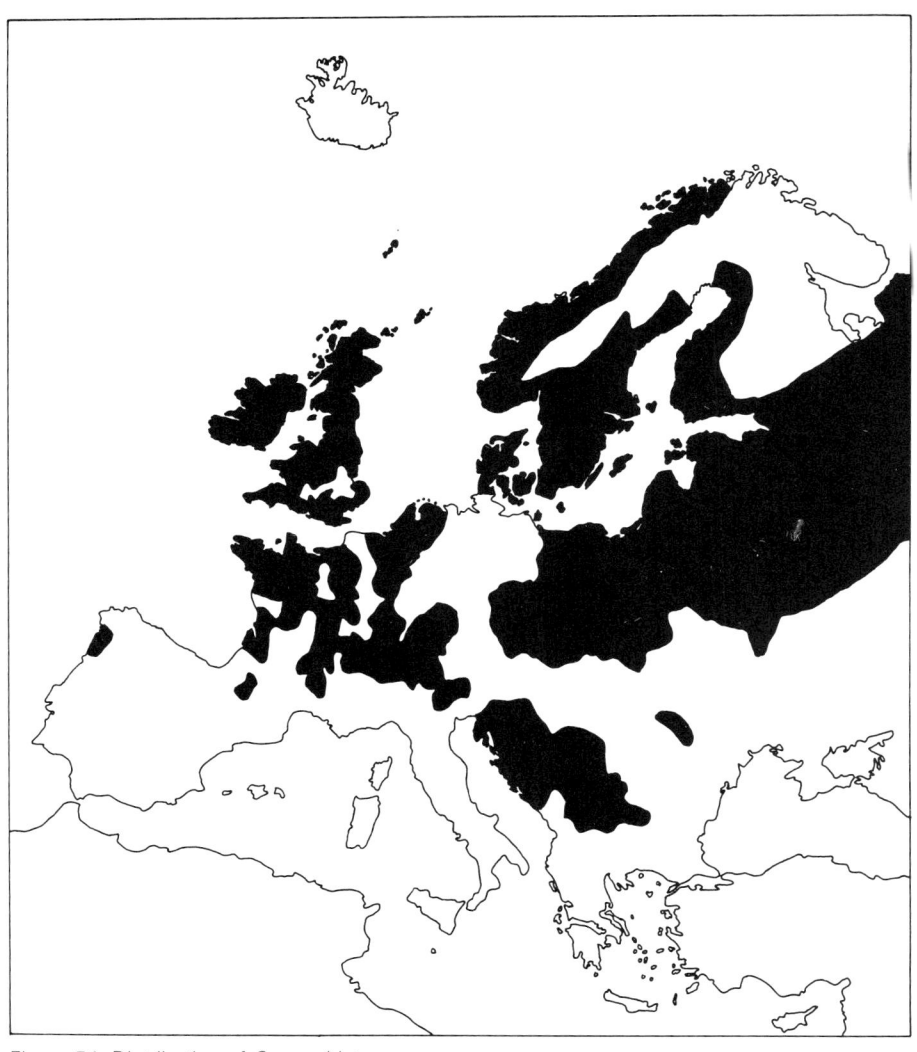

Figure 54. Distribution of *S. cuspidatum*

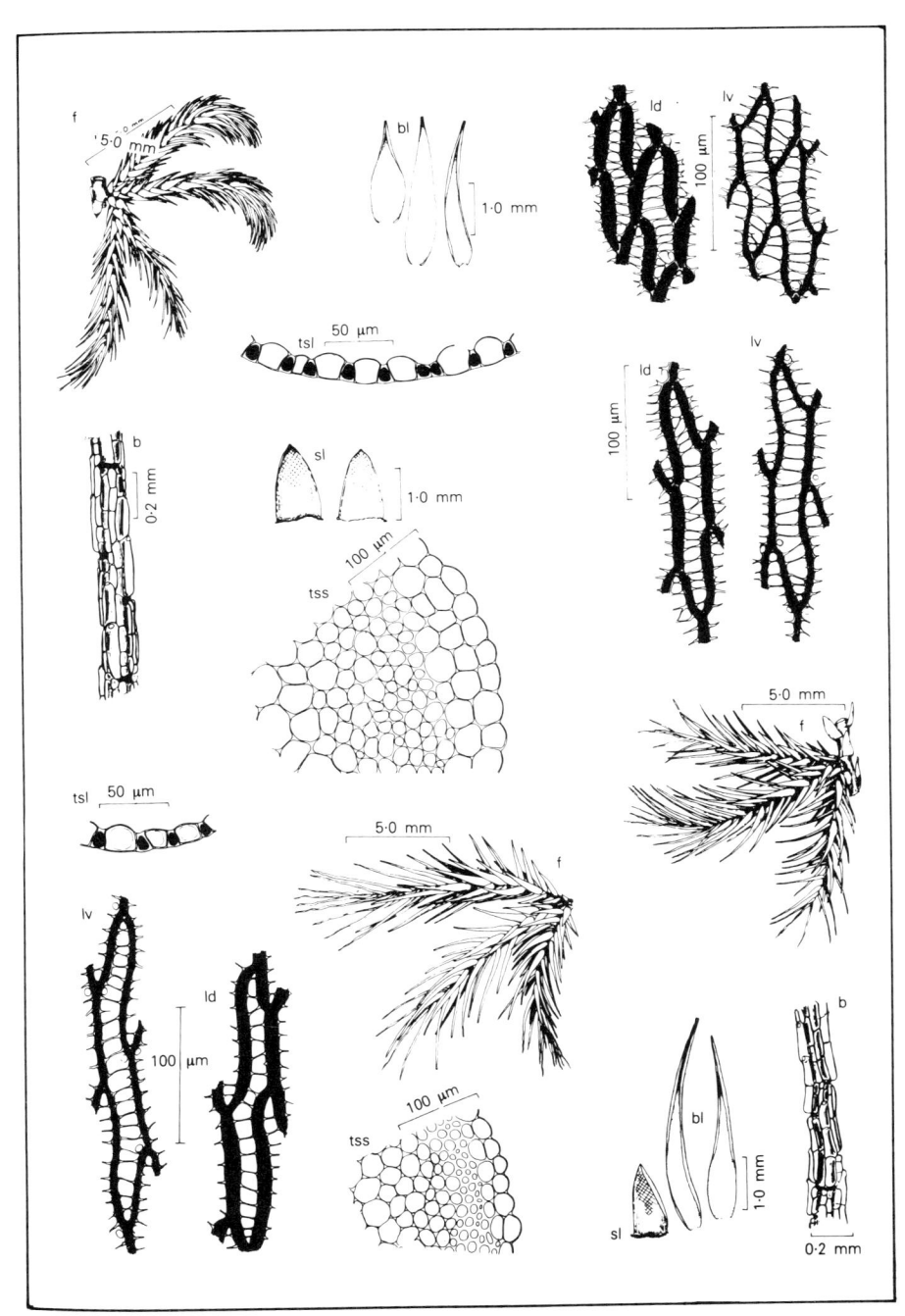

Figure 55. *Sphagnum cuspidatum*

distinctly and remotely dentate leaf margins (caused by the lateral projection of the upper cell ends of the outer series of border cells), and the partial or complete replacement of fibrillose cells by undifferentiated prosenchymatous cells near the leaf apex. An extreme, and rare, form has its fascicles reduced to single branches which may continue to grow indefinitely to produce plants of very un-*Sphagnum*-like appearance. (This form is probably the *S. monocladum* of Warnstorf *et al.*) Although many of these variants have been accorded varietal, or even specific, names, there are no clear distinctions which allow the application of such names with any consistency. Some of the forms have been identified with non-European taxa, eg *S. torreyanum* Sull. and *S. serratum* Aust., both of which are North American, but it is now generally accepted that these names have been applied in error (see Isoviita 1966).

Although dioecious, many species of the section Cuspidata are commonly fertile and a degree of genetic heterogeneity within and between populations is only to be expected. It is quite often easy to distinguish different clones within the same population due to slight variations in colour or habit, which may or may not be sex-linked. Flatberg (1988b) considers one of the more marked and widespread variants of the *S. cuspidatum* complex to be sufficiently distinct to be accorded species rank: he has named it *Sphagnum viridum* Flatberg. As the name implies, the brownish colours that are often present in the capitulum of *S. cuspidatum* are virtually absent (except at the proximal ends of the branches) in *S. viridum.* Other features distinguishing *S. viridum* from *S. cuspidatum* are the relatively shorter and proportionately broader branch leaves, and shorter hyaline and photosynthetic cells in the apical portions of these leaves. However, quoting Flatberg (1988b), 'The general morphologic similarity of *S. cuspidatum* and *S. viridum* in combination with their considerable phenoplasticity and the occurrence of sexual leaf dimorphism, will cause many identification problems. To solve these requires long experience both in the field and laboratory'.

The ecological and phytogeographical ranges of the above 2 taxa coincide so closely that there appears to be little practical advantage in pursuing such a narrow species concept in this work. *S. viridum* is, therefore, considered to be a form of *S. cuspidatum*. Students wishing to investigate further infraspecific variation within *S. cuspidatum* should refer to Flatberg's paper.

26. SPHAGNUM RIPARIUM

Sphagnum riparium Ångstr. (*Öfvers. K. VetenskAkad. Förh.*, **21**, 198. 1864)

PLANTS: Robust, tall or short, with tumid branches and well-developed capitula; green or, occasionally, yellowish or brownish. **Fascicles:** Of 4(–5) dimorphic branches; spreading branches 2(–3); 15.0–25.0 mm long, strong, often more or less julaceous. Pendent branches 2(–3); weaker, variable in length, shorter or longer than the spreading. **Stem:** Strong, 0.7–1.1 mm diameter; green to pale yellowish; *cortex not, or scarcely, distinct* of thick-walled cells which merge into the internal cylinder; internal cylinder green or yellowish, of slightly to moderately thickened cells (in section similar to that of *S. flexuosum*). **Branch anatomy:** Retort cells moderately distinct, in linear groups of 2(–3); internal cylinder pale green or yellowish. **Stem leaves:** Hanging; ovate-lingulate to triangular-lingulate; large, 1.4–1.7 mm long, 1.0–1.3 mm across insertion; border strong, widely expanded below into large patches of prosenchymatous tissue which are often almost confluent across the leaf base; apex rounded-obtuse, *deeply notched or torn* due to apical resorption (often very deeply bifid, as though pulled apart from the tip); hyaline cells near apex conspicuously enlarged; hyaline cells without fibrils, septa frequent. **Branch leaves:** Erect to slightly patent, 5-ranked (at least in part); *large*, 2.0–2.6 × 0.8–1.2 mm; lanceolate, widest at quarter to half-way from insertion; border rather thin, becoming confluent above to give narrow, *more or less acute apex of prosenchymatous tissue* (except in leaves of weak specimens), which can extend as much as 0.3 mm or more below the leaf tip. Pendent branch leaves lanceolate, less acute and usually lacking the prosenchymatous apices. **Hyaline cells:** Variable; in upper mid-leaf small, ca 70 µm × 10.0–15.0(–18.0) µm; abaxial surface without pores or, usually, with a small resorption gap in the apical and/or basal angle, occasionally with 1–5 additional small pores or thinnings; adaxial surface with 3–6 faint to clearly defined pores near the cell angles or near the cell mid-line (mostly clearly separate from the commissures). Hyaline cells of lower lateral areas of leaves much larger, 120–170 × 20–30 µm, *apical end conspicuously wider than basal end*, both abaxial and adaxial surfaces with a large resorption gap occupying the whole of the apical cell end; abaxial surface otherwise lacking pores; adaxial surface with additional large, unringed pores. Hyaline cells of pendent branch leaves 70–100 × 18–30 µm, *apical end much wider than basal end;* abaxial surface with a large resorption gap in the apical angle, with or without an additional smaller gap in one of the upper lateral angles; adaxial surface with 2–4 large pores, the apical pore much larger than the others. **Leaf TS:** Hyaline cells almost plane on abaxial surface, strongly convex on the

Figure 56. Distribution of *S. riparium*

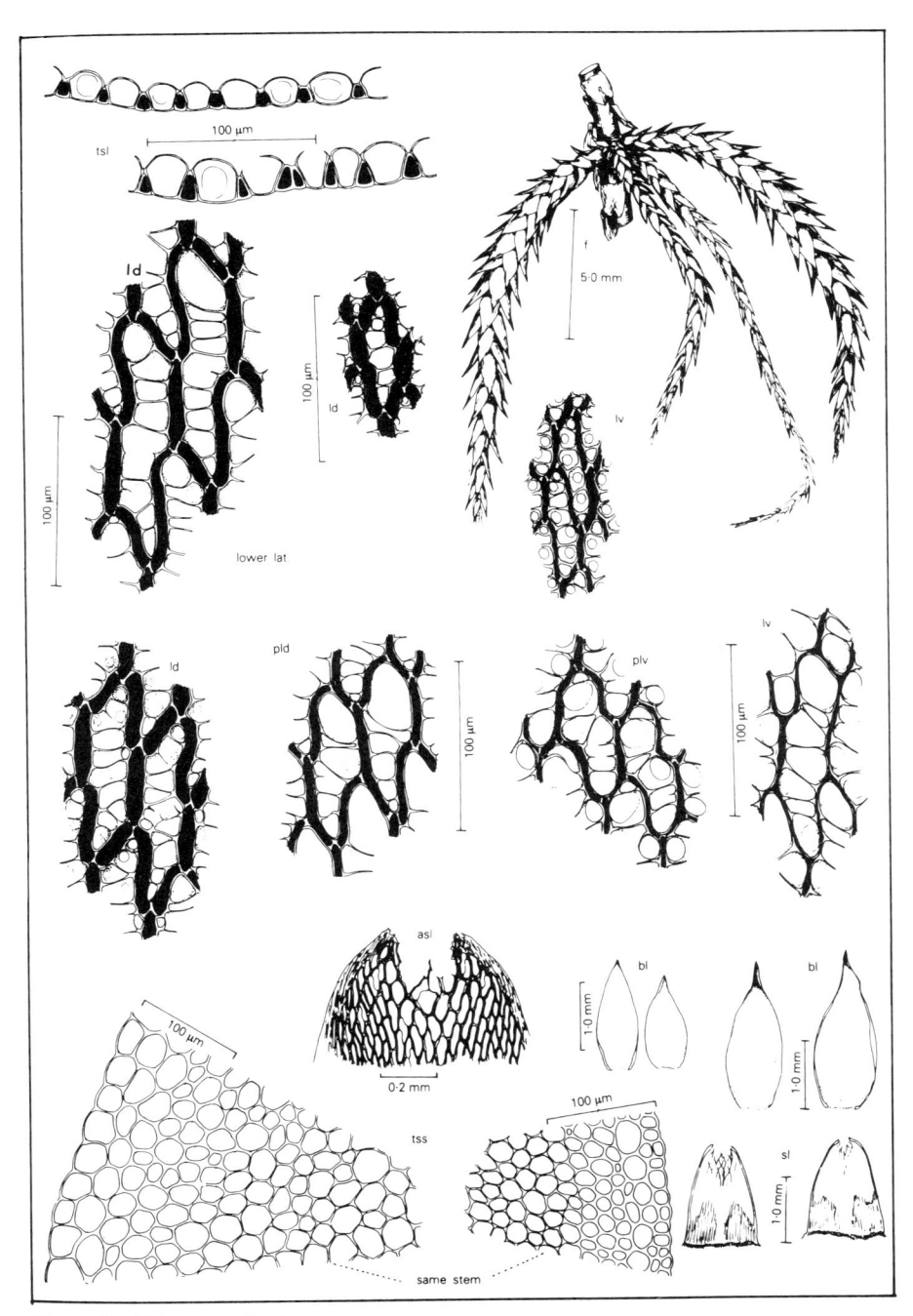

Figure 57. *Sphagnum riparium*

adaxial. Photosynthetic cells triangular to trapezoid, mostly *reaching or narrowly exposed on the adaxial leaf surface*, occasional cells shallowly enclosed. **Fertile plants:** Dioecious. Antheridial branches brownish. Inner perichaetial bracts very large, ca 5.0 × 2.0 mm, concave; apices widely obtuse to retuse; *tissue almost entirely prosenchymatous* or with small differentiated zone below apex. Female plants occasional to frequent, but capsules rare. Spores papillose, 24–26 μm.

HABITAT: Forming loose mats in wet mesotrophic to somewhat oligotrophic mires, in ditches or along stream margins, usually under scattered scrub or more mature fen woodland. *S. riparium* is confined to very wet areas, often being found in shallow water, and may have as associates *S. lindbergii, S. jensenii* or small, raised tufts of *S. angustifolium. Carex canescens* is also a frequent companion species, especially in Scandinavia. In the southern part of its range, *S. riparium* appears to occur in more mesotrophic sites so that in Britain, for example, it is frequently found amongst *Juncus* species in areas with mineral water enrichment.

DISTRIBUTION: Circumpolar in the sub-arctic and the northern part of the boreal zone of Europe, Asia and North America. In Europe, its distinct northerly and slightly continental tendency is shown by the abundance of the species in northern Scandinavia, where it is one of the dominant *Sphagnum* species, and its increasing rarity further south. It is uncommon in southern Scandinavia and rare in north-central Europe, where it is mostly confined to montane areas. It is rare and scattered in parts of north-east Britain and found mostly in upland areas.

S. riparium is one of the most robust species of section Cuspidata in Europe, but relatively weak plants, which are by no means rare, may be confused in the field with other members of the section, eg *S. flexuosum, S. obtusum* and *S. recurvum*. The characteristic stem leaves can usually be seen easily with a hand lens, and only rarely lack the conspicuous apical notch or tear. (Stem leaves of *S. flexuosum* or *S. obtusum* can suffer mechanical damage, eg during removal from stems, and appear similar to those of *S. riparium*. However, stem leaves of these species lack the enlarged hyaline cells in the central apical region of the leaf.) The large branch leaves have a more finely acute appearance than those of related species, even in the field, and the leaves themselves may be more or less erect and convolute, giving the branches a cylindrical (at times almost 'rat-tailed') appearance not often found among other species of this section. Under the microscope, *S. riparium* is easily identified. The large apical resorption gaps in at least the pendent branch leaves distinguish it from *S. flexuosum. S. obtusum* has a very different pore morphology, and size alone should prevent confusion with *S. angustifolium*.

Morphological details suggest that *S. riparium* is probably most closely related to *S. angustifolium*, despite the disparity in size.

27. SPHAGNUM OBTUSUM

Sphagnum obtusum Warnst. (*Bot. Ztg,* **35,** 478. 1877)

PLANTS: Medium-sized to, usually, robust (comparable to *S. riparium*), with rather large capitula; green or yellowish, occasionally brownish, especially in the capitulum. **Fascicles:** Closely set to more or less distant; of 4–5 scarcely dimorphic branches; spreading branches 2(–3), (14.0–)16.0–20.0(–30.0) mm long; often appearing stout due to size and arrangement of leaves; pendent branches (when distinct) 2–3; variable in length, from shorter than the spreading and not differing in form, to as long as or longer than spreading, and then much attenuated distally. **Stem:** Strong, 0.7–1.3 mm diameter; cortex distinguishable but inconspicuous, of 3–4 layers of slightly to strongly enlarged, thick-walled cells; internal cylinder green or yellowish. **Branch anatomy:** Retort cells distinct, in linear groups of 2–3, apertures not strongly protuberant; internal cylinder yellowish or green. **Stem leaves:** Hanging, occasionally more or less spreading; 1.1–1.3 mm long, 0.9–1.2 mm across insertion; triangular-lingulate, tapering to *broadly truncate, eroded* and more or less fringed, but not bifid (seldom notched) apices (0.2–0.3 mm across tip); border strong, widely expanded below. Hyaline cells without fibrils, (occasionally a few, very faint, fibrils present near the apex); adaxial surfaces more or less completely resorbed. **Branch leaves:** Distinctly 5-ranked (sometimes conspicuously so); *large,* commonly 2.2 mm, rarely under 1.8 mm long, at least on the stronger branches; lanceolate and widest about a quarter to one third from base; apices truncate-dentate; border rather narrow, 2–3(–4) cells wide, occasionally confluent at the apex but the latter not mainly prosenchymatous. Leaves of pendent branches smaller, though seldom less than 1.1 mm long at a point one third of the way along the branch from its insertion; broadly lanceolate. **Hyaline cells:** In upper mid-leaf 70.0–110.0 µm long × 11.0–13.0 µm wide on abaxial surface, 15.0–18.0–µm on the adaxial. Abaxial surface without pores, or with a small, inconspicuous resorption gap in the apical, occasionally also in the basal, angle. Adaxial surface variable, normally *without or with few, rather large* (7.0–10.0 µm diameter), *usually inconspicuous pores or thinnings* (usually visible only after staining). In the *lower lateral parts of leaves, hyaline cells with rather numerous, small* (ca 2.0 µm diameter), *faint pores, in or at either side of the cell mid-line on one or, usually, both leaf surfaces* (visible only after heavy staining and best seen in the 2–3 intra-marginal cell series about one quarter of the way above leaf insertion). Hyaline cells of pendent branch leaves similar to those of the spreading branches, not conspicuously enlarged in the apical half, but occasionally with a moderately large resorption gap in the apical angle. **Leaf TS:** Hyaline cells more or less plane on the abaxial surface, shallowly convex on the adaxial.

Photosynthetic cells triangular with slightly bulging sides, widely exposed abaxially, but mainly shallowly enclosed on the adaxial. **Fertile plants:** Dioecious. Antheridial bracts rather shorter than branch leaves and usually orange-brown. Inner perichaetial bracts large, up to 4.5 mm long; apices broad, retuse; hyaline cells clearly differentiated, except at extreme tip and lower third to sixth of bract, without fibrils or fibrils thin and scattered in an occasional cell. Capsules uncommon to locally frequent; spores strongly papillose, 23.0–27.0 μm diameter.

HABITAT: As mats in wet, mesotrophic to eutrophic mires, often partly immersed in pools, or along stream or lake margins. As a species able to grow in eutrophic habitats, it may be found associated with S. warnstorfii or S. teres, or with S. squarrosum in less eutrophic areas. It may also be found at the margins of more oligotrophic peatlands and under sparse Salix scrub.

DISTRIBUTION: A circumpolar species with continental tendencies. In Europe, it is scattered throughout northern, eastern and central parts but becomes rare in the extreme north and in montane areas. Absent from Britain, though recorded formerly from Lancashire.

S. obtusum, when well grown, is one of the most robust species of the Sphagnum recurvum group, comparable in vigour with S. riparium and the most robust forms of S. flexuosum. In many of its characters, it seems to lie between S. riparium and S. flexuosum, and is probably more closely related to the latter. The minute pores in the branch leaf hyaline cells are easily overlooked in casual examination. They vary in number and may even be absent from some specimens. In such cases, the 5-ranked branch leaves and adaxially enclosed photosynthetic cells will separate S. obtusum from S. flexuosum. S. riparium, a plant of similar dimensions, has apically resorbed-bifid stem leaves, typically entirely prosenchymatous branch leaf apices and different pores (although minute pores, of the S. obtusum type, are by no means rare in S. riparium). In S. jensenii and S. majus, pores similar to those in S. obtusum are found, but in the former species (which are normally brown and have fibrillose stem leaves) these pores are much more clearly defined and occur throughout the leaves.

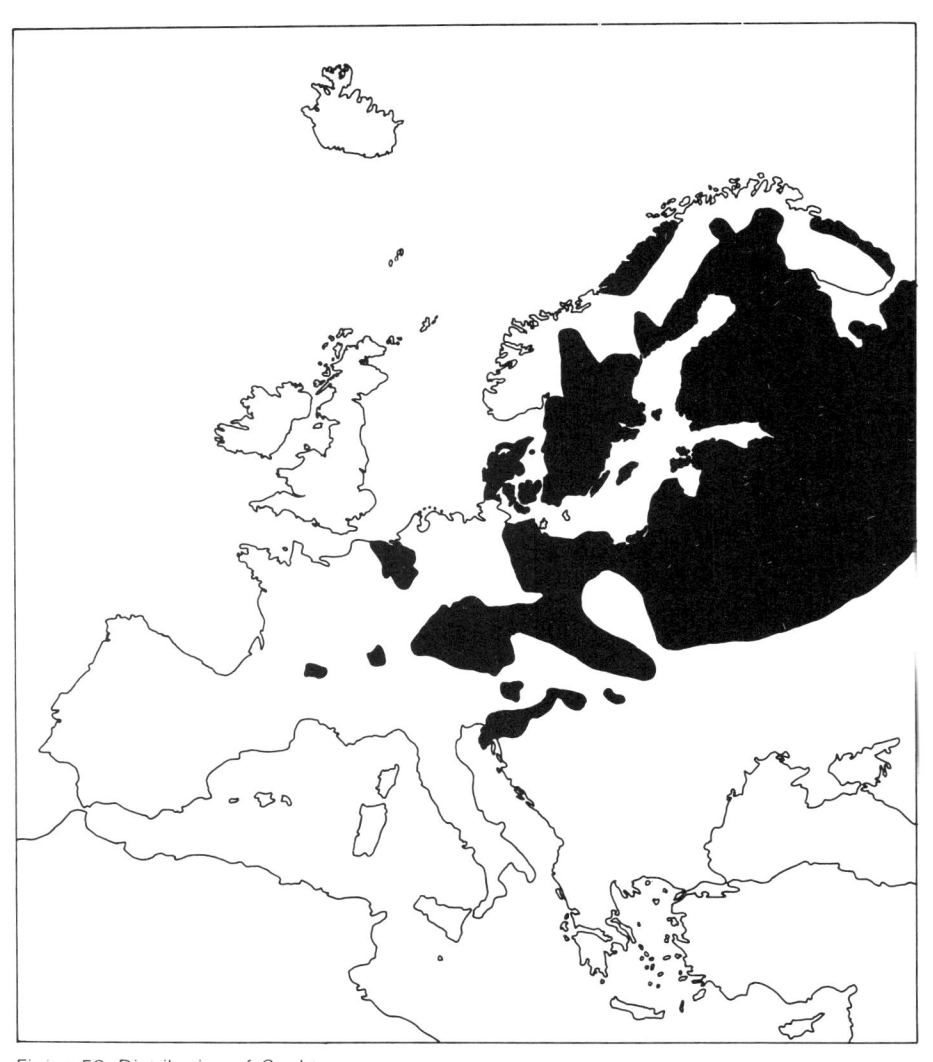

Figure 58. Distribution of *S. obtusum*

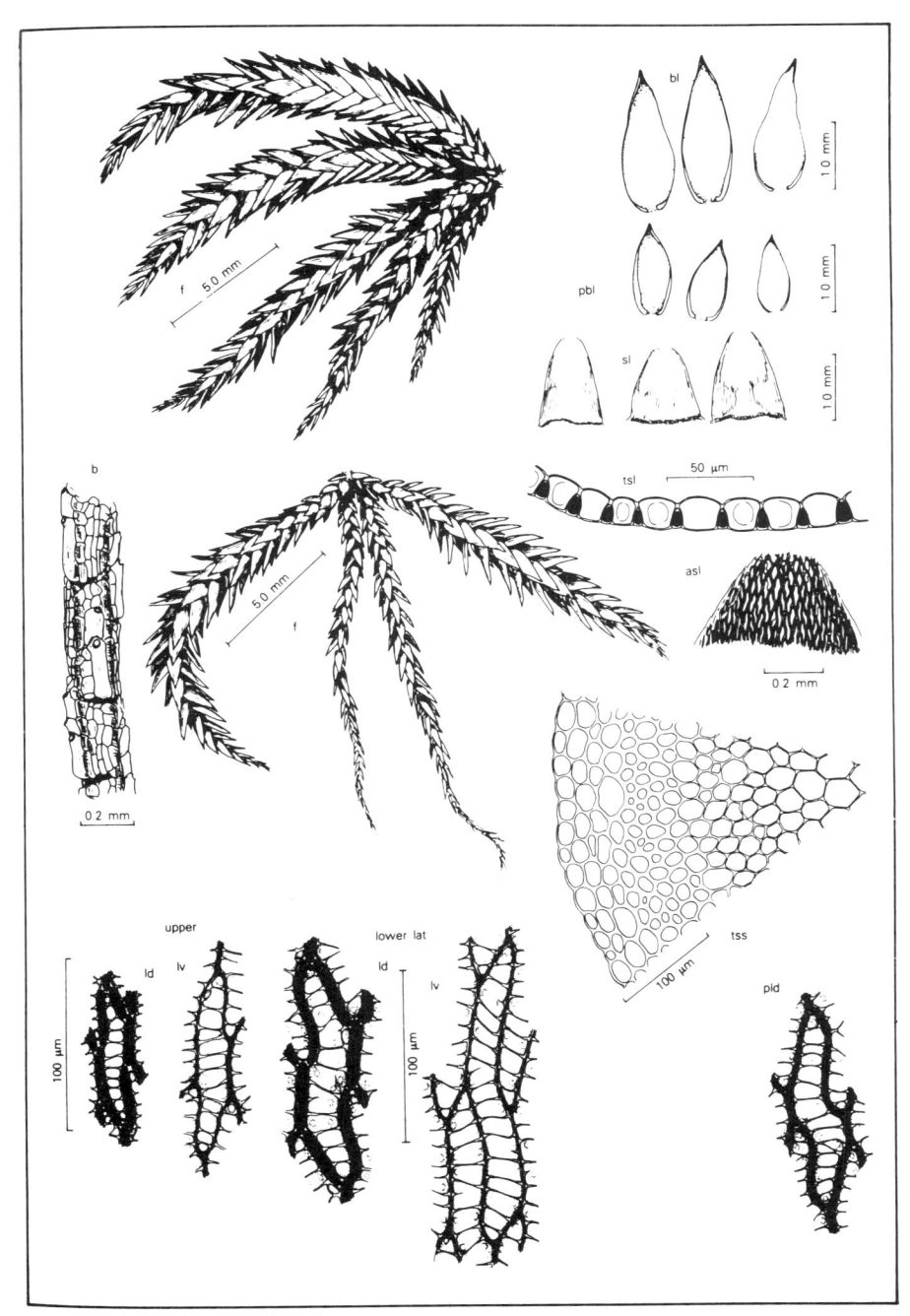

Figure 59. *Sphagnum obtusum*

28. SPHAGNUM FLEXUOSUM

Sphagnum flexuosum Doz. & Molk. *(Prodr. fl. Batav.,* 76, 106. 1851)
S. amblyphyllum (Russ.) Zickend. *(Byull. mosk. Obshch. Ispyt. Prir. Otd. Biol.,* n.s. **14,** 278. 1900)
S. recurvum var. amblyphyllum (Russ.) Warnst. (*Bot. Gaz.,* **15,** 219. 1890)

PLANTS: Medium-sized to rather robust, lax to rather compact; capitula well developed, often rather large and hemispherical; green to pale ochre. **Fascicles:** Mostly well spaced, occasionally dense; branches usually 5, *not or weakly dimorphic;* spreading 2(–3), long and tapering (12.0–25.0 mm or more), arched spreading; pendent (2–)3, weaker. **Stem:** Strong, 0.6–1.3 mm diameter; *cortex not distinguishable from internal cylinder* (except sometimes adjacent to leaf insertion; internal cylinder of slightly to moderately thickened cells, pale green to pale yellow-green. **Branch anatomy:** Retort cells distinct, in linear pairs; internal cylinder pale yellowish green, rarely faintly brown or pink. **Stem leaves:** Hanging and more or less appressed to stem; 0.7–1.3 × 0.7–1.0 mm; triangular-ovate to shortly triangular-lingulate; *apices obtuse and rounded-truncate, finely fringed,* usually with a *small notch at the tip;* border strong, widely expanded below. Hyaline cells without fibrils. **Branch leaves:** *Not consistently 5-ranked;* lanceolate, more or less uniform or, commonly, *in the distal parts of branches, rather abruptly becoming linear-lanceolate to linear.* Spreading branch leaves 1.5–2.4 mm long, the linear leaves, if present, up to 3.0 mm or more. Pendent branch leaves similar to those of spreading branches. **Hyaline cells:** Very narrow, in upper mid-leaf 70–170 µm long × 9.0–13.0 µm on the abaxial surface, up to 22.0 µm on the adaxial. Abaxial surface without pores, or with a small and inconspicuous resorption gap in the apical angle; with or without pseudopores in some lateral angles. Hyaline cells of lower pendent branch leaves proportionately wider, usually with 1–4 resorption gaps in the upper half of the abaxial surface; adaxial pores up to 12.0 µm diameter, often distinctly ringed. **Leaf TS:** Hyaline cells more or less plane on the abaxial surface, convex on the adaxial. Photosynthetic cells large, thin-walled, *trapezoid and exposed on the adaxial surface as well as the abaxial,* wider on the abaxial (in the upper half of leaf, often as wide as the hyaline cells. **Fertile plants:** Dioecious. Antheridial branches similar to sterile branches, but usually more highly coloured, yellowish to orange-brown. Inner perichaetial bracts large, up to 4.7 mm long; apices blunt and somewhat retuse. Hyaline cells distinct in the upper part of the bract, lacking fibrils. Capsules apparently rather rare; spores pale brown, slightly papillose to almost smooth, 23.0–26.0 µm diameter.

HABITAT: As mats or wider carpets in wet, mesotrophic to slightly oligotrophic mires, or along their margins, and in flushed grassland over mineral soil with a high organic content. *S. flexuosum* is able to tolerate a range of chemical conditions and some degree of shade, but is restricted in relation to water level: it is normally confined to intermediate zones and is absent from the wettest hollows and the tops of hummocks. It may be found in some wooded areas but is more typical of at least partially open communities. At the more oligotrophic end of its range, associates may include *S. magellanicum* or *S. papillosum,* whilst it may be found with *S. squarrosum* or *S. teres* in mesotrophic situations. Under intermediate conditions, it can occur with *S. russowii, S. lindbergii* and *S. riparium,* or in mixed stands with the closely related *S. recurvum* var. *mucronatum* and *S. angustifolium.*

DISTRIBUTION: Circumboreal with oceanic tendencies. Widespread in Europe, but with a somewhat oceanic and southern tendency. It is rare in the northern part of Scandinavia (and mainly confined to coastal areas) but more abundant further south. It is present in the lowlands and extends into the sub-alpine and alpine zones, and is restricted to montane areas in the southern part of its range in Italy, Yugoslavia and Bulgaria. Although its British distribution is not fully known (because of past confusion with *S. recurvum* var. *mucronatum* and, to a lesser extent, with *S. angustifolium*), it appears to be widespread but scattered throughout.

S. flexuosum is similar in appearance to *S. recurvum,* especially the lax '*fallax*' form of the latter, and the 2 species have often been confused, in Britain at least. However, the anatomical differences are sufficiently numerous and consistent to support treatment of *S. flexuosum* as a species in its own right. Within the '*recurvum* complex', there may be inconsistencies in some characters in some specimens, but the overall balance of differences should make distinction possible without undue difficulty. The distinctive leaf section, much more like that of *S. cuspidatum* than *S. recurvum* (with trapezoid photosynthetic cells which are adaxially exposed) and the obtuse, apparently never mucronate, stem leaves devoid of fibrils are the best distinguishing features of *S. flexuosum* which separate it from *S. recurvum. S. obtusum* has similar stem leaves but has a different leaf section, and the numerous, though faint, pores in the lower lateral hyaline cells have no counterpart in *S. flexuosum.* Attenuated forms from rather dry woodland habitats can bear a strong superficial resemblance to *S. fimbriatum,* but are readily distinguished by the lack of a large terminal stem bud, and small hanging, rather than large erect, stem leaves. Such forms may also closely resemble robust forms of *S. angustifolium,* and microscopic examination may be necessary to separate them: *S. angustifolium* is readily distinguished by the presence of large (over 12 μm diameter) pores in the apical angles of the

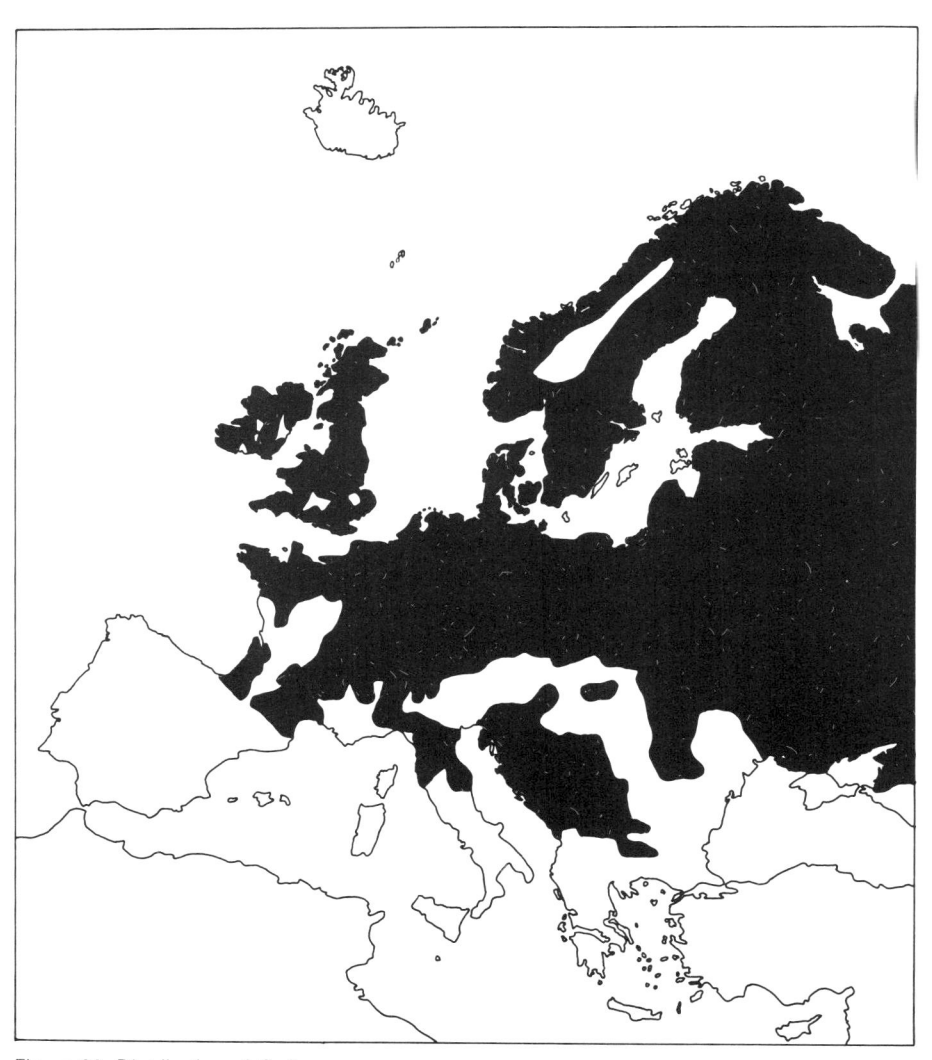

Figure 60. Distribution of *S. flexuosum*

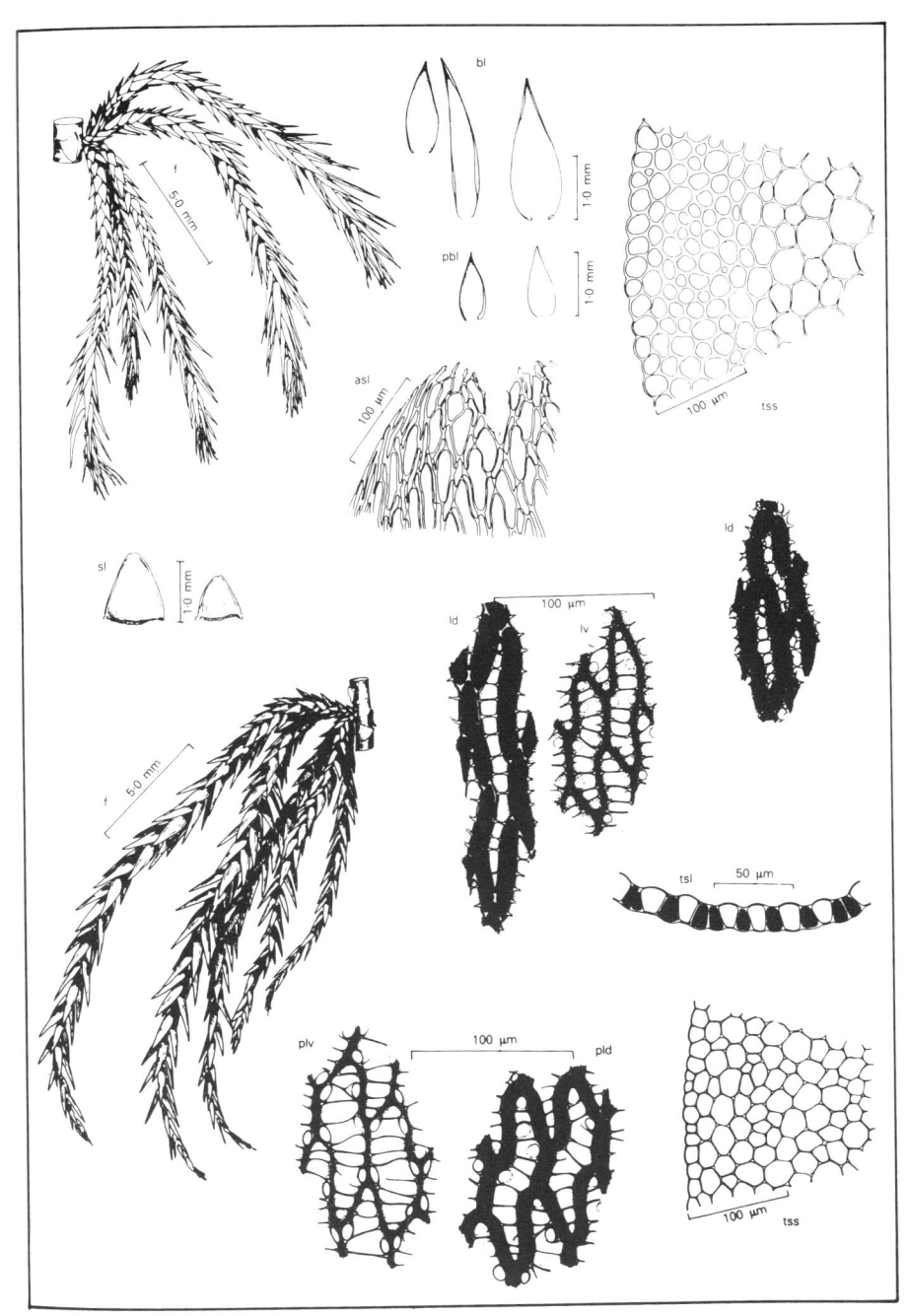

Figure 61. *Sphagnum flexuosum*

abaxial faces of hyaline cells in the lower lateral parts of hanging branch leaves and by the adaxially enclosed photosynthetic cells.

29. *SPHAGNUM RECURVUM*

Sphagnum recurvum P. Beauv. (*Prodr. fam. aeth.*, 88. 1805)
var. *recurvum* (The type variety probably does not occur in Europe)

29a. *S. recurvum* var. *mucronatum (Russ.)* Warnst. (*Bot. Gaz.* 1890)
S. recurvum subsp. *mucronatum* Russ. (*Sber. naturfGes. Univ. Dorpat,* **9,** 99. 1889)
S. fallax (Klinggr). Klinggr. (*Flora Westpreuss.,* 128. 1880)

PLANTS: Medium-sized, rarely compact; growing in deep, soft tufts, or pure to somewhat mixed carpets; deep green, yellow-green, ochre to orange-brown. **Fascicles:** Regularly well spaced along stems; of (4–)5 *moderately to strongly dimorphic branches;* spreading branches 2, rather rigid, widely divergent; 16.0–20.0 mm long; pendent branches 2–3, mostly more or less as long as, or appreciably shorter than, spreading branches (9.0–18.0 mm), sometimes considerably elongated in older fascicles. **Stem:** Relatively strong, rigid and rather brittle, 0.6–1.1 mm diameter; green to pale yellowish, rarely with areas of translucent pale brown; *cortex moderately distinct* of (2–)3–4 layers of thin-walled or, usually, slightly thick-walled hyaline cells, 15.0–30.0 μm diameter; internal cylinder strongly developed, green to pale yellowish. **Branch anatomy:** Retort cells distinct, 2–3 times as large as normal cortical cells, mostly in linear pairs; internal cylinder pale, rarely translucent pink. **Stem leaves:** Hanging and appressed to stem; rather small, 0.8–1.1 × 0.6–1.0 mm; equilateral to short isosceles triangular with slightly rounded sides; apices abruptly tapered to *apparently acute, or somewhat mucronate* due to inrolling of the upper leaf margins (normally only flattened with difficulty, then rounded-obtuse and obsoletely dentate); border strong, several cells wide to apex, strongly expanded below into wide patches of prosenchymatous tissue, sometimes almost confluent across leaf base, moderately to rather weakly expanded in the forma *fallax*. Hyaline cells typically without fibrils, or at most with a few cells near the apex containing fibrils on the abaxial surface; adaxial surface entirely resorbed. (In forma *fallax*, fibrillose cells, resembling those of branch leaves, sometimes present almost to mid-leaf.) **Branch leaves:** Numerous and densely arranged; almost always *clearly 5-ranked*, rarely spirally arranged; *uniform in size and shape; except near branch tips where they are shorter*, in mid-branch 1.4–3.0 mm long (sometimes longer in forma *fallax*) × 0.5–0.6 mm wide; lanceolate, widest one quarter to one third from insertion; apices truncate and 5–7 dentate, but often appearing acute due to inrolled upper leaf margins; border strong, 3–4 cells wide. Pendent branch leaves small, 0.8–1.2 mm long (to 1.5 mm in forma *fallax*), ovate to ovate-lanceolate (at least in lower parts of branches), concave.

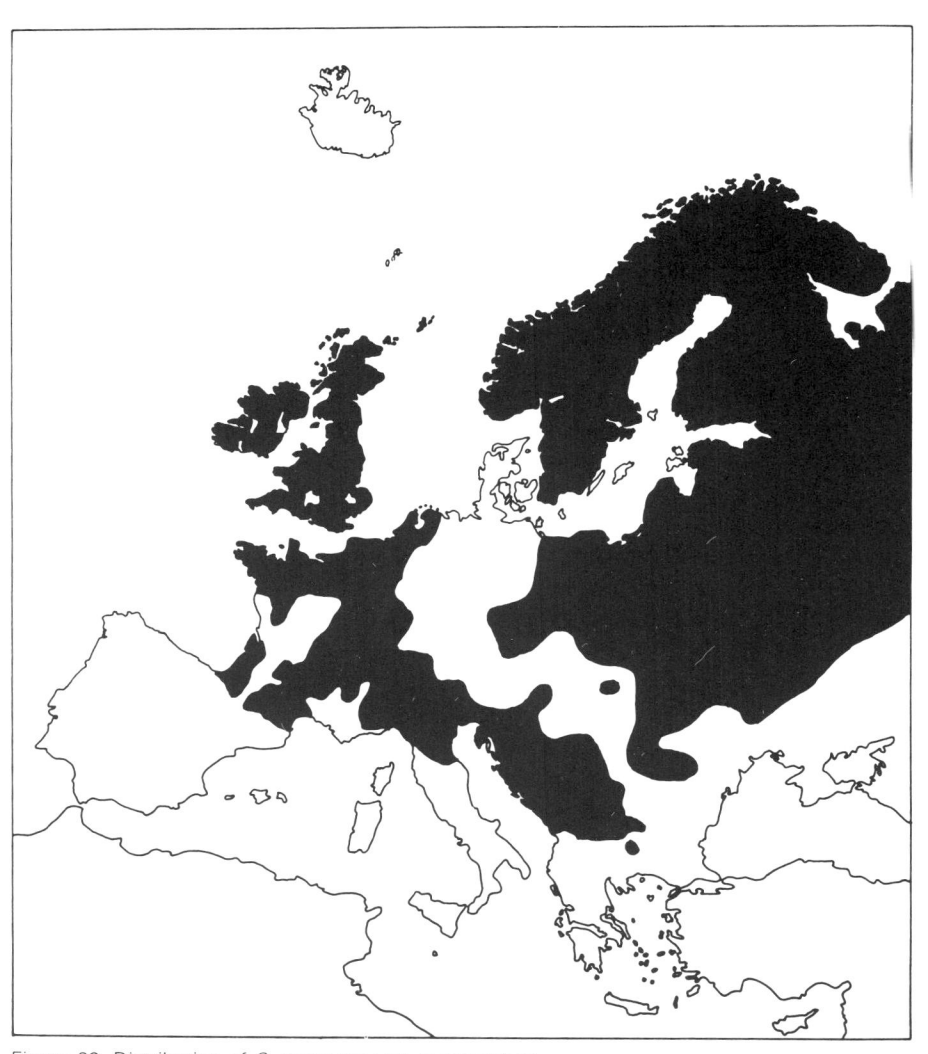

Figure 62. Distribution of *S. recurvum* var. *mucronatum*

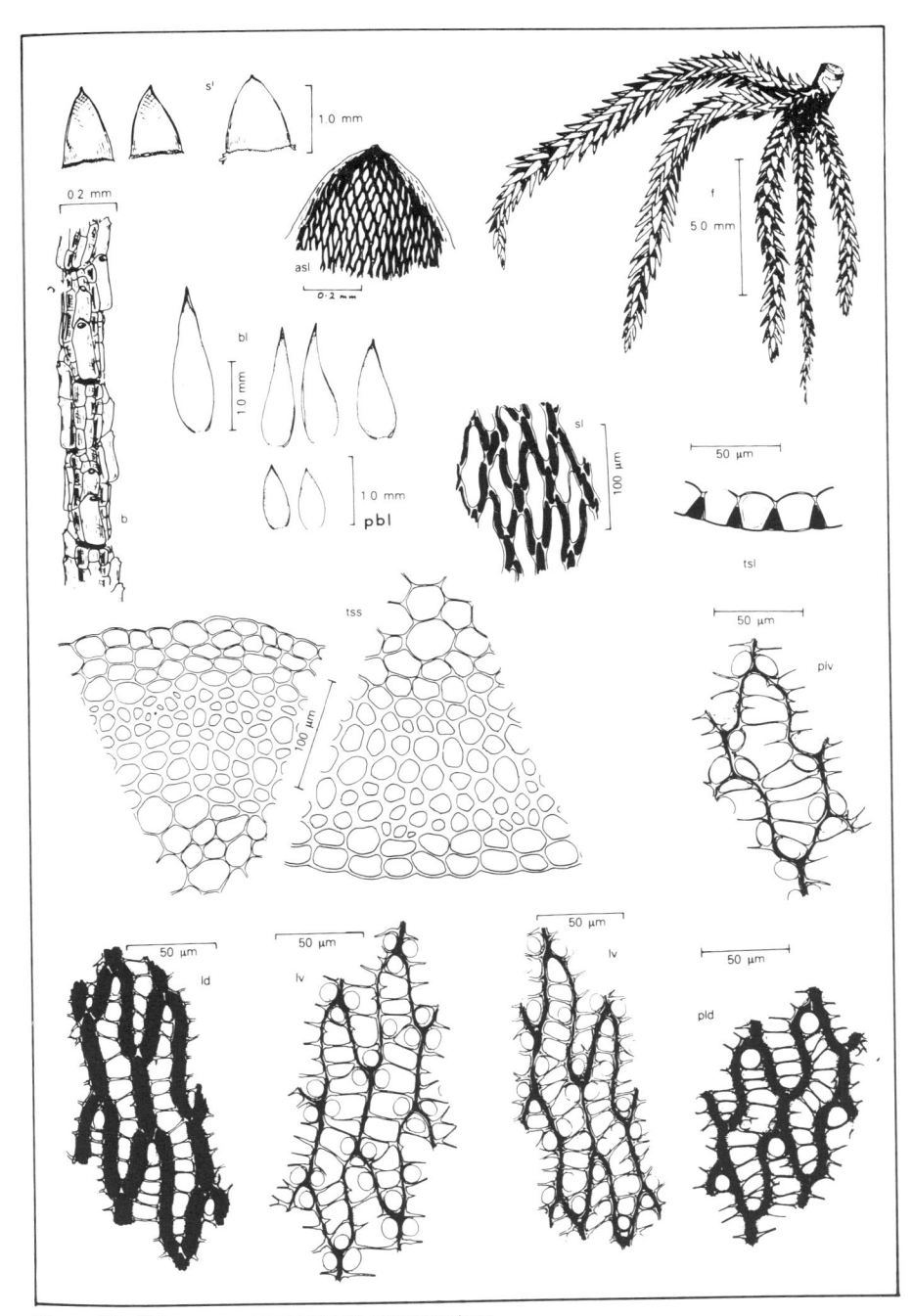

Figure 63. *Sphagnum recurvum* var. *mucronatum*

Hyaline cells: Narrow, in upper mid-leaf 120–160 × 10.0–15.0 μm on abaxial surface, 15.0–20.0(–23.0) μm on the adaxial. Abaxial surface without or, usually, with an inconspicuous resorption gap in the apical angle, with or without an occasional pseudopore in one or more lateral angles; true pores absent. Adaxial surface with numerous (usually 5–8) *rather large* (7.0–10.0 μm), faint to distinct, unringed or, occasionally, faintly ringed circular pores or thinnings, mostly adjacent to the cell angles. Hyaline cells of pendent branch leaves shorter and wider, 80–120 × 20.0–30.0 μm; apical angles not conspicuously wider than basal ones. Abaxial surface with medium-sized to rather large pores (8.0–12.0 μm) in the apical angle of each cell, otherwise without pores. Adaxial surface with large, unringed to thinly ringed pores which may appear elliptical because of the convexity of the cell surface. **Leaf TS:** Hyaline cells plane on the abaxial surface, convex on the adaxial. Photosynthetic cells rather thin-walled, *isosceles triangular* with broad abaxial exposure, *mostly distinctly but shallowly enclosed on adaxial surface* (occasionally with a few cells trapezoid in TS which may be narrowly exposed on adaxial surface). **Fertile plants:** Dioecious. Antheridial branches strongly orange or brown, otherwise similar to sterile branches. Perichaetial bracts 3.0–3.5 × 2.3–2.6 mm; widely ovate, strongly concave, convolute; apices broadly rounded but, like stem leaves, usually with pinched apices, and so appearing more or less mucronate; with differentiated hyaline cells in upper mid-bract, but these little, or no, larger than the photosynthetic cells and sometimes only discernible after staining, almost never with fibrils; rest of bract of prosenchymatous cells. Capsules uncommon to locally abundant; spores rather pale, almost smooth to lightly papillose, 25.0–28.0 μm diameter.

HABITAT: In oligotrophic to somewhat mesotrophic areas, often forming wide lawns in moderately wet, open peatlands. Exceptionally, it may also be found in more distinctly minerotrophic sites or drier localities, being tolerant of a fairly wide range of chemical and hydrological conditions. It is usually found in ombrotrophic or geotrophic mires as an intermediate between strict hollow species, eg *S. cuspidatum*, and main hummock-builders, eg *S. papillosum* and *S. magellanicum*. In the north, it extends into increasingly mesotrophic locations so that, in northern Scandinavia, it may be found in the lagg zone of mires, with scattered *Salix*. In central Sweden and Finland, it is a somewhat mesotrophic species, showing a preference for sites with slight mineral influence, and in northern Germany it is more distinctly a species of ombrotrophic mires. A characteristic associate of *S. recurvum* var. *mucronatum* is *Eriophorum angustifolium* which grows as scattered shoots at various densities within a carpet of *Sphagnum* composed largely or entirely of this species. In Britain, it grows mainly in situations similar to those in which it is found in southern Scandinavia, ie it is generally weakly minerotrophic.

DISTRIBUTION: Circumboreal with sub-oceanic tendencies. Widespread in Europe but somewhat oceanic, being more abundant in the north and west. Common throughout most of Britain.

S. recurvum var. *mucronatum* is perhaps the most common member of a group of closely related taxa which have variously been regarded by different authors as distinct species, distinct varieties, or simply as forms of a single species. The 3 taxa, treated here as distinct species, which are most frequently confused with *S. recurvum* are *S. flexuosum*, *S. angustifolium* and *S. pulchrum*. *S. flexuosum* is usually a fairly robust plant with much the habit of *S. recurvum* forma *fallax* but lacking the basically 5-ranked leaf arrangement normally detectable in the latter. The lack of a distinct stem cortex and the adaxially exposed photosynthetic cells of *S. flexuosum* are found only in highly modified forms of *S. recurvum* (which would then have fibrillose stem leaves and not the blunt, non-fibrillose leaves of *S. flexuosum*). *S. angustifolium* is a small plant with proportionately narrow branch leaves and very much larger apical pores on the abaxial surface of the pendent branch leaves. *S. pulchrum* is often a more brightly coloured plant, gold to deep gold-brown or chestnut, with rather short, stubby branches and a slightly shaggy appearance. The stem is darker than in *S. recurvum* and the branch leaves have a different outline (widest at about half-way, rather than a quarter to a third as in *S. recurvum*). In TS, the photosynthetic cells of *S. pulchrum* are more deeply immersed.

There is some controversy over the correct name to be applied to this species in Europe. British authors over recent years have used the name *S. recurvum*, but doubt has been cast on the conspecificity of European plants with the American one to which P. de Beauvoir gave the name. If the name *S. recurvum* is to be rejected, then *S. fallax* should be used (Isoviita 1966). Here, we follow the usage adopted by Hill (Smith 1978) and retain *S. recurvum* var. *mucronatum*, recognizing that the type variety probably does not occur in Europe.

30. SPHAGNUM ANGUSTIFOLIUM

Sphagnum angustifolium (Russ.) C. Jens. in Tolf (*Bih. K. svenska VetenskAkad. Handl.*, ser. 3, **16,** 9, 48. 1891)
S. recurvum var. *tenue* Klinggr. (*Schr. phys.-ökon. Ges. Köningsb.*, **13,** 5 1872)
S. parvifolium (Warnst.) Warnst. (*Bot. Zbl.*, **82,** 67. 1900)

PLANTS: Rather small, compact or attenuated; green or yellowish (resembling delicate forms of *S. recurvum, S. fimbriatum* or *S. capillifolium* in the field); capitula well developed. **Fascicles:** Well spaced, of 4 distinctly, and sometimes strongly, dimorphic branches; spreading branches 2, rather short (in open habitats) or long and tapering (shaded habitats), 6.0–10.0 (–17.0) mm long; pendent branches normally longer than the spreading, sometimes very long and attenuated, (8.0–)12.0–18.0 mm long. **Stem:** Thin, 0.4–0.8 mm diameter; green, pale yellow or faintly translucent pink near the capitulum; *cortex not distinct,* except immediately adjacent to stem leaves where there are local swellings of 2–4 layers of enlarged, thin-walled cells. **Branch anatomy:** Retort cells very distinct, mostly occurring singly, occasionally in linear pairs, distinctly protuberant; internal cylinder pale or, frequently, translucent pink. **Stem leaves:** Hanging; small, 0.7–0.9(–1.0) mm long; shortly isosceles to equilateral triangular; apices obtuse and eroded, mostly slightly concave, rarely with pinched or pseudomucronate tips; border strong, widely expanded below. Hyaline cells mostly without fibrils, rarely a few near apex with rudimentary fibrils on the abaxial surface; adaxial surfaces more or less completely resorbed. **Branch leaves:** Rather densely imbricated (except in shaded plants), in the lower parts of the branches, 5-ranked or not, suberect to slightly patent; rather small, 1.1–1.6 × 0.3–0.5 mm, narrowly lanceolate. Pendent branch leaves small, 0.7–1.0 mm long, narrowly ovate to lanceolate. **Hyaline cells:** In upper mid-leaf 90–100 μm long × 10.0–15.0 μm wide on abaxial surface, 15.0–22.0 μm on the adaxial. Abaxial surface often with a small pore or pseudopore in some lateral angles, and usually with a rather indistinct small to medium-sized resorption gap in the apical, occasionally also in the basal, angle. Adaxial surface with 4–6 relatively large (8.0–10.0 μm), mainly unringed circular pores adjacent to cell angles. Hyaline cells of pendent branch leaves, especially in the lower lateral parts of the leaves, *conspicuously wider at the apical end than at the basal end* and, on the abaxial surface, *with large resorption gaps or pores* (12.0–18.0 μm diameter) in the apical angles. Adaxial pores as in spreading branch leaves. **Leaf TS:** Hyaline cells much wider than photosynthetic cells; almost plane on abaxial face, shallowly convex on adaxial. Photosynthetic cells isosceles ovate-triangular, with relatively thin or slightly thickened abaxial walls; distinctly but rather shallowly enclosed on the adaxial surface. **Fertile plants:** Dioecious. Antheridial bracts often brightly coloured yellow or pale orange, and sharply contrasted with the green branch leaves. Inner perichaetial bracts large and conspicuous, 3.5–4.5 mm long; lacking fibrils; apex obtuse and minutely retuse, mainly of prosenchymatous cells; hyaline cells differentiated in the upper half of bract. Capsules locally frequent, sometimes abundant; spores yellow, slightly papillose 22.0–24.0 µm diameter.

HABITAT: Has a fairly wide tolerance of trophic status, but is usually found in wet, mesotrophic mires, where it grows as pure mats or as scattered shoots among other *Sphagnum* species, eg *S. riparium, S. flexuosum, S. fuscum, S. subnitens, S. recurvum* var. *mucronatum, S, russowii* or *S. magellanicum*. It is usually present as an intermediate between hummock and hollow species, but may occasionally be found in partially submerged *Sphagnum* lawns or soaks with, for example, *S. riparium* or *S. jensenii,* or on rather drier hummocks with *S. magellanicum. S. angustifolium* is also able to tolerate moderate shade and may be dominant under some areas of wooded mire, although it can also be common in *Carex*-dominated communities. In Britain, it is found most usually towards the most mesotrophic part of its chemical range, in sites with a distinct inflow of minerotrophic water.

DISTRIBUTION: Circumboreal with distinct continental tendencies. It is the only member of the *S. recurvum* complex (including *S. recurvum* itself and *S. flexuosum,* as well as *S. angustifolium*) to be found in the continental interior of North America. In Europe, it is widespread from the lowlands to the sub-alpine, but has slight continental tendencies and is comparatively uncommon in coastal areas. Scattered and uncommon in Britain, but its distribution has not yet been investigated fully.

S. angustifolium is fairly stable in its microscopic characters, but is rather variable in gross morphology. In the field, it can superficially resemble. *S. recurvum* or *S. flexuosum,* but differs mainly in its smaller dimensions and proportionately narrower branch leaves. These species, as well as being more robust, usually have at least some fascicles with 5 rather than a maximum of 4 branches per fascicle: weak forms may not be distinguishable without microscopic examination. In exposed, wet habitats, particularly in the hilly districts of northern and western Britain, it may resemble terrestrial forms of *S. cuspidatum,* but it differs from that species by the possession of small-leaved pendent branches and the lack of linear distal leaves on the spreading branches. Some authors emphasize the translucent pink (often cited as red) colour commonly present in the internal cylinders of branches, but this feature can also be found in *S. recurvum* and, less frequently, in *S. flexuosum* and *S. cuspidatum. S. angustifolium* is most readily distinguished

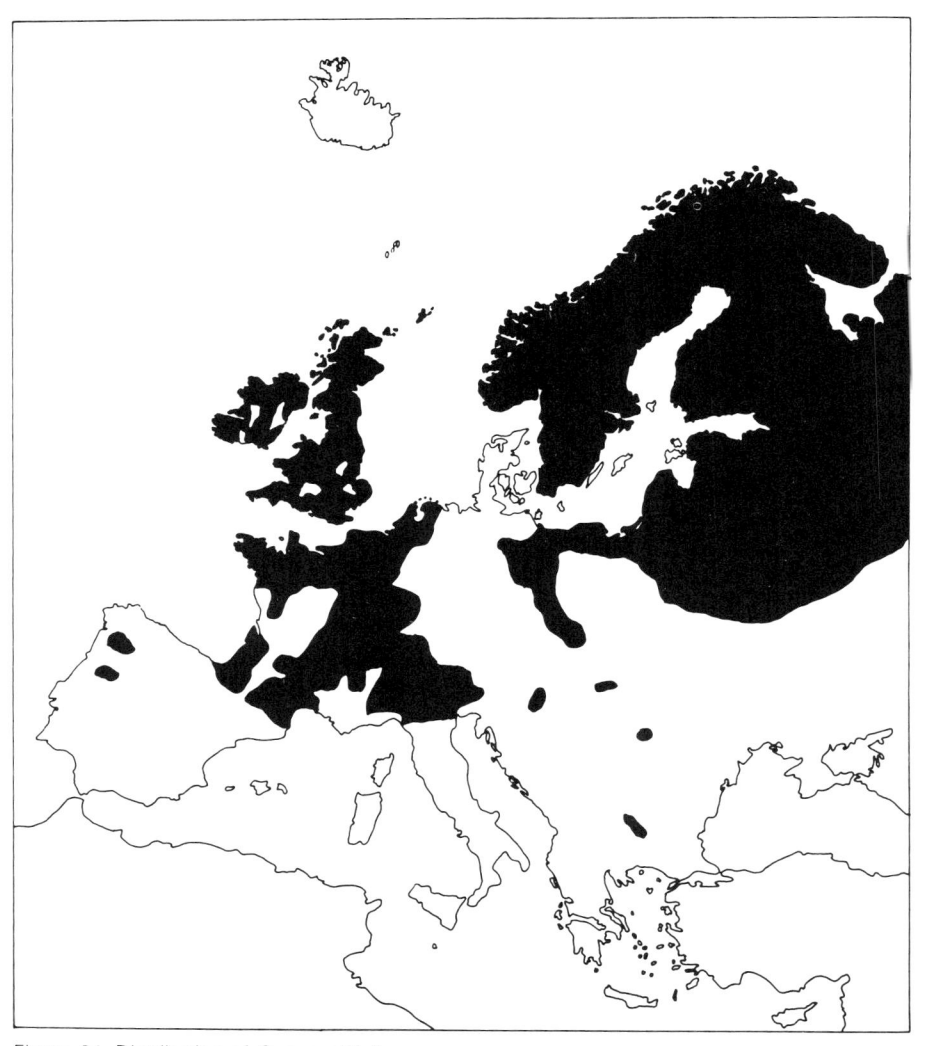

Figure 64. Distribution of *S. angustifolium*

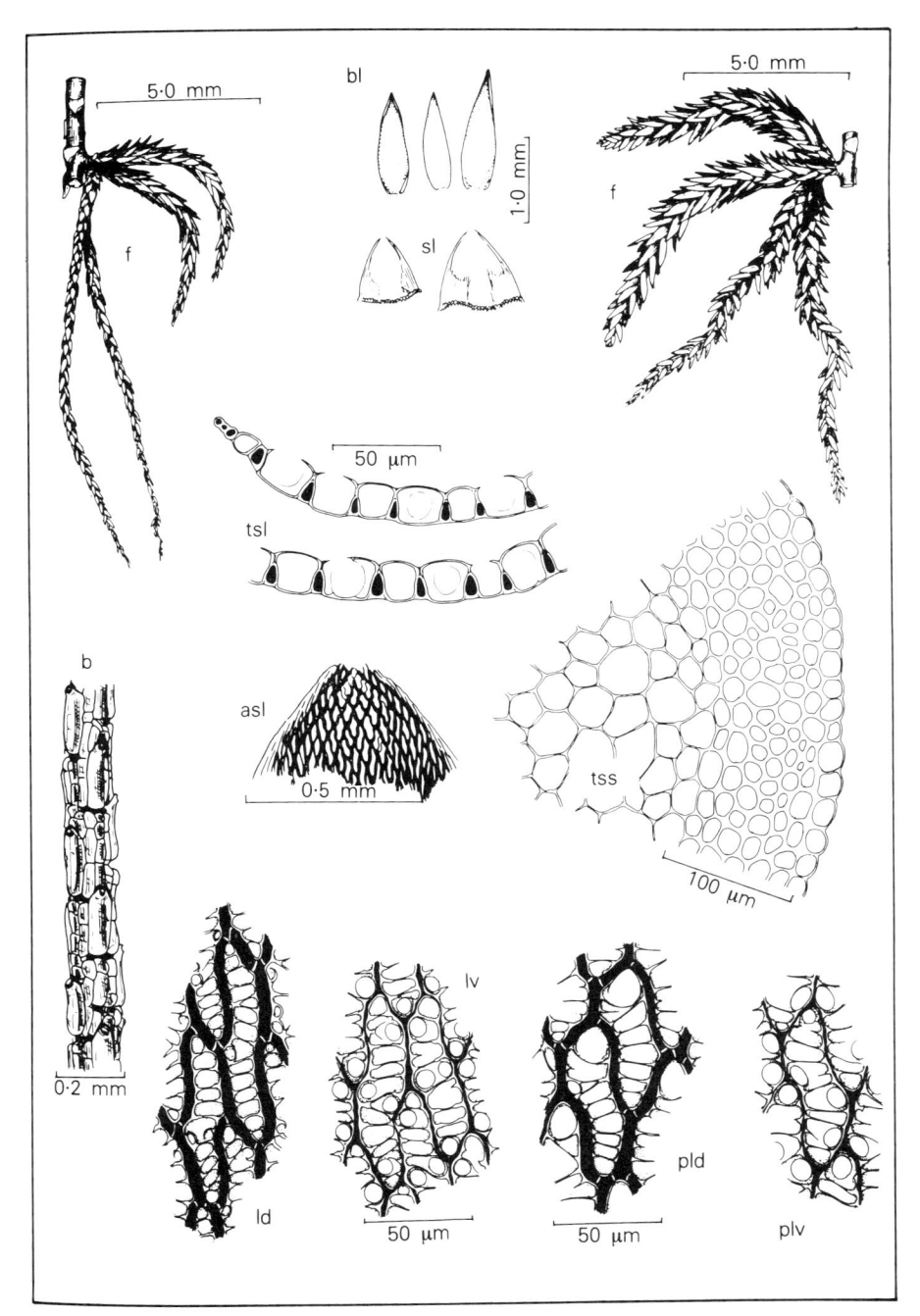

Figure 65. *Sphagnum angustifolium*

from related species (except sometimes *S. balticum*) by the hyaline cells of the pendent branch leaves with their large abaxial, apical pores.

31. SPHAGNUM BALTICUM

Sphagnum balticum (Russ.) C. Jens. (*Festskr. Bot. Kjøb. halvhundre-daarsf.*, 100, 116. 1890)

PLANTS: *Small*, slender; yellow-brown to dark brown, green only in shade (resembling in habit and stature, but not in colour, *S. angustifolium*). **Fascicles:** Commonly with 3, at most 4, dimorphic branches; spreading branches 2, short, 4.0–8.0 mm, rarely attenuated and exceeding 9.0 mm, densely foliate; pendent branches 1(–2), weaker than, but more or less as long as, spreading branches, less densely foliate. **Stem:** 0.4–0.6 mm diameter; cortex well developed, of 2–3 layers of thin-walled (or the inner layers slightly thickened) cells; cells in section several times larger than those of the strongly developed, green-yellow to pale brown internal cylinder. **Branch anatomy:** Retort cells distinct, highly inflated, *usually solitary,* moderately protuberant at apertures; internal cylinder yellowish to brownish. **Stem leaves:** Spreading to suberect, but not appressed to stem; *relatively large,* 0.8–1.1 × 0.6–0.8 mm; triangular-ovate to shortly lingulate or slightly spatulate, usually distinctly concave; apices rounded-obtuse, often notched; border strong, widely expanded below. Hyaline cells below leaf apex fibrillose, at least on abaxial surface, rarely almost lacking fibrils; abaxial surface with few or no complete fibrils, but numerous fibril stumps; interfibrillar spaces with indistinct thinning (visible only after heavy staining) or an occasional small pore; adaxial surfaces of hyaline cells almost completely resorbed. **Branch leaves:** Small, 1.2–1.6 × 0.4 mm, not consistently 5-ranked, rather densely arranged; lanceolate, often slightly curved and secund. **Hyaline cells:** 80.0–110.0 × 9.0–15.0 μm abaxially, up to 22.0 μm wide adaxially, in upper mid-leaf. Abaxial surface variable, typically with *few to several* pores or *ringed and unringed pores and pseudopores* (irregular in size, but rarely more than 5.0 μm diameter) *along the commissures* and a small resorption gap in the apical angle. Adaxial surface with medium-sized (5.0–9.0 μm) circular, unringed pores, mainly adjacent to cell angles. Hyaline cells of lower lateral parts of leaves with few small pores (sometimes none), but with a large resorption gap in the apical angle on the abaxial, and sometimes the adaxial, surface. Hyaline cells of pendent branch leaves somewhat wider (up to 30.0 μm) and with larger apical resorption gaps (8.0–13.0 μm), otherwise similar to those of spreading branches. **Leaf TS:** Hyaline cells plane on the abaxial surface, shallowly convex adaxially. Photosynthetic cells oval-triangular, the abaxial walls often thickened considerably, mostly rather deeply enclosed on adaxial surface. **Fertile plants:** Dioecious. Antheridial branches distinct or not, the bracts deep brown. Inner perichaetial bracts to 4.0 mm long, broad and more or less retuse at apex; subapical tissue similar to upper stem leaf tissue, fibrillose. Capsules

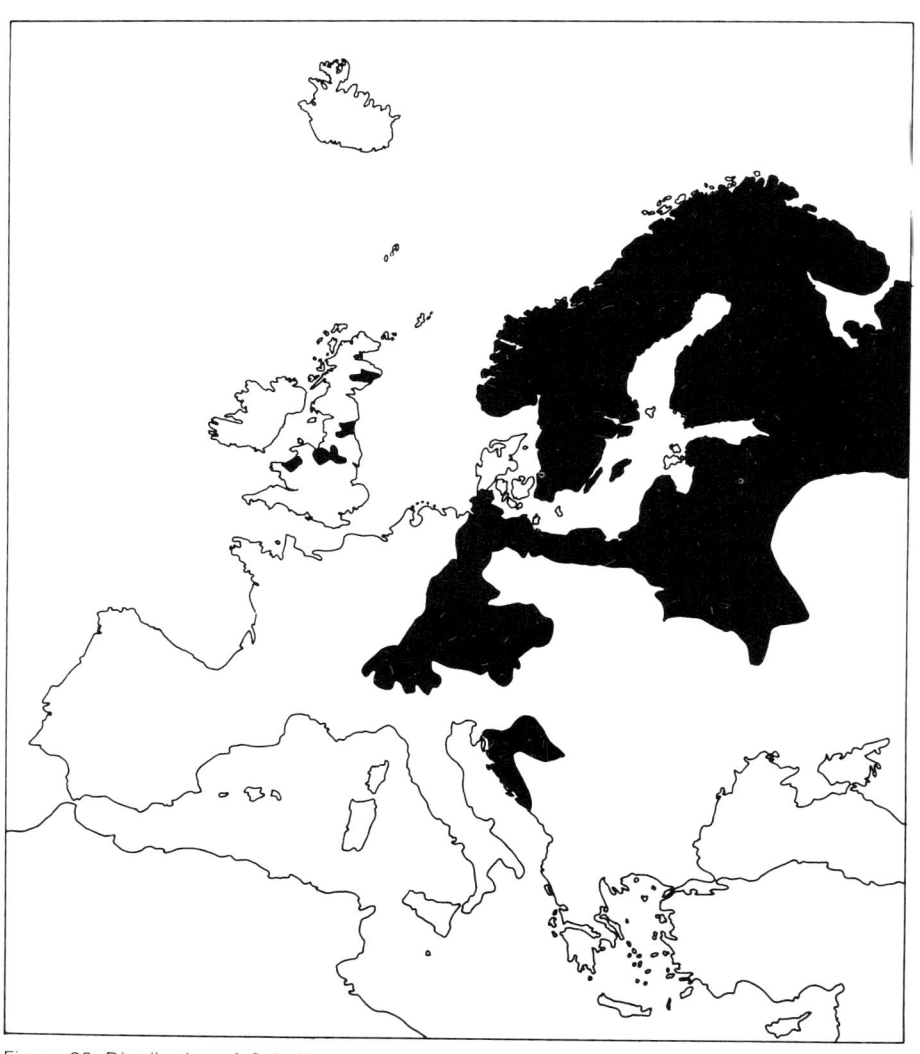

Figure 66. Distribution of *S. balticum*

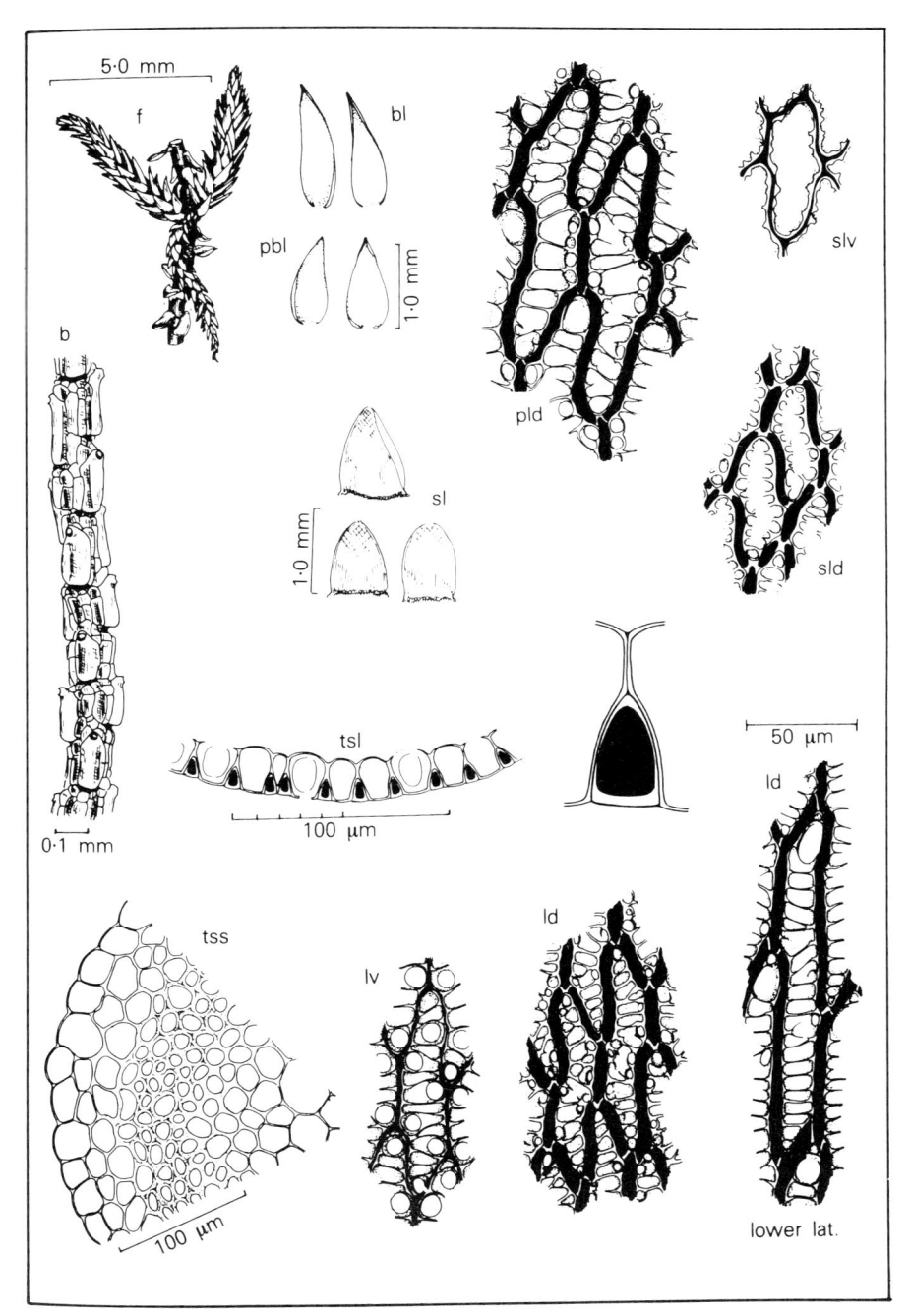

Figure 67. *Sphagnum balticum*

rare (unknown in Britain), but locally frequent in some arctic localities; spores yellow, lightly papillose to almost smooth, 27–31 µm diameter.

HABITAT: In wet, oligotrophic to slightly mesotrophic areas. It forms cushions or carpets in the wetter parts of open, ombrotrophic *Sphagnum*/ dwarf shrub mires, either as the major *Sphagnum* species, or in association with *S. majus, S. fuscum* or *S. cuspidatum* in the wetter parts, or with *S. papillosum* or *S. magellanicum* in the somewhat drier parts. Vascular plants associated with these lawns may include *Erica tetralix, Eriophorum angustifolium, Carex limosa, C. pauciflora* or *Scheuchzeria palustris.* In the north, it may also be associated with *S. lindbergii,* particularly between dwarf shrub tussocks.

DISTRIBUTION: Circumpolar in northern Europe, northern Asia, North America and Greenland. In Europe, it is mainly a lowland species (although reaching 1250 m in northern Sweden), with northern and somewhat continental tendencies. It is present in most of Scandinavia, except the south-west, and extends as far south as the Alps, but becomes increasingly rare in the sub-alpine zones. Rare and scattered in north-east Britain.

Sphagnum balticum is rather like a brown version of *S. angustifolium* or weak plants of *S. annulatum,* and in many ways seems to form a connecting link between these species. *S. angustifolium* is a paler plant (never deep brown), with non-fibrillose stem leaves and mostly indistinct stem cortex. However, some plants of *S. balticum* from central Europe are structurally very close to *S. angustifolium. S. annulatum* is usually distinctly larger, has some pores on or near the hyaline cell mid-line (abaxial side), and lacks the large apical resorption gaps in the lower lateral cells of the spreading branch leaves and the pendent branch leaves. *S. annulatum* has not been found in Britain, but in northern Europe it may occur in the same areas as *S. balticum.* It is generally unsafe to rely solely on field identifications in regions where both species occur.

Sphagnum fuscum (Acutifolia) could be mistaken for this species, in the field, by beginners, but can be separated easily by its dark stems, thin-textured, larger stem leaves, and strongly dimorphic branches. There is no possibility of confusion under the microscope.

32. *SPHAGNUM ANNULATUM*

Sphagnum annulatum Warnst. (*Bot. Zbl.*, **76,** 422. 1898)

PLANTS: Medium-sized to small, rather lax to more or less compact; yellow-brown to brown, rarely green (resembling a robust *S. balticum* or a weak *S. jensenii*). **Fascicles:** Distant or rather close set, slightly dimorphic; of 3, occasionally 4, tapered branches; stronger branches 8.0–14.0 mm long. **Stem:** 0.4–0.8 mm diameter; cortex distinct, 2–3 layers of rather narrow cells with distinctly thickened walls; internal cylinder yellowish. **Branch anatomy:** Retort cells distinct, mainly in linear groups of 2–3, slightly protuberant at apertures; internal cylinder yellowish to pale brown. **Stem leaves:** Erect-spreading; 1.0–1.2 mm long, triangular-ovate, concave; apices broadly rounded-obtuse, eroded; border expanded below into wide areas of prosenchymatous tissue. Hyaline cells below apex more or less strongly fibrillose (leaves indistinguishable from those of *S. balticum*). **Branch leaves:** 5-ranked or not, moderately large, 1.5–2.0 × 0.8 mm on stronger branches, broadly lanceolate. Leaves of pendent branches (in more dimorphic fascicles) smaller, ovate to ovate-lanceolate in lower third of branch, 0.8–1.2 mm long. **Hyaline cells:** Rather small in upper mid-leaf, 70–90 × 11.0–15.0 μm on abaxial surface, 12.0–20.0 μm on adaxial. Abaxial surface variable but mostly with some pores and pseudopores on, or near, the commissures and usually with one or a few ringed and/or unringed pores in, or near, the cell mid-line (these pores not as numerous, or in regular rows, as in *S. jensenii*, although, in some lower lateral cells, they may be in interrupted rows). Adaxial surface often without true pores, or pores few (rarely more than 4 per cell), sometimes with numerous pseudopores along the commissures and, particularly, adjacent to the lateral angles; unringed, circular pores, when present, seldom larger than 5.0 μm diameter. Hyaline cells of pendent branch leaves with slightly larger pores, otherwise identical to those of the spreading branch leaves, normally lacking large, apical resorption gaps on the abaxial surface. **Leaf TS:** Hyaline cells almost plane on both surfaces. Photosynthetic cells oval-triangular with more or less oval lumina and distinctly thickened abaxial walls; mostly distinctly, but shallowly, enclosed on the adaxial surface. **Fertile plants:** Probably dioecious, but no fertile material seen.

HABITAT: Like the closely related *S. jensenii*, this is a species of wet, unshaded, mesotrophic habitats, though it tends to be more aquatic and is more or less confined to areas of open, moving water, eg soaks and ditches within minerotrophic mires. Its most common associate is *S. jensenii*, though it may also be found in *S. majus*.

Figure 68. Distribution of *S. annulatum*

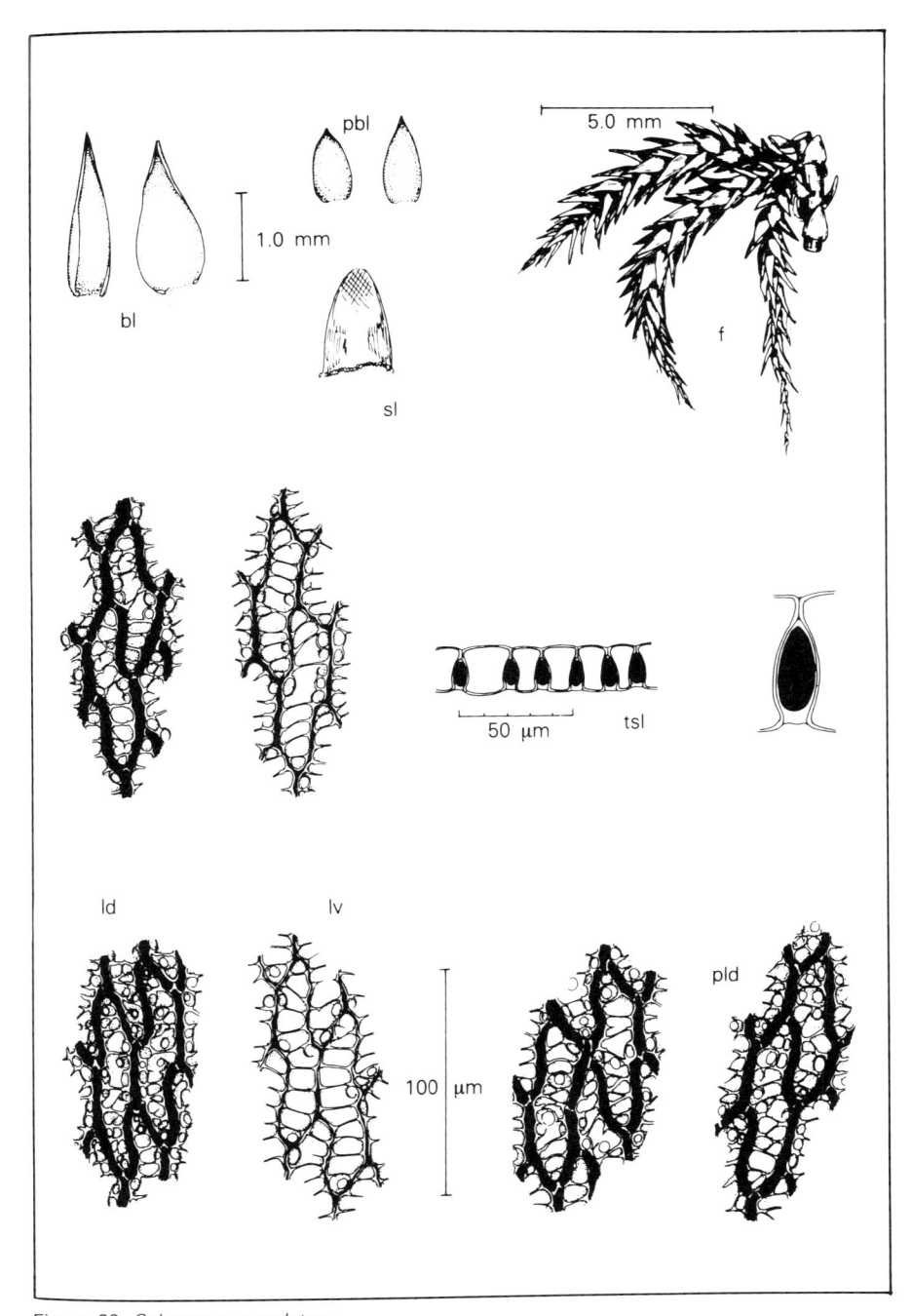

Figure 69. *Sphagnum annulatum*

DISTRIBUTION: Discontinuously circumboreal, being found in parts of northern Fennoscandia and USSR, and in north-eastern Canada (Ontario and Labrador).

There has been a great deal of confusion about the identity and status of *S. annulatum*, especially in its relationship to *S. jensenii* and, in North America, *S. mendocinum* Sull. It has been treated as a variety of *S. jensenii* (or *vice versa*) by many authorities. We follow the nomenclature of Isoviita (1966). As here interpreted, *S. annulatum* is intermediate between *S. jensenii* and *S. balticum*. The geographical ranges and ecological preferences of these 2 species and of *S. majus* are similar, and possibly *S. annulatum* is of hybrid origin.

We have been unable to find fertile material from Europe, the only putative fruiting plant from the British Museum herbarium being a misidentified *S. majus* mixed with sterile *S. jensenii*.

In the field, *S. annulatum* resembles a weak *S. jensenii* or a rather robust *S. balticum*, but can really only be identified with certainty on the microscopic characters of the branch leaf pores. The absence or paucity of adaxial pores and the lack of large, apical resorption gaps distinguish *S. annulatum* from *S. balticum*. *S. jensenii* has minute pores, either in the cell mid-line or partially in two rows, one either side of the mid-line, and none, or very few, along the commissures.

In a recent appraisal of *S. annulatum* and its relatives, Flatberg (1988c) has proposed a different demarcation between this species and *S. jensenii* from that presented above. In effect, he dismisses *S. annulatum*, as typified in the sense of Lindberg (1899) and understood by us in this treatment, as an aberrant plant that does not deserve taxonomic recognition. Instead, he redefines the species to include the later-described *S. propinquum* Lindb. ex Warnst., which has hitherto been regarded by most authors as synonymous with *S. jensenii*. The characters which separate the 2 taxa are concisely presented in the key (Flatberg 1988c, p.349) thus:

> Capitulum glossy chestnut-brown to sometimes (in spring fen vegetation) dark red-brown; outer branches of the capitulum, and the divergent branches below the capitulum, distinctly arcuato-decurved in their distal part; leaves of the divergent branches markedly concave and distinctly smaller near the proximal end than the middle part of the branches; terminal bud of the capitulum conspicuous, usually of about equal length or only slightly shorter than the inner branches, and always visible. ....*S. annulatum*

Capitulum dull yellowish brown to brown, outer branches of the capitulum, and divergent branches below the capitulum not distinctly arcuato-decurved in their distal part; leaves of the divergent branches not markedly concave and not distinctly smaller near the proximal end than in the middle part of the branches; terminal bud of the capitulum moderately large, exceeded in length by the inner branches, and sometimes concealed by them. ....*S. jensenii*

According to Flatberg (1988c), *S. annulatum* is predominantly a species which prefers somewhat richer, less wet sites than the oligotrophic to weakly minerotrophic mires preferred by *S. jensenii*. Their geographical ranges are largely coincident, but *S. annulatum* shows more pronounced arctic-alpine tendencies.

33. *SPHAGNUM JENSENII*

Sphagnum jensenii Lindb. (*Acta Soc. Fauna Flora fenn.*, **18,** 13. 1899)

PLANTS: Medium-sized to robust rather lax, sometimes rigid; brown, green-brown or yellow-brown, rarely all green. **Fascicles:** Distant or close set, not, or slightly, dimorphic; of 3–4 branches; spreading branches 2, short or much elongated (8.0–20.0 mm), often appearing stout; pendent branches weaker and attenuated distally, or almost as vigorous as spreading and scarcely differing (if 4 branches, then one usually intermediate in attitude and form between the 2 spreading and the weakest pendent branch). **Stem:** Rather thin, 0.4–0.7 mm diameter; cortex developed but not conspicuous, sometimes indistinct, in TS of 2–4 layers of rather narrow cells with distinctly thickened walls; internal cylinder yellowish. **Branch anatomy:** Retort cells distinct, mainly in linear groups of 3, slightly protuberant at apertures; internal cylinder relatively thick, yellowish to pale brown. **Stem leaves:** Mainly erect-spreading; 1.0–1.3 × 1.0 mm, oval-triangular, concave; apices rounded-obtuse; border strong, expanded below. Hyaline cells below leaf apex on abaxial side strongly fibrillose, lacking pores, or with an occasional small and inconspicuous pore; adaxial surface with or without scattered complete fibrils, cell wall almost completely resorbed, but with numerous, strongly ringed, small pores or pseudopores, often in short series along the commissures. **Branch leaves:** Large, mostly over 2.0 mm long, erect-spreading, more or less 5-ranked, lanceolate. Leaves of pendent branches (where distinct) broadly lanceolate, 1.3–1.8 mm long in basal third of branch. **Hyaline cells:** Long, 11.0–20.0 × 100–150 µm. Abaxial surface with frequent to abundant

Figure 70. Distribution of *S. jensenii*

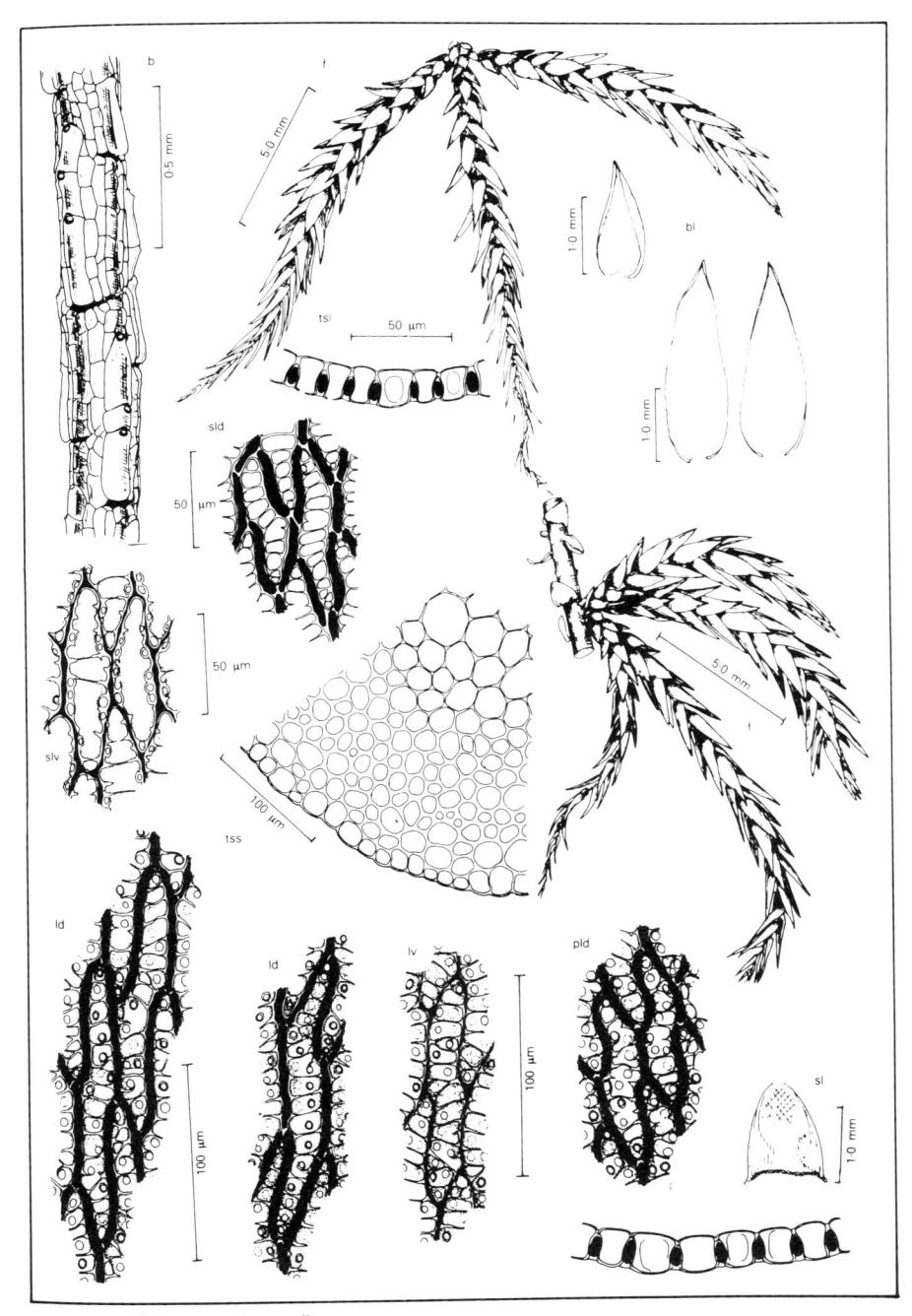

Figure 71. *Sphagnum jensenii*

small, ringed and unringed pores in one or 2 rows in, or on either side of, the cell mid-line; with no, or only an occasional, pore or pseudopore on or near the commissures. Adaxial surface (at least in cells from the lower lateral parts of the leaf, usually throughout the leaf) with small pores more or less identical to those on the abaxial surface. Hyaline cells of pendent branch leaves similar to those of spreading branch leaves. **Leaf TS:** Hyaline cells almost plane on both surfaces. Photosynthetic cells triangular-oval with oval lumina and, usually, strongly thickened abaxial walls; mostly shallowly enclosed on adaxial surface. **Fertile plants:** Dioecious. No fertile material seen by the authors.

HABITAT: In open, wet, mesotrophic mires, usually partly submerged in pools or in hollows subject to seasonal flooding: a typical habitat is in the elongated pools, or 'flarks' of aapa fens (string bogs). It prefers open habitats without shade, and either forms mats of rather prostrate individuals or occurs in association with, most commonly, *S. majus*, though it may also be found with *S. riparium*, *S. lindbergii* or *S. annulatum*. Other associates may include *Drepanocladus exannulatus*, *Carex limosa* and *Menyanthes trifoliata*.

DISTRIBUTION: Circumpolar in the boreal and sub-arctic regions, but extending as far south as the Rocky Mountains of Wyoming (at an altitude of over 3000 m) and Honshu, Japan. In Europe, it is distinctly northern-continental and is found throughout much of northern and central Scandinavia, in northern USSR as far as west Siberia, in northern Poland and, reportedly, as an isolated population in the Ukraine.

The pores of *S. jensenii* are much more clearly defined than those of *S. obtusum* and occur throughout the leaf. In depauperate specimens, the pores may be less numerous, and it is not always possible to separate such plants from *S. annulatum* (assuming them to be distinct species). *S. majus* has similar pores, but these are usually larger and confined to the abaxial leaf surface. *S. majus* can also be distinguished by its leaf section, which shows the photosynthetic cells exposed on the adaxial leaf surface, not enclosed as in *S. jensenii*.

The close relationship between *S. jensenii* and *S. annulatum* is discussed in the notes under the latter species. In North America, the position is further complicated by the occurrence of another, apparently closely related, taxon, *S. mendocinum* Sull.

34. *SPHAGNUM MAJUS*

Sphagnum majus (Russ.) C. Jens. (*Festskr. Bot. Kjøb. halvhundre-daarsf.*, 106. 1890)
S. dusenii Warnst. (*Hedwigia*, **29**, 214. 1890)

PLANTS: Semi-rigid to very lax (resembling robust forms of *S. cuspidatum* or lax *S. jensenii*); nearly always at least partly *brown,* and often the whole plant brown, green only in shade. **Fascicles:** Of 3–4, often vigorous, branches (15.0–20.0 mm or more) which are at most only slightly dimorphic (though varying in vigour). **Stem:** 0.6–0.8 mm diameter, pale yellow-green; cortex moderately distinct, of 2–3 layers of enlarged hyaline cells with slightly thickened walls; internal cylinder well developed, yellow. **Branch anatomy:** Retort cells in linear pairs, distinct, slightly protuberant at apertures; internal cylinder greenish yellow to faintly brown. **Stem leaves:** Hanging or spreading, often loosely appressed to stem; triangular-ovate, concave; apices rounded-obtuse, eroded across the tips; border strong, expanded below into, usually large, patches of prosenchymatous tissue; hyaline cells near apex strongly fibrillose on abaxial surface; adaxial surface almost completely resorbed, but with numerous, rather irregular, thick-ringed pores or pseudopores along the commissures, sometimes with occasional intact fibrils, especially towards cell ends (stem leaves like those of *S. jensenii*). **Branch leaves:** Lanceolate to linear-lanceolate (distal leaves somewhat reduced and linear), often curved and secund; 2.0–2.8 mm long on the more vigorous branches; mostly obscurely 5-ranked. **Hyaline cells:** Long, 100–220 × 13.0–22.0 µm in upper mid-leaf; abaxial surface with *numerous (rarely few) medium-sized mainly unringed (never strongly ringed) pores* (ca 5.0 µm diameter), more or less scattered over the surface, or in one or 2 irregular rows; adaxial surface without pores or with few (usually 1–3 at most) rather large, unringed pores. **Leaf TS:** Hyaline cells almost plane on abaxial face, shallowly convex on adaxial. Photosynthetic cells trapezoid, moderately thick-walled, with oval-triangular lumina; reaching, and often appreciably exposed on, adaxial leaf surface, but with wider exposure on abaxial surface. **Fertile plants:** Dioecious. Antheridial bracts brown; shorter and proportionately wider than branch leaves. Inner perichaetial bracts spatulate; about 4.5 mm long; apices broad and rounded, tissue near apex strongly fibrillose, in lower third of bract more or less uniformly prosenchymatous. Capsules occasional; spores pale, coarsely papillose, 33–35 µm diameter.

HABITAT: An extremely hydrophilous species, submerged in pools or forming carpets in very wet parts of open, oligotrophic to weakly mesotrophic mires. Its range overlaps with those of *S. cuspidatum* and *S. jensenii,* and it

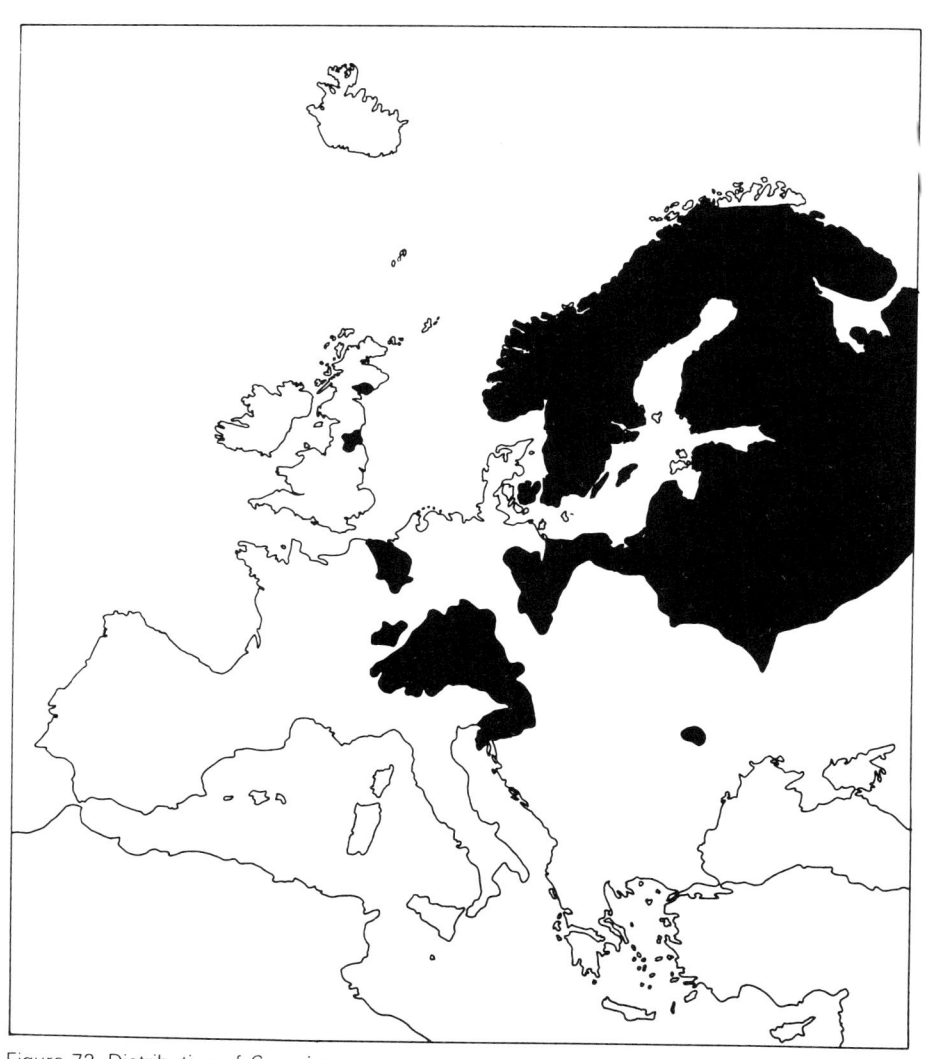

Figure 72. Distribution of *S. majus*

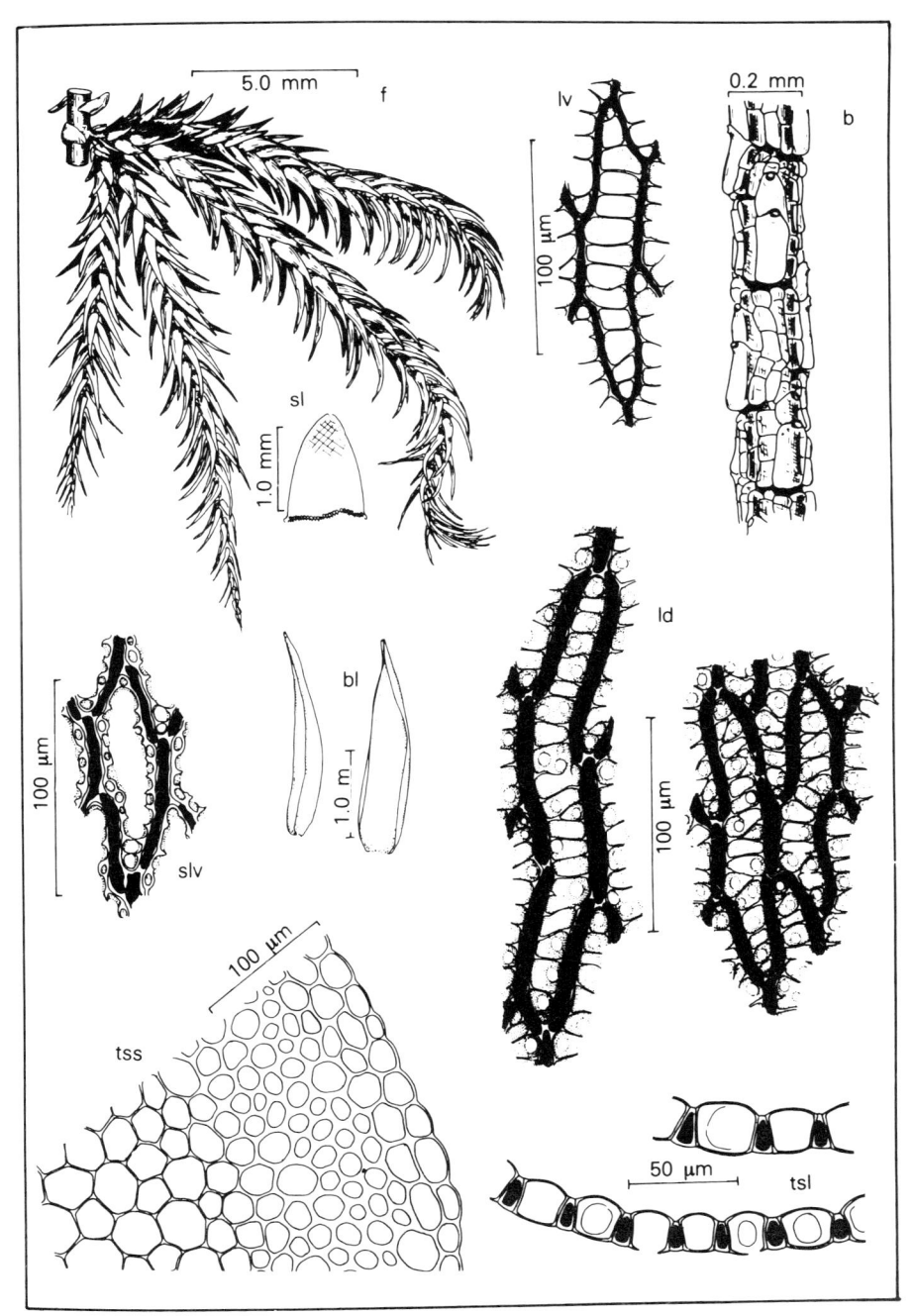

Figure 73. Sphagnum majus

may be found in association with either of these species, or forming monospecific carpets beneath scattered sedges, *Eriophorum angustifolium* or *Scheuchzeria palustris*. It may be found in ombrotrophic bogs, and in the north it may replace *S. cuspidatum* even in distinctly oligotrophic pools, or in those with a limited amount of mineral water influence.

DISTRIBUTION: Circumpolar in the boreal and sub-arctic zones of Europe, northern Asia and eastern North America, but rare in western North America. The European distribution is somewhat continental, though it is more oceanic in North America: the species is common through most of Fennoscandia but decreases in frequency towards the south and west. It extends to the Vosges and the Alps, but is rare above the sub-alpine zone. It is found in 2 localities in north-east Britain.

With its rather elongated branch leaves and lax habit, *S. majus* bears a superficial resemblance to *S. cuspidatum*, a resemblance enhanced further in weaker or shade forms. However, the two species are not closely related, *S. majus* belonging to the species group that contains *S. jensenii*, *S. annulatum* and *S. balticum*. *S. cuspidatum* has narrower, more widely truncate branch leaves and lacks the abaxial pores of *S. majus*. Lax forms of *S. jensenii* can be identified by the smaller pores which usually extend to the adaxial as well as the abaxial leaf surface, and by the adaxially enclosed photosynthetic cells.

35. *SPHAGNUM PULCHRUM*

Sphagnum pulchrum (Braithw.) Warnst. (*Bot. Zbl.*, **82,** 42. 1900)
S. intermedium var. *pulchrum* Braithw. (*Sphag. Eur. N. Am.*, 81. 1880)

PLANTS: Medium-sized to rather robust, often growing in wide carpets or smaller mats; yellow or yellow-brown to chestnut. (Of similar appearance to *S. lindbergii,* but with a different range, or to a less slender, more shaggy *S. recurvum.*) **Fascicles:** Rather close set, rarely distant; of 4 slightly to (usually) rather strongly dimorphic branches; spreading branches 2, relatively *short* (10.0–13.0 mm, rarely over 15.0 mm long) *and thick*, shortly tapered; pendent branches 2, weaker and often shorter, mostly less than 12.0 mm long. **Stem:** 0.5–0.9 mm diameter; cortex distinct, 2–3 layers of enlarged cells with slightly thickened walls; at least the outer layer of cortex with *some brown pigment;* internal cylinder well developed, *brown to yellow-brown,* rarely pale, green. **Branch anatomy:** Retort cells distinct, in linear pairs, the lower with relatively strong protuberant apertures; internal cylinder yellowish to brown. **Stem leaves:** Hanging or spreading; small, 0.8–1.0 mm long and about the same width; more or less equilateral ovate-triangular; apices rounded-obtuse, but mostly appearing mucronate due to inrolling of leaf margins; border strong, expanded below into large, almost confluent, patches of prosenchymatous tissue. Hyaline cells in upper half of leaf, or less, *weakly fibrillose on the abaxial surface,* occasionally without fibrils; adaxial surfaces partially to completely resorbed. **Branch leaves:** Erect-spreading, *uniformly and conspicuously 5-ranked,* densely arranged and branches appearing prismatic; 1.4–1.8 × 0.6–0.9 mm; broadly lanceolate, *widest at, or just below, mid-leaf,* often shallowly inrolled above. Pendent branch leaves small (in dimorphic forms), 0.9–1.1 × 0.5 mm, ovate, concave. **Hyaline cells:** In upper mid-leaf 100–150 × 13.0–18.0(–20.0) μm abaxially, up to 25.0 μm adaxially. Abaxial surface sometimes without pores, usually with resorption gaps in the apical angles and *often 1(–2) pores* (6.0–9.0 μm diameter) *in the upper lateral angles;* lateral angles with or without pseudopores. Adaxial surface with medium-sized (ca 8.0 μm) unringed pores, mainly adjacent to cell angles. Hyaline cells of pendent branch leaves proportionately shorter, but otherwise similar to those of spreading branches. **Leaf TS:** Hyaline cells plane on abaxial surface, shallowly convex on adaxial. Photosynthetic cells *equilateral, or shortly isosceles, triangular,* with wide abaxial exposure but deeply enclosed on adaxial surface (the fused walls of the hyaline cells on either side more or less equalling in length the height of the photosynthetic cell). **Fertile plants:** Dioecious. Antheridial branches rather inconspicuous, especially in brown plants; antheridial bracts orange-brown. Inner perichaetial bracts large, 4.5 mm long, ovate; apices broadly rounded-obtuse, often minutely retuse;

Figure 74. Distribution of *S. pulchrum*

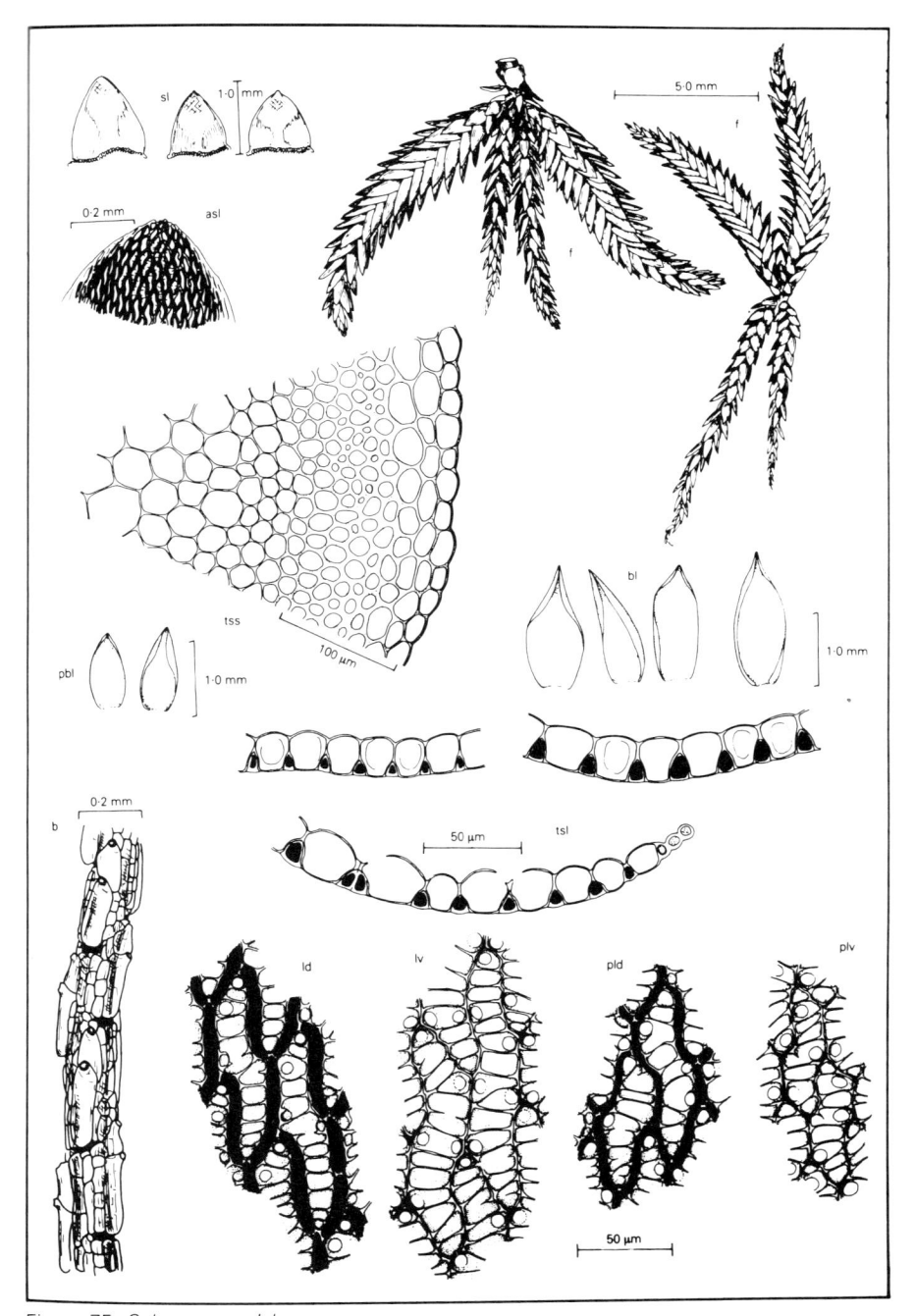

Figure 75. *Sphagnum pulchrum*

upper hyaline cells intact, lacking fibrils, or a few cells with irregular fibrils; hyaline cells differentiated to below mid-bract; basal tissue more or less uniformly prosenchymatous. Capsules very rare (a single capsule seen on British material); spores yellow-brown, slightly papillose, ca 28.0 μm diameter.

HABITAT: As loose mats or dense carpets in wet oligotrophic mires. Occasionally, it may be submerged in shallow pools. In north-western Britain and Ireland, it may form extensive carpets on ombrotrophic peatlands, but further east it is increasingly confined to oligotrophic sites with a throughflow of water (eg valley mires of Dorset in southern England) or the margins of soakways or pools within the mire expanse. Although forming pure mats or carpets in many locations, it may also be mixed with other *Sphagnum* species (eg *S. papillosum, S. magellanicum, S. tenellum*) and leafy hepatics, as well as associated with scattered individuals of such vascular plants as *Rhynchospora alba, Drosera rotundifolia, D. intermedia, Andromeda polifolia, Erica tetralix* and *Narthecium ossifragum*.

DISTRIBUTION: A lowland, sub-oceanic species of north-west Europe, eastern Asia (Japan and Kamchatka) and eastern North America. It has a scattered distribution through the north-western parts of Europe, extending as far east as south-central Finland and the Baltic coast of Russia, and south as far as the Black Forest, although it is also recorded from the Ukraine. In Britain, it has a disjunct distribution in western areas from Dorset, through west Wales and north-west England, to Galloway, Argyll and Skye. It also occurs in parts of western Ireland.

Sphagnum recurvum is the most likely species to be confused with *S. pulchrum*. In the field, the shorter, relatively fatter branches, tapering fairly sharply both towards the insertion and the distal end, together with the wider set of the leaves on the branch and the dark stem, should distinguish *S. pulchrum*. The branch leaf outline is also different, being widest near the middle of the leaf in *S. pulchrum* and widest a little above the insertion to about a third of the leaf length in *S. recurvum*, whilst under the microscope the more shortly triangular and more deeply immersed photosynthetic cells characterize *S. pulchrum*. The dark stem of *S. pulchrum* separates it from forms of *S. jensenii* with distinctly 5-ranked branch leaves, and the triangular stem leaf outline contrasts with the more or less rectangular stem leaf with a tattered apex found in *S. lindbergii*, but, in any case, field confusion should be no real problem as the geographical ranges of these 2 species are very different from that of the oceanic, western to somewhat south-western *S. pulchrum*.

36. SPHAGNUM LINDBERGII

Sphagnum lindbergii Schimp. ex Lindb. (*Öfvers. K. VetenskAkad. Förh.,* **14,** 126. 1857)

PLANTS: Robust and usually tall (somewhat resembling *S. pulchrum,* but with longer branches); brown, often growing in extensive light to dark brown mats. **Fascicles:** Of 4(–5) more or less dimorphic branches; spreading branches 2, (10.0–)15.0–25.0 mm long, prismatic, strong; pendent branches 2(–3), weaker and usually slightly shorter than spreading. **Stem:** Rigid; 0.6–0.8 mm diameter; *dark brown to almost black,* except in etiolated plants; *cortex well developed and distinct,* of 3–4 layers of *inflated thin-walled cells;* internal cylinder dark brown, thick, of strongly thickened cells. **Branch anatomy:** Retort cells variable in length; in linear pairs or threes or, often, *in clusters 2–3 cells wide and 2–3 cells* long; internal cylinder brown. **Stem leaves:** Hanging and appressed to stems; large, 1.3–1.6 × 0.9–1.5 mm, rectangular to broadly spatulate; *apices wide, truncate, conspicuously fimbriate;* border inconspicuous above, merging below mid-leaf into large areas of prosenchymatous tissue. Hyaline cells much enlarged in upper mid-leaf, without fibrils, rarely with a few fibrils below; broadly rhomboid near apex; both surfaces resorbed in upper part of leaf. **Branch leaves:** Densely imbricated in 5 ranks; 1.6–2.0(–3.5) × 0.7 mm; broadly lanceolate, in submerged plants sometimes lanceolate to linear-lanceolate; with a minute spur at either side of insertion (formed by projection of the lowest hyaline cells); border 3–5 cells wide. Pendent branch leaves ovate to ovate-lanceolate, the lower about 1.0 mm long. **Hyaline cells:** In upper mid-leaf 80–150 μm long × 10.0–15.0 μm on abaxial surface, 18.0–25.0 μm adaxially; in lower lateral areas of leaf, up to 200 μm long or more. Abaxial surface mostly without pores, except for a small resorption gap in the apical angle, usually with a few pseudopores in one or more lateral angles; cells in lower lateral parts of leaf usually with 1–4 circular pores adjacent to cell angles. Adaxial surface with 5–8 rather small, 3.0–5.0 μm diameter, unringed pores mainly near the cell angles. Hyaline cells of pendent branch leaves similar to those of the spreading branches. **Leaf TS:** Hyaline cells plane on the abaxial surface, shallowly convex on the adaxial. Photosynthetic cells ovate-triangular to somewhat ovate-trapezoid, with oval lumina and strongly thickened abaxial walls; moderately widely exposed on the abaxial surface, enclosed to varying degrees on the adaxial. **Fertile plants:** Dioecious or autoecious. Antheridial branches similar to sterile branches. Perichaetial bracts large; lower bracts ovate, entire, the middle and upper ones fimbriate like the stem leaves. Inner bracts with 3 types of tissue: near base, lax; in mid-bract, thick-walled, prosenchymatous; near apex, differentiated, with enlarged, non-fibrillose

Figure 76. Distribution of *S. lindbergii*

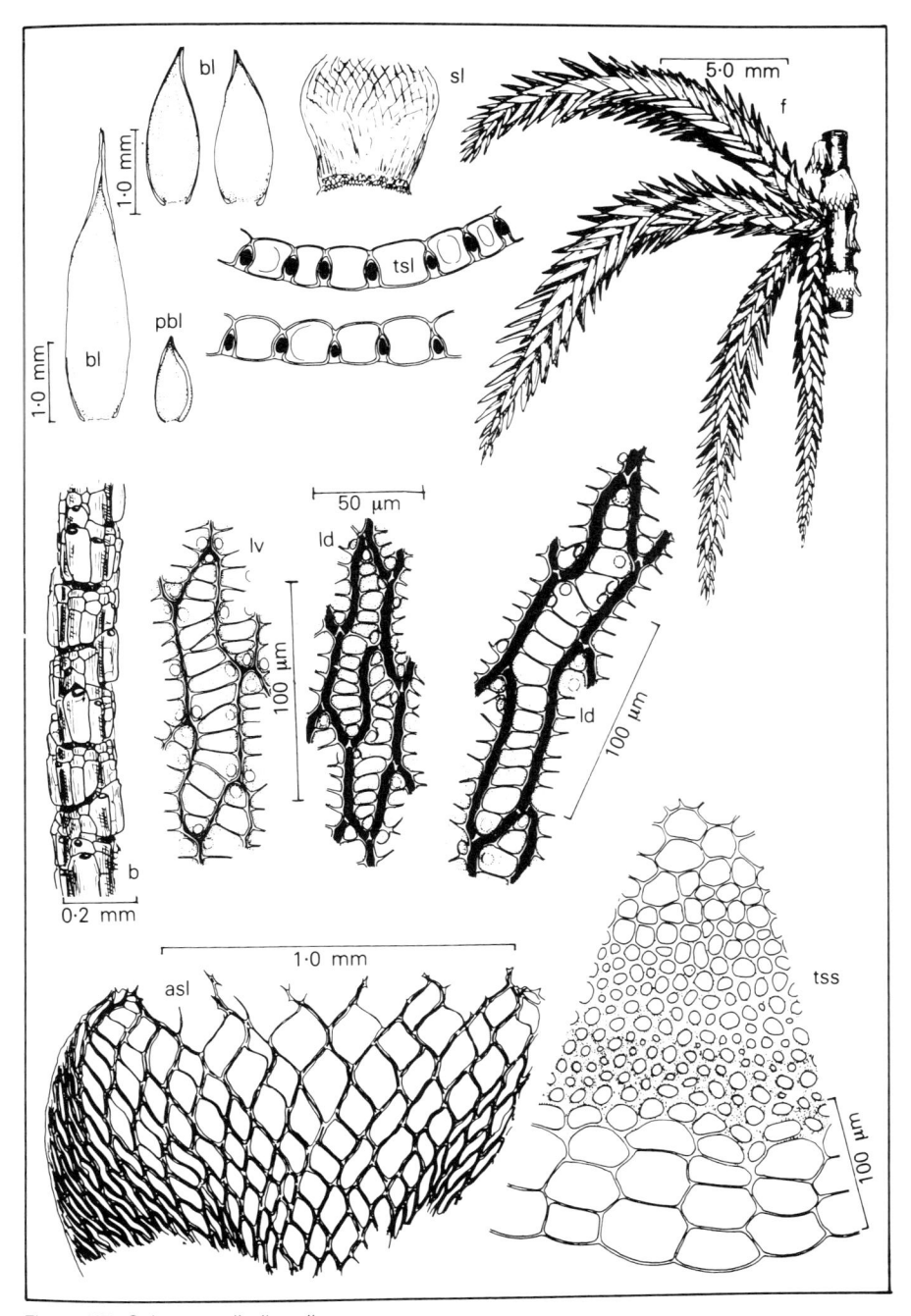

Figure 77. *Sphagnum lindbergii*

hyaline cells. Capsules rather rare; spores pale, coarsely papillose, ca 32 μm diameter.

HABITAT: A hydrophilous species forming brown carpets in oligotrophic mires, beside lakes and streams and in montane flushes. In the north, these carpets may become extensive and form pure stands or include other species of *Sphagnum*, eg *S. jensenii* or *S. riparium*. Like a number of other species, *S. lindbergii* appears to be restricted to more oligotrophic conditions near the southern limit of its range whereas in the north it is found in rather more mesotrophic locations. In Britain, it is restricted to montane flushes.

DISTRIBUTION: Circumpolar in the arctic, sub-arctic and boreal zones. Common in the arctic of northern Scandinavia and Russia, but becoming increasingly rare, and progressively confined to montane areas, further south. In Britain, it is rare and confined to flushes above about 600 m in the Scottish highlands.

This handsome species bears only a superficial resemblance in the field to other brown members of the section Cuspidata and, with its dark stems and strongly fimbriate stem leaves, should pose no problems in identification. *S. lindbergii* is only closely related to the rare, and possibly non-European, *S. lenense,* but it is a much larger plant with more widely fimbriate stem leaves than its rarer relative.

The stem leaves of *S. lindbergii* show an interesting parallel with those of *S. fimbriatum* and may lead to speculation about convergent adaptation, but these unrelated species differ in so many other respects that confusion between them is highly unlikely.

37. *SPHAGNUM LENENSE*

Sphagnum lenense H. Lindb. in Pohl. (*Acta Horti. Petropolit.*, **33**(1), 14. 1915)

PLANTS: Lax or compact; brown (resembling *S. fuscum*, but with 5-ranked branch leaves). **Fascicles:** Well spaced or closely set: of 4–5, often strongly dimorphic, branches; spreading branches 2, *rarely exceeding 10.0 mm;* pendent branches thinner and sometimes considerably longer than spreading, but rarely exceeding 14.0 mm, attenuated. **Stem:** Thin, 0.3–0.6 mm diameter; dark brown to almost black; cortex well developed and distinct, of 3–4 layers of inflated, thin-walled cells; internal cylinder thick, dark brown, of strongly thickened cells. **Branch anatomy:** Retort cells distinct, sometimes solitary, commonly in linear pairs, occasionally with an additional perforate cell on one or both sides; internal cylinder brown. **Stem leaves:** Hanging; rectangular, not much wider above than at insertion; about 0.8 × 0.7 mm; apex strongly resorbed-fimbriate (as in *S. lindbergii*, but more narrowly so and sometimes more reminiscent of *S. riparium*); border intact, except at apex, expanded below into conspicuous patches of prosenchymatous tissue. **Branch leaves:** 5-ranked; *small, rarely exceeding 1.5 mm in length;* lanceolate. Pendent branch leaves ovate to ovate-lanceolate, less than 1.0 mm long. **Hyaline cells:** 90–150 × 12.0–20.0 μm in upper mid-leaf. Abaxial surface without pores, or with 1(–2) faint, small pores, mainly in the apical angles. Adaxial surface with medium-sized, 6.0–10.0 μm, unringed circular pores, mainly near (but not confined to) the cell angles. **Leaf TS:** Hyaline cells plane on the abaxial surface, shallowly convex on the adaxial. Photosynthetic cells ovate-triangular to ovate-trapezoid, with oval lumina and strongly thickened abaxial walls; moderately widely exposed on the abaxial surface, enclosed to varying degrees on the adaxial. **Fertile plants:** Probably dioecious, but no fertile material seen.

HABITAT: In dense brownish cushions or mats, resembling those of *S. fuscum*, in mesotrophic mires or on moist soil, often under scrub. It favours somewhat drier locations than the closely related *S. lindbergii*.

DISTRIBUTION: Incompletely circumpolar in arctic Asia, Alaska, northern Canada and Greenland. Reported from the Kola peninsula, but the nearest confirmed record is from the Archangel area.

S. lenense is essentially a miniature version of *S. lindbergii*, from which it differs in its much smaller dimensions and more narrowly fimbriate stem leaves. The size difference is clear, and even weak specimens of *S. lindbergii* should not, even on size measurements alone, be mistaken for *S. lenense*. Compact forms of *S. balticum* differ in their concave, fibrillose stem leaf

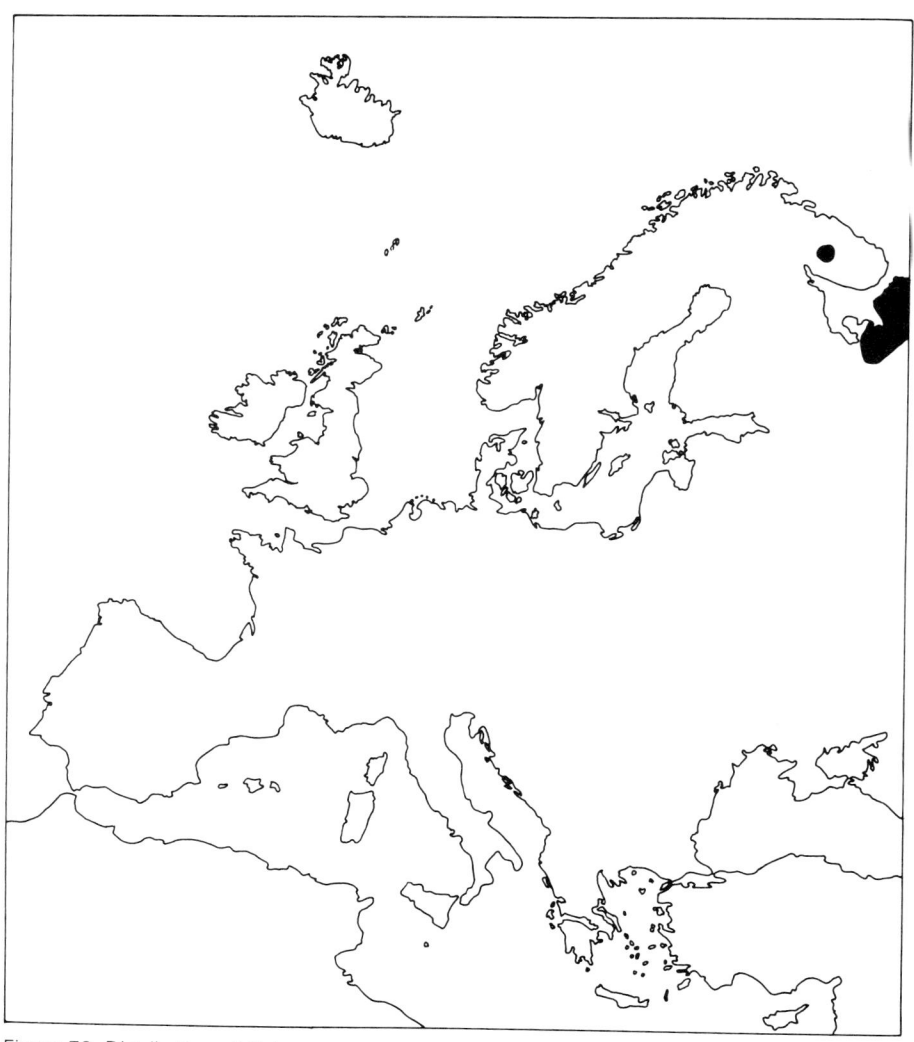

Figure 78. Distribution of *S. lenense*

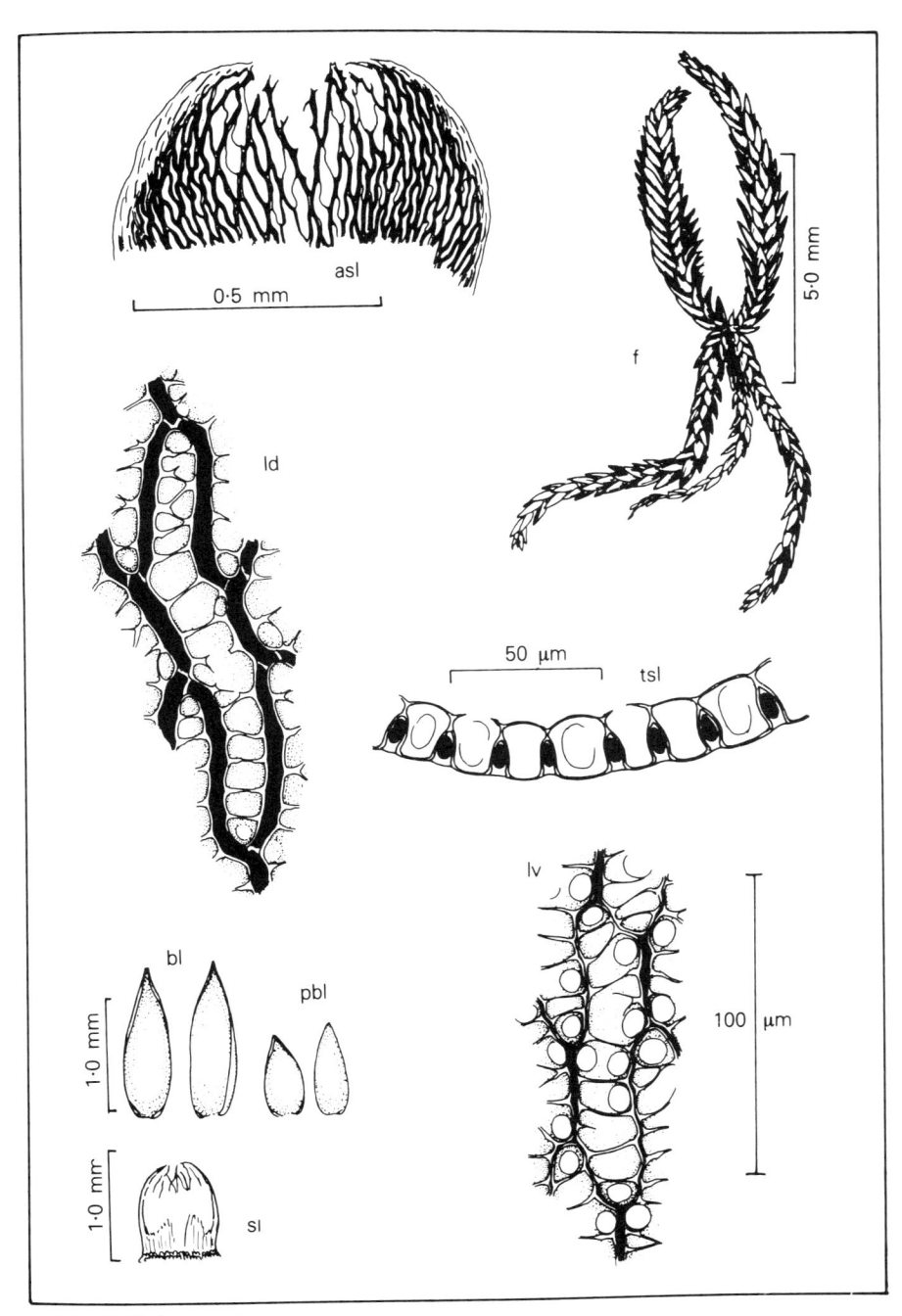

Figure 79. *Sphagnum lenense*

apices. In the field, *S. fuscum* is distinguishable by its non-ranked branch leaves and larger, lingulate stem leaves.

SECTION MOLLUSCA

Sphagnum sect. *Mollusca* Schlieph. ex Casares-Gil (*Mems R. Soc. esp. Hist. nat.,* **13,** 51. 1925)

This section, formerly included in section Cuspidata, is represented by a single species, *S. tenellum*. The combination of italicized characters given in the species description may be taken as diagnostic of the section. Most recent authors have been content to include *S. tenellum* in the section Cuspidata, but most have accepted that it is the most divergent species of that section, and Isoviita (1966), for example, recognized that it deserved, at least, the rank of subsection. The view adopted here is that, phylogenetically, *S. tenellum* is at least as divergent from section Cuspidata as *S. aongstroemii*, for example, is from section Squarrosa, and the resurrection of 'Mollusca' as a separate section to contain it seems appropriate.

38. *SPHAGNUM TENELLUM*

Sphagnum tenellum (Brid.) Brid. (*Musc. recent. suppl. IV,* 1, 1819)
S. molluscum Bruch (*Flora, Jena,* **8,** 633–635. 1825)

PLANTS: Small, delicate; green or yellow, sometimes tinged brown or orange; tufted or low growing with small capitula. **Fascicles:** Rather distant, occasionally close-set; of 2–3(–4) branches; spreading branches, usually 2, short or rather elongated, (4–)6–10(–12) mm long, not tapering; pendent branches, normally one, *not different from,* or rarely barely different from, spreading branches. **Stem:** Pale, thin, rarely exceeding 0.5 mm diameter; *cortex very well developed,* of (1–)2–3(–4) layers of *highly inflated* hyaline cells; outer surfaces of cells without pores; internal cylinder well defined, yellow. **Branch anatomy:** Retort cells *very well developed,* in linear pairs; several times larger than the small imperforate cells, pores set on long protuberances; internal cylinder yellow-green. **Stem leaves:** Spreading; large, almost as large as branch leaves, 1.0–1.4 × 0.5–0.7 mm; ovate-lingulate, concave above; tapered abruptly to eroded or obtuse and dentate apices, often somewhat hooded; border fairly strong above, distinctly expanded below into pseudoprosenchymatous patches in which the hyaline cells and photosynthetic cells remain distinct (when stained); strongly fibrillose in upper half and there identical to branch leaves. **Branch leaves:** 5-ranked, but very obscurely so in mature branches; laxly imbricated; *broadly ovate, concave* (giving branches a beaded appearance); 1.0–1.5 × 0.7–0.9 mm; border narrow, usually 2 cells wide; apices truncate-dentate. **Hyaline cells:** *Short and wide,* 50–110 × 20–35 μm; abaxial surface without

Figure 80. Distribution of *S. tenellum*

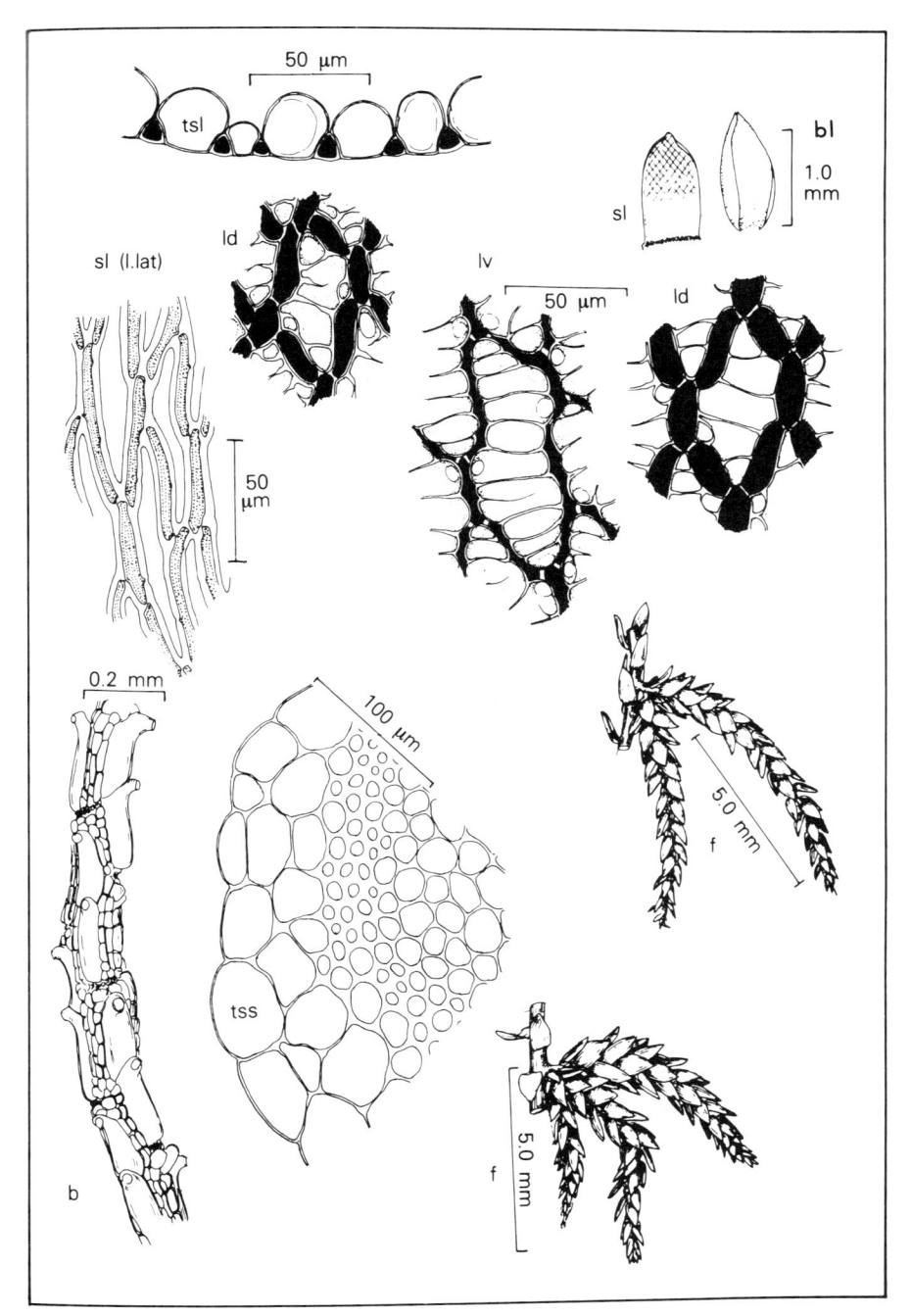

Figure 81. *Sphagnum tenellum*

pores, or with a single obscure or clearly defined pore in the apical angle; usually with one or more pseudopores in some of the lateral angles; adaxial surface variable, usually with 2–5 small to medium-sized, faint or well-defined pores, more or less confined to the cell angles. **Leaf TS:** Hyaline cells plane on abaxial face, *strongly inflated on the adaxial;* internal commissural walls smooth. Photosynthetic cells more or less equilateral triangular; widely exposed on abaxial surface and just reaching the adaxial (sometimes apparently immersed due to close proximity of walls of adjacent hyaline cells). **Fertile plants:** Monoecious, but individual plants often apparently unisexual. Antheridial bracts more or less identical to branch leaves, or orange to brown. Inner perichaetial bracts large, ovate, *tapered above to more or less acute or narrowly obtuse apices;* hyaline cells differentiated almost, or quite, to insertion; in the upper part, *mostly extensively, but often erratically, fibrillose.* Capsules very common (maturing earlier in the year than most other species); spores almost smooth to lightly papillose, 33–42 µm diameter.

HABITAT: A common species of open, damp, oligotrophic habitats where it frequently grows in mixtures with other 'unaggressive' species of *Sphagnum*. It is found as single shoots or small, loose cushions on the central parts of active raised bogs, where it is associated with *S. subnitens, S. papillosum* or *S. capillifolium*. A particularly characteristic associate is *S. compactum,* and both species occur together on peat bared by burning, along wet heath-bog margins, in upland oligotrophic flushes or on bare, peaty soils. When found by itself in, for example, damp hollows in compacted peat, it forms small, neat cushions or mats.

DISTRIBUTION: A sub-oceanic species of Europe, eastern Asia, and both the Atlantic and Pacific coastal areas of North America. It is also present in Ecuador. In Europe, it is widespread, but scattered in the lowlands and mountains from Scandinavia to north-west Spain, but decreases in abundance towards the south-east. Present throughout most of Britain, except for parts of central England, but more widespread in the north and west.

Sphagnum tenellum is a highly distinctive species, usually recognized readily by its small size, delicacy and pale colour. It often appears almost translucent and 'sparkling', especially when squeezed. The ovate, concave branch leaves and steam leaves, which are similar in shape, and the lack of distinction between spreading and pendent branches also help to characterize this species. No other European species has the combination of short, broad branch leaves with large pores and markedly abaxially displaced photosynthetic cells. Although some members of the Subsecunda have a similar leaf shape, they are normally more opaque due to the presence of narrow hyaline cells, and usually have dark stems. In the Arctic, attenuated forms of *S. aongstroemii* are not unlike *S. tenellum,* but they differ in their widely truncated branch leaves (at least in the lower half of the branch).

SECTION RIGIDA

Sphagnum sect. 4. *Rigida* (Lindb.) Schlieph. (*Verh. zool.-bot. Ges. Wien*, **15,** 413. 1865)
Sphagnum sect. *Malacosphagnum* C. Müll. Hal. (*Flora, Jena,* **70,** 404. 1887)

PLANTS: Usually moderately robust but low-growing in Europe, although often very robust in the tropics and southern hemisphere; generally forming compact, low cushions varying in colour from pale straw to deep orange-brown, very occasionally purplish; capitulum often concealed by upwardly directed branches so that the plant may resemble a *Leucobryum* rather than a *Sphagnum*. **Fascicles:** Usually closely set; of 5 branches, 2 spreading and 3 pendent, but frequently one or 2 branches suppressed or, rarely, a fourth, much reduced pendent branch present (occasionally there also appears to be a diminutive extra erect 'branch'); branches strongly dimorphic, spreading short, rarely tapering distally; pendent thin, colourless, finely tapering, varying in length but usually not, or only slightly, longer than the spreading branches. **Stem:** Up to 0.9 mm diameter (up to 1.2 mm in extra-European material); in European species, usually more or less hidden by the branches; cortex well developed, but of uneven thickness (1–3, occasionally 4, cell layers) but never as wide and conspicuous as in section Sphagnum; outer surface of external cortical cells without fibrils, but usually with a single large pore; internal cylinder thick, pale yellowish to dark brown. **Branch anatomy:** Spreading branches commonly less than 10 mm long; pendent branches 2.0–15 mm or more; cortex a single layer of more or less *uniform, large, hyaline cells, most or all of which have a large pore at the distal end* (ie lacking distinct retort cells); internal cylinder pale, or dark brown. **Stem leaves:** Hanging; *very small, rarely exceeding 0.8 mm* in the European species; roughly triangular with more or less eroded and, usually, slightly hooded apices; fibrils normally absent; septa few or none; border more or less distinct and usually slightly widened above the insertion (to 4–8 cells), vanishing at or near the apex; abaxial face of hyaline cells intact, with or without fibrils; adaxial surface completely or partially resorbed; septa few or absent. **Branch leaves:** Large, often exceeding 3.0 mm, widely ovate to lanceolate, frequently squarrose and usually narrowed abruptly at the point of flexion; apex usually broadly truncate with conspicuous teeth, occasionally eroded and concave (then resembling members of the section Sphagnum, but not roughened on the abaxial surface), usually with a subapical hump which does not obscure the apical teeth; border very narrow, composed of a single cell series in which the outer lateral wall is resorbed (ie with a resorption furrow as in section Sphagnum). Pendent branch leaves, except for a few, short, basal ones, elongate-lanceolate to linear, narrow and relatively delicate, frequently lacking a border (bounded by fibrillose hyaline cells) and narrowed to a slightly, or not, resorbed apex. **Hyaline cells:** Of spreading branch leaves relatively short and wide; pores variable, but usually more numerous on the abaxial surface and frequently absent from the adaxial, except for *pseudolacunae (triple pores) which are present at each conjunction of basal and 2 lateral angles of the hyaline cells,* and which have triangular or triradiate common openings. **Leaf TS:** Hyaline cells not strongly inflated, usually almost flat on the adaxial surface; triple pores conspicuous. Photosynthetic cells elliptical, enclosed or narrowly emergent on one (abaxial) or both leaf surfaces, lumina oval; internal commissural walls smooth or papillose. **Fertile plants:** Monoecious. Antheridia borne mainly or exclusively on the pendent branches; outermost female bracts small, resembling stem leaves, often without fibrils; innermost bracts large, up to 7.0 mm, or more, resembling branch leaves, fibrillose throughout, or lacking fibrils towards the base; apex only slightly, or not, resorbed and sometimes with a considerable area of the tip composed of prosenchymatous tissue; border broad.

The European species of this section are nearly always short, but in other parts of the world, particularly New Zealand, species such as *S. australe* Mitt. reach imposing dimensions and are among the largest known *Sphagna*. Like section Sphagnum, this section is clearly circumscribed and should offer few difficulties in field identification and none under the microscope. Occasionally, plants from other sections may bear a superficial resemblance, but no European species from other groups has the same combination of anatomical features; section Sphagnum has large stem leaves and abaxially roughened branch leaf apices, together with abaxial, not adaxial, triple pores; large-leaved members of the section Subsecunda have larger, at least partially fibrillose, stem leaves, narrow hyaline cells, and no resorption furrow.

39. SPHAGNUM STRICTUM

Sphagnum strictum Sull. *(Musci alleghan.,* 49. 1845)

PLANTS: Low-growing in rather loose mats (resembling pale, lax forms of *S. compactum*); pale green or yellow-green to straw-coloured, never orange or brown. **Fascicles:** Moderately closely spaced and more or less hiding the stem; upper branches erect; branches dimorphic; spreading branches typically 2, short, not tapering; pendent branches (1–)3 finely tapering, pale, thin and terete. **Stem:** Thin relative to the other dimensions of the plant, 0.5–0.8 mm diameter: cortical cells in 2 or 3 layers, in places reduced to a single layer, especially near the insertion of a stem leaf; exposed face of cortex with a large pore in most cells; internal cylinder green or pale brown, *never dark brown.* **Branch anatomy:** Spreading branches usually less than 10 mm; cortex of *uniform hyaline cells, most, or all, of which have a large pore at the distal end.* **Stem leaves:** Hanging; *minute, rarely exceeding 0.8 mm in mature plants;* triangular with rounded, somewhat cucullate apices which are often eroded; abaxial faces of hyaline cells intact, lacking pores or fibrils; adaxial faces partly or wholly resorbed (when partly resorbed, giving large 'membrane gaps' which may resemble pores); septa few or absent; border well defined to just below apex, 2–4 cells wide except just above insertion where it is expanded to 6–8 cells wide. **Branch leaves:** Large, often exceeding 2.8 mm long, ovate, typically subsquarrose; margins strongly inrolled; apices widely truncate, dentate, not eroded or hooded (but with the 'hump' characteristic of the section); border narrow (a single cell width) with outer lateral wall partly or wholly resorbed to give a resorption furrow. **Hyaline cells:** *Large and relatively short* in relation to width, rather uniform and only elongated near insertion; abaxial surface variable, with none or a few ringed pores and/or pseudopores, and usually with a single large, unringed resorption gap or pore in the upper angle; adaxial surface with few or no pores, except for the pseudolacunae; 1–2 series of lateral hyaline cells porose on both faces; internal commissural walls *minutely papillose,* rarely smooth (best seen in oblique leaf sections). **Leaf TS:** Hyaline cells large but not strongly inflated, the adaxial face slightly more convex than the abaxial. Photosynthetic cells elliptical, normally *narrowly exposed on the abaxial leaf surface* (usually via a thickened cell wall); sometimes also exposed on the adaxial surface. **Fertile plants:** Monoecious. Antheridia borne on pendent branches, fertile branches not differing markedly from sterile ones; inner female bracts very large, often more than 4.0 mm long; apex more or less acute and often consisting of purely prosenchymatous tissue with little or no trace of apical and marginal resorption; tissue otherwise more or less the same as that of the branch leaves, except near insertion, where some cells may lack fibrils. Capsules frequent; spores papillose 28–32 μm diameter.

HABITAT: *S. strictum* grows in similar habitats to, and may be associated with, *S. compactum*, although it appears to be slightly more shade-tolerant and less resistant to drought conditions. It forms loose mats or more compact tussocks on blanket bog or wet moorland, especially amongst *Molinia caerulea* or *Narthecium ossifragum*. It may also be found with *Molinia* and *Myrica gale* on wet, flushed slopes, or with *Molinia*, *Calluna vulgaris* and *Trichophorum cespitosum* in wet heath communities.

DISTRIBUTION: *S. strictum* is really a species complex which has its main area of distribution in the warmer regions of the world. European plants belong to the subspecies *strictum* which is not very tolerant of sub-zero temperatures. This subspecies also occurs in eastern North America, from Newfoundland to Florida and Mexico. It has a distinctly oceanic distribution in Europe and is confined to north-western coastal areas, although a few isolated records from further east do exist. In the British Isles, it is virtually restricted to the north-west coastal areas of Scotland and western Ireland, although it has been recorded from Galloway and north Wales.

With a little experience, *S. strictum* can be identified with some confidence in the field, confusion only being likely with forms of the related *S. compactum*. Typically, however, *S. compactum* has a dark stem. Atypical plants may have rather paler stems and widely spreading branch leaves (particularly when growing in shade). In a number of unrelated taxa, spreading leaves are found as a response to poor illumination. Final determination depends upon branch leaf anatomy, as seen in TS. The papillae on the internal commissural walls are very small in European plants and are seldom, if ever, as conspicuous as those in *S. papillosum*.

Care should be taken to use only fully developed plants for examination of stem leaves (ie stems which have at least 4 branches per fascicle), as the stem leaves of juvenile shoots may be large and resemble the branch leaves to varying degrees. Such juvenile shoots may often be present in gatherings taken from sites where there has been a check to normal growth, as often happens with this species. The European range may be marginal for the species as a whole, and both growth and reproductive performance may suffer in colder winters.

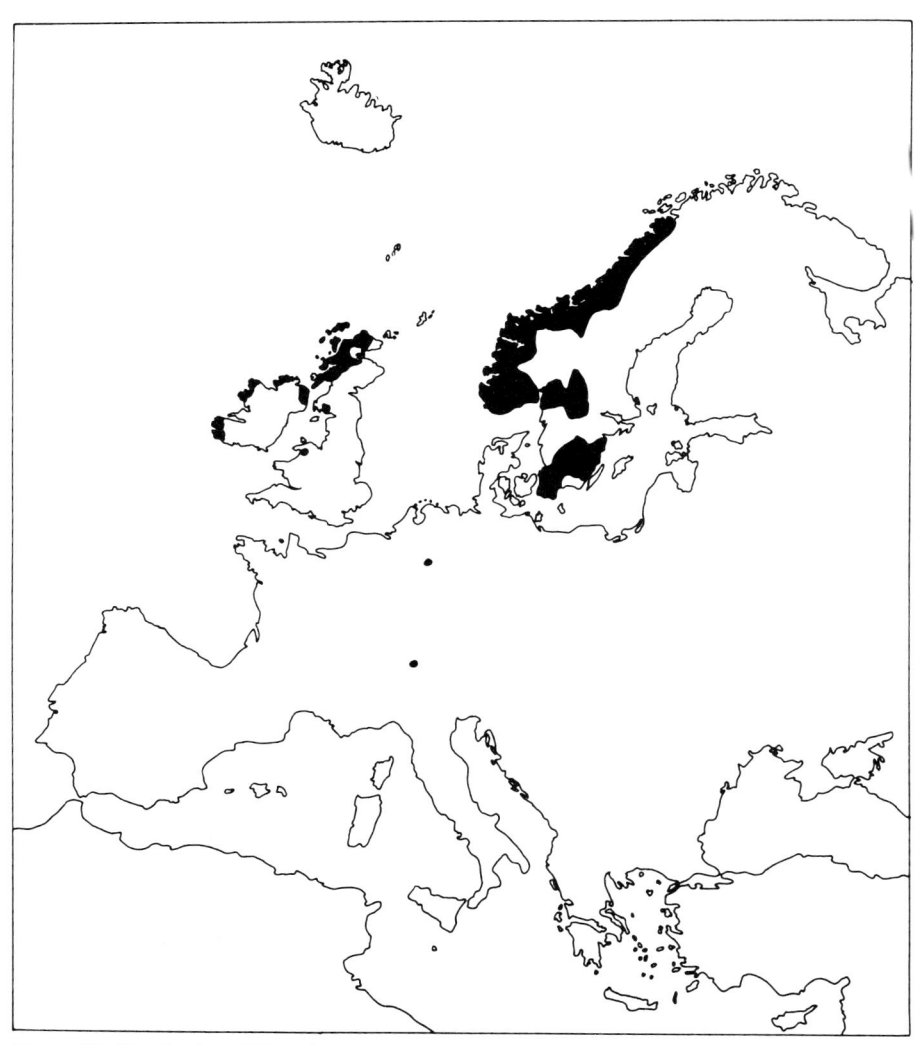

Figure 82. Distribution of *S. strictum*

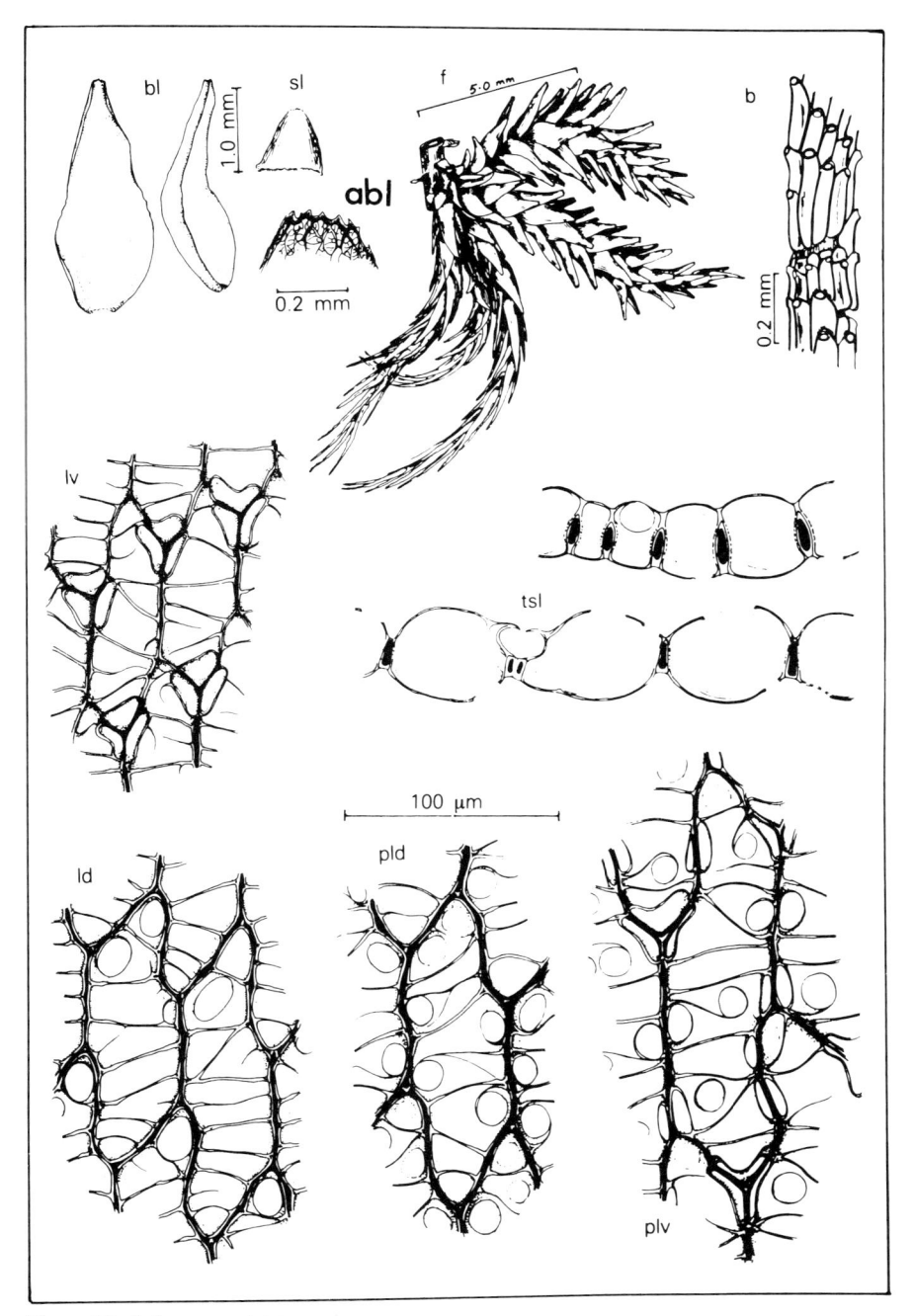

Figure 83. *Sphagnum strictum*

40. SPHAGNUM COMPACTUM

Sphagnum compactum DC in Lamarck & De Candolle (*Flore franc.*, 443. 1805)

PLANTS: Low-growing, forming dense cushions or compact mats; whitish green, ochre or orange-brown, occasionally with some purple coloration, but never wine-red; in shade, sometimes dull straw or yellow-green, rarely in dense shade deeper green; *capitula small and usually concealed by the upper branches* (so that the plant resembles *Leucobryum*). **Fascicles:** Closely spaced and concealing the stems, forcing the upper branches to stand erect; branches strongly dimorphic; (1–)2 short spreading branches not tapering distally; (2–)3 pendent branches varying in length, not much longer than the spreading branches (often very short), pale, thin and terete, finely tapering. **Stem:** Thin relative to the other dimensions of the plant, 0.5–0.8 mm diameter; cortical cells in 2 or 3 layers (variable even on a single stem), in places reduced to a single layer, especially near the insertion of a stem leaf; exposed face of cortex with a large pore in most cells; internal cylinder dark brown to almost black in surface view. **Branch anatomy:** Spreading branches usually less than 10.0 mm; cortex of *uniform hyaline cells most, or all, of which have a large pore at the distal end.* **Stem leaves:** Hanging; *minute, rarely exceeding 0.8 mm in length in mature plants;* more or less triangular with rounded, often eroded, apices; abaxial faces of hyaline cells intact, lacking pores or fibrils; adaxial faces partly or completely resorbed (when partly resorbed, giving large 'membrane gaps' which sometimes resemble pores); septa few or absent; border well defined to just below apex, 2–4 cells wide, wider above insertion, 6–8 cells wide. **Branch leaves:** Slightly spreading or imbricate, only in dense shade becoming distinctly subsquarrose; large, 1.8–3.0 mm long, widely ovate or ovate-rectangular, rarely ovate-lanceolate, concave, occasionally slightly narrowed about one third from insertion; margins inrolled; apex occasionally hooded (but never abaxially roughened) or eroded, but, more typically, truncate with 5–7 conspicuous teeth (such leaves, in profile, have a characteristic hump below the apex); border narrow (a single cell width) with outer lateral wall partly or wholly resorbed to give a resorption furrow. Pendent branch leaves, except for a few short, basal ones, narrow, linear to linear-lanceolate, delicate; border often absent, but, where present, similar to that of spreading branch leaves. **Hyaline cells:** *Large, relatively short in relation to width*, 110–130 × 30–40 μm in upper mid-leaf, rather uniform and only elongated near insertion; abaxial surface variable, usually with rather numerous (4 or more) pores and/or pseudopores in varying proportions; true pores more or less circular, mostly ringed, 9.0–12.0 μm diameter, set 2.0–4.0 μm from the commissures; sometimes with a larger, unringed resorption pore near the upper angle and one or more free pores in the cell mid-line (the latter more numerous in the pendent branch leaves); adaxial surface with few or no pores, except for the pseudolacunae which are a constant and regular feature; 1–2 series of lateral hyaline cells porose on both faces. **Leaf TS:** Hyaline cells large but not strongly inflated, the adaxial face slightly more convex than the abaxial. Photosynthetic cells oval, *thin-walled, completely enclosed by the hyaline cells, though more shallowly so on the abaxial surface of the leaf* (there occasionally 'exposed' via a thickened wall); internal commissural walls smooth. **Fertile plants:** Monoecious. Antheridia borne mostly on pendent branches; inner female bracts very large, often more than 4.0 mm long; apex more or less acute and often consisting entirely of prosenchymatous tissue and with little or no trace of apical and marginal resorption; tissue otherwise more or less the same as that of branch leaves, except near the insertion, where some cells may lack fibrils. Capsules common; spores pale brown, papillose, 32.0–36.0 μm diameter.

HABITAT: Open, oligotrophic areas: it is intolerant of even moderate shade and incapable of competing with more vigorous plants. Typically, it is found on thin, compacted peats of wet heathland (where it is usually associated with other 'unaggressive' species such as *S. tenellum* and *Campylopus* spp.) on low hummocks on wet raised bogs or blanket peat. It also forms extensive lawns in the 'flarks' of aapa mires or may be submerged in shallow pools on western ombrotrophic bogs. At high altitude it may occur on damp, stony ground, often with *S. fuscum,* or may be a component of oligotrophic flush vegetation with *S. tenellum, S. auriculatum* and *S. fuscum.* It can withstand drought much more successfully than most other species of *Sphagnum* and is often one of the first colonizers of peat bared by burning.

DISTRIBUTION: Extensive in the northern temperate zones of Europe, Asia and North America, and extending southwards as far as Madeira, Mexico and Hawaii. Widespread in Europe in both lowland and upland areas (reaching 2500 m in central Europe) but largely confined to sub-alpine zones in the southern part of its range. Frequent to common throughout the British Isles, though absent from parts of central Ireland and most of south-east and midland England. It ascends to moderately high altitudes in western Britain, but is most common below 500 m.

The compact habit, coupled with the large branch leaves and often strong coloration, makes *S. compactum* fairly easy to identify in the field. Some forms may be much paler and have widely spreading branch leaves, and may then be difficult to separate, with certainty, from the related *S. strictum.* The latter species does, however, have pale stems, rather than dark ones: under the

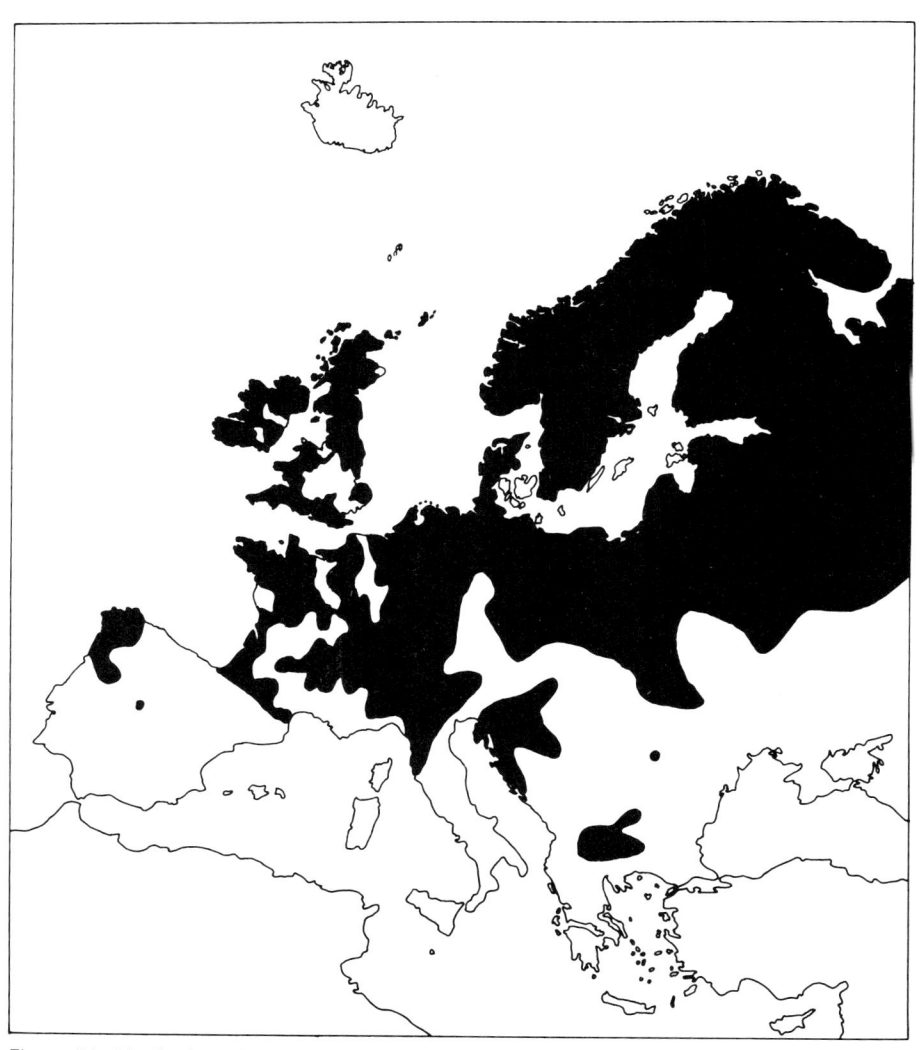

Figure 84. Distribution of S. *compactum*

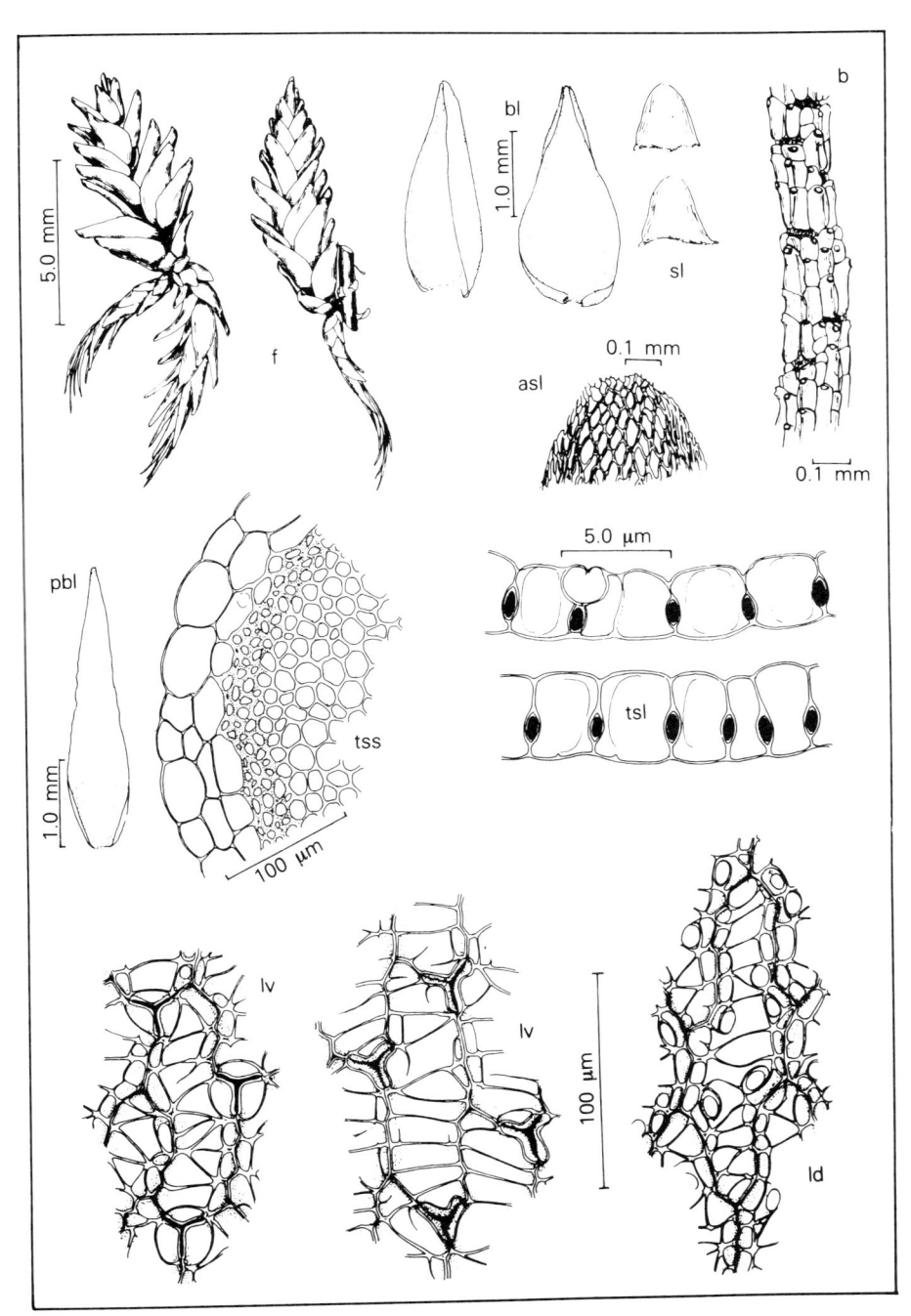

Figure 85. *Sphagnum compactum*

microscope, it may be distinguished by the branch leaf photosynthetic cells which are narrowly exposed on the abaxial leaf surface and the normally minutely papillose commissural walls of *S. strictum*. In any case, *S. strictum* is a much rarer species, confined to the mild areas adjacent to western coasts, so that doubtful specimens from eastern, central or southern Britain or Europe will almost certainly be *S. compactum*. Forms with cucullate branch leaves can be separated from members of the section Sphagnum by their dense habit, small and mostly hidden capitula, and minute stem leaves. Squarrose-leaved forms, which are rare, can be similarly distinguished from *S. squarrosum*. (The minute stem leaves will also separate this species from similar forms of *S. molle*.) The pattern of hyaline and photosynthetic cells in the leaves will distinguish this species from *Leucobryum*.

GLOSSARY

Abaxial	surface of leaf away from the axis = convex surface
Adaxial	surface of leaf nearest axis = concave surface
Anterior	(of leaf) adaxial surface
Antheridium	stalked oval organ producing male gametes
Antheridial branch	branch bearing antheridia
Archegonium	flask-shaped organ producing female gametes and retaining the zygote
Archegonial branch	branch bearing archegonia
Arcuato-decurved	arched backwards
Areolation	pattern of cells as seen in surface view
Autoecious	with antheridial and archegonial branches on separate stems of the same plant
Blanket bog	ombrotrophic and acid peat deposit developed over level or gently undulating mineral ground and covering it almost completely
Bog	ombrotrophic peat deposit, ie one in which the sole source of water supply is from precipitation: bogs are poor in plant nutrients and their waters are distinctly acid in reaction
Border	(of a leaf) margin consisting of a row or rows of prosenchymatous cells
Bract	modified leaf surrounding antheridium or archegonium
Capitulum	head formed of crowded branches around the stem apex
Chlorocyst	photosynthetic cell in leaf
Clavate	club-shaped: narrowed towards the base and broader near the apex
Comb fibril	lamella extending into the lumen of a branch leaf hyaline cell (in Europe only present in *S. imbricatum*)
Commissure	junction of a hyaline cell and a photosynthetic cell
Commissural pores	pores lying along the commissures
Concave surface	adaxial surface (of a leaf)
Concolorous	of the same colour
Convex surface	abaxial surface (of a leaf)
Convolute	sheathing: rolled together
Coronate	having corona: with a cup-shaped ring of leaves
Cortex	outer layer or layers of stem or branch, consisting of comparatively large, thin-walled cells
Crest	= lamella

Cucullate	hooded at the apex by incurving of the margins (of leaves)
Dentate	toothed
Dimorphic	having 2 forms
Dioecious	with antheridia and archegonia on separate plants
Efibrose	lacking fibrils (also efibrillose)
Eporose	lacking pores
Erect	with the distal end pointing towards apex of axis (used of branches or leaves)
Erose	irregularly notched
Eroded	worn away
Eutrophic	with abundant mineral ions and hence of high nutrient status; usually with a pH reaction which is ± neutral or slightly basic
Evanescent	disappearing abruptly or quickly
Exposure	(of photosynthetic leaf cells) reaching the surface of the leaf and not occluded by adjacent hyaline cells
Fascicle	tuft or group of branches originating at the same point on a stem
Fibrils	spiral bands of thickening on the walls of hyaline cells of leaves: also present in the cortical cells of stems and branches in section Sphagnum
Fibrillose	bearing fibrils
Fibrose	= fibrillose
Flaccid	flabby: lying loosely and lacking stiffness
Fruit	spore capsule and stalk
Geotrophic	fed by groundwater
Green cells	photosynthetic cells of leaves, ie those containing chloroplasts
Hanging branches	those branches within a fascicle which hang downwards from the point of insertion and are ± appressed to the stem
Hyaline cells	large empty cells found between the network of photosynthetic cells in leaves
Immersed	(of photosynthetic cells) enclosed by adjacent hyaline cells so that they do not reach the leaf surface
Imperforate	without perforations: lacking pores or resorption gaps
Incumbent	lying upon (of leaves lying closely along the axis and not spreading)
Inflated	(of hyaline cells seen in cross-section) having a somewhat rounded or swollen shape
Internal cylinder	central part of stem or branch consisting of, at least in the outermost layers, smaller, thicker-walled cells than the surrounding cortex

Isophyllous	with stem and branch leaves of similar shape
Lacerate	deeply divided into an irregular number of segments; appearing torn
Lamella	elongated projection from the cell wall projecting into the lumen of a hyaline cell (seen only in *S. imbricatum*, in Europe)
Lamellate	having lamellae
Lanceolate	lance-shaped: tapering to a ± long point
Leucocyst	= hyaline cell
Lingulate	tongue-shaped
Lumen	cell cavity
Lyrate	lyre-shaped: waisted
Membrane gap	hole formed by resorption of part of hyaline cell wall: usually rather irregular in shape but sometimes more regular and resembling a large pore
Mesotrophic	with intermediate concentrations of mineral ions, and hence of moderate nutrient status: usually weakly acid
Minerotrophic	supplied by groundwater draining from mineral soil or rock
Mire	actively growing peatland
Monoecious	with archegonia and antheridia on the same plant, often on the same stem
Oligotrophic	with few mineral ions and so of low nutrient status: usually distinctly acid
Ombrotrophic	with water and mineral ions supplied solely by atmospheric precipitation
Ovate	egg-shaped: widest near the base and narrowed above
Pallid	pale
Papilla (plural papillae)	small elongated projection from the internal surface of the cell wall of a hyaline cell
Papillose	having papillae
Parenchymatous cells	± hexagonal cells with end walls perpendicular to main axis
Pectinate	comb-like
Pendent branches	= hanging branches
Perichaetium	female "inflorescence": structure formed around the archegonium by modified leaves
Perichaetial bracts	modified leaves surrounding the archegonium and forming the perichaetium
Photosynthetic cells	cells containing chloroplasts
Plano-convex	(of hyaline cells seen in cross-section) having one outer wall almost flat and the other inflated or convex

Pore	round or oval opening on the outer wall of a cortical cell or a hyaline leaf cell
Porose	having pores
Posterior	(of a leaf) abaxial surface
Prosenchymatous	elongated tapering cells lying parallel to the axis: not differentiated into hyaline and green cells
Protonema (plural protonemata)	filamentous (sometimes thalloid) structure produced by germinating spore and giving rise to gametophyte
Pseudolacuna	triple pore formed at the junction of 3 hyaline cells (one basal and 2 upper lateral hyaline cell angles)
Pseudopore	round or oval thinning of the cell wall of a hyaline cell: like a pore which retains a thin membrane across the 'hole'
Quinquefarious	in 5 ranks
Raised bog	domed ombrotrophic peat deposit
Recurved	(of leaves) with the upper part curved back away from the axis
Reflexed	(of leaves) bent backwards so that the apex points away from the axis apex
Resorption	erosion or breakdown of parts of cell walls
Resorption furrow	channel along the margin of a leaf, formed by erosion of side walls of the outermost cells
Resorption gap	hole formed by erosion of part of hyaline cell wall (usually less regular in shape than a pore)
Retort cell	large cell of stem or, more usually, branch cortex in which the distal end is perforated by a single pore; it is also often narrowed and turned outwards at the distal end
Ringed pore	pore surrounded by a ring of thickened cell wall
Rostrum	a beak (hence rostrate; with a beak)
Scabrid	roughened: rather like a cheese grater
Septate	having septa
Septum (plural septa)	partition across a hyaline cell: it gives the appearance of 2 hyaline cells without a photosynthetic cell between them
Shadow pore	round or oval thinning of the cell wall of a hyaline cell (only revealed by heavy staining)
Spatulate	paddle-shaped: narrowest at the base and wider above
Spreading branches	those branches of a fascicle which are, at least near their insertion, pointing upwards or horizontal: longer branches appear arched
Squarrose	(of leaves) with the upper part bent back abruptly
Subacute	acute but not sharply so
Suberect	(of branches or leaves) with distal end pointing + towards apex of axis

Subsquarrose	(of leaves) with the upper part bent back abruptly, but not to quite the same extent as in squarrose examples
Terete	rounded in cross-section: without ridges or angles
Tetrahedral	shaped like a 3-sided pyramid
Triradiate	having 3 limbs diverging from a central point
Truncate	cut off abruptly at the distal end
Tumid	swollen
Ventral	(of leaf) concave or adaxial surface
Vitta	a central band of cells in a leaf which are of different shape, size or colour from those on either side.

INDEX OF SECTION, SPECIES AND VARIETY NAMES

Acisphagnum Andr.		= Acutifolia	65
Acisphagnum C. Müll.		= Cuspidata	164
Acutifolia Wils.			65
acutifolium Ehrh. ex Schrad.		= S. capillifolium	92
acutifolium	var. fuscum Schimp.	= S. fuscum	84
	var. quinquefarium Lindb. ex Braithw.	= S. quinquefarium	88
	var. robustum Russ.	= S. russowii	102
	var. subtile Russ.	= S. capillifolium	92
	var. tenellum Schimp.	= S. capillifolium var. rubellum	96
affine Ren. et Card.		= S. imbricatum	56
amblyphyllum (Russow) Zick.		= S. flexuosum	180
americanum Warnst. ex Paul		= S. angermanicum	76
angermanicum Melin			76
angustifolium (C. Jens. ex Russow) C. Jens.			190
annulatum H. Lindb. ex Warnst.			199
annulatum	var. porosum (Schlieph. & Warnst.) Maas & Isov.	= S. jensenii	203
antarcticum Mitt.		= S. australe	229
aongstroemii C. Hartm.			126
apiculatum H. Lindb.		= S. recurvum	185
arcticum Flatb. & Frisvoll		= S. girgensohnii	110
auriculatum Schimp.			150
auriculatum	var. auriculatum	= S. auriculatum	150
	var. inundatum (Russ.) M. O. Hill	= S. subsecundum subsp. inundatum	143
austinii Sull. ex Aust.		= S. imbricatum	56
balticum (Russow) Russow ex C. Jens.			195
bavaricum Warnst.		= S. auriculatum	150
camusii (Card.) Warnst.		= S. auriculatum	150
capillaceum (Weiss) Schrank		= S. capillifolium	92
capillifolium (Ehrh.) Hedw.			92
capillifolium	var. **capillifolium**		96
	var. **rubellum** (Wils.) A. Eddy		96
	var. tenerum (Sull. & Lesq. ex Sull.) Crum		97
centrale C. Jens.		= S. palustre var. centrale	50
Compacta Lindb.		= Rigida	228
compactum D.C.			234
compactum Brid.		= S. compactum	234
contortum K. F. Schultz			160
contortum	var. platyphyllum (Lindb. ex Braithw.) Åberg.	= S. platyphyllum	156
Cuspidata Lindb.			164
cuspidatum Ehrh. ex Hoffm.			166
Cymbifolia Lindb.		= Sphagnum	44
cymbifolium Hedw.		= S. palustre	46
dusenii Warnst.		= S. majus	207

244

fallax (Klinggr.) Klinggr.		= *S. recurvum* var. *mucronatum*	185
fallax	var. *angustifolium*	= *S. angustifolium*	190
	var. *flexuosum*	= *S. flexuosum*	180
fimbriatum Wils.			111
flavicomans (Card.) Warnst.			81
flexuosum Dozy & Molk.			180
flexuosum	var. *angustifolium*	= *S. angustifolium*	190
	var. *fallax* (Klinggr.) M. O. Hill ex A. J. E. Smith	= *S. recurvum* var. *mucronatum*	185
	var. *flexuosum*	= *S. flexuosum*	180
	var. *tenue* M. O. Hill ex A. J. E. Smith	= *S. angustifolium*	190
fuscum (Schimp.) Klinggr.			84
girgensohnii Russow			107
gravetii Russow		= *S. auriculatum*	150
hakkodense Warnst. & Card. in Card.		= *S. papillosum*	52
Hemitheca Lindb. ex Braithw.			136
imbricatum Hornsch. ex Russow			56
imbricatum	var. **affine** (Ren. & Card.) Warnst.		60
	var. **cristatum**		60
	var. **sublaeve**		60
Inophloea Russow		= Sphagnum	44
Insulosa Isov.			126
insulosum Ångstr. ex Schimp.		= *S. aongstroemii*	126
intermedium Hoffm.		= *S. recurvum*	185
intermedium Lindb.		= *S. recurvum*	185
intermedium (Warnst.) Russow ex Warnst.		= *S. palustre*	40
Inundata Russow		= Subsecunda	140
inundatum Russow		= *S. subsecundum* subsp. *inundatum*	143
isophyllum Russow		= *S. platyphyllum*	156
jensenii H. Lindb.			203
jensenii	var. *annulatum* (H. Lindb. ex Warnst.) Warnst.	= *S. annulatum*	199
junghuhnianum Dozy & Molk.			75
laricinum (Wils.) Spruce ex Ångstr.		= *S. contortum*	160
laricinum	var. *platyphyllum* Lindb. ex Braithw.	= *S. platyphyllum*	156
lenense H. Lindb.			219
lindbergii Schimp.			215
lindbergii	var. *microphyllum* Warnst.	= *S. lenense*	219
Litophloea Russow		= Acutifolia	65
macrophyllum Bernh. ex Brid.			16
magellanicum Brid.			61
majus (Russow) C. Jens.			207
Malacosphagnum C. Müll.		= Rigida	228
medium Limpr.		= *S. magellanicum*	61
mendocinum Sull. & Lesq.			202/206
Mollia Schimp.		= Acutifolia	65
molle Sull.			67

Mollusca Schlieph. ex Cas.-Gil.			223
molluscum Bruch		= *S. tenellum*	223
monocladum (Klinggr. ex Warnst.)		= *S. cuspidatum*	170
mucronatum (Russow) Zick.		= *S. recurvum* var. *mucronatum*	185
nemoreum Scop.		= *S. capillifolium*	92
nemoreum var. *tenerum*		= *S. capillifolium*	97
obesum (Wils.) Warnst.		= *S. auriculatum*	150
Obtusifolia Wils.		= Sphagnum	44
obtusifolium Ehrh. ex Hoffm.		= *S. palustre*	50
obtusum Warnst.			176
Palustria Lindb.		= Sphagnum	44
palustre L.			46
palustre var. *centrale* (C. Jens.)			50
Papillosa Russow		= Squarrosa	116
papillosum Lindb.			52
papillosum var. *laeve* Warnst.			53
var. *sublaeve* Limpr.			53
parvifolium (Warnst.) Warnst.		= *S. angustifolium*	190
platyphyllum (Lindb. ex Braithw.) Sull. ex Warnst.			156
plumulosum Röll		= *S. subnitens*	71
Polyclada C. Jens.			131
portoricense			57
pulchrum (Lindb. ex Braithw.) Warnst.			211
Pycnoclada Russow		= Polyclada	131
pycnocladum Ångstr.		= *S. wulfianum*	131
pylaesii Brid.			136
pylaesii var. *sedoides* (Brid.)			137
pylaiei Braithw.		= *S. pylaesii*	136
quinquefarium (Lindb. ex Braithw.) Warnst.			88
recurvum P. Beauv.			185
recurvum var. *amblyphyllum* (Russow) Warnst.		= *S. flexuosum*	180
subsp. *balticum* Russow		= *S. balticum*	195
var. ***mucronatum*** forma *fallax* (Russow) Warnst.			185
var. *parvifolium* Sendtner ex Warnst.		= *S. angustifolium*	190
var. *tenue* Klinggr.		= *S. angustifolium*	190
Rigida Lindb.			228
rigidum (Nees & Hornsch.) Schimp.		= *S. compactum*	234
riparium Ångstr.			171
robustum (Warnst.) Card.		= *S. russowii*	102
rubellum Wils.		= *S. capillifolium* var. *rubellum*	96
russowii Warnst.			102
sericeum C. Müll.			16
serratum Aust.			170
skyense Flatberg			75
Sphagnum L.			44
Squarrosa Russow			116
squarrosum Crome			122
strictum Sull.			230
strictum subsp. ***pappeanum***			21

subbicolor Hampe		= *S. palustre* var. *centrale*	50
subfulvum Sjörs			80
subnitens Russow & Warnst.			71
Subsecunda Lindb.			140
subsecundum Nees in Sturm			142
subsecundum	var. *auriculatum* (Schimp.) Lindb.	= *S. auriculatum*	150
	subsp. *inundatum* (Russ.) C. Jens.		143
	subsp. *subsecundum*		143
subtile (Russow) Warnst.		= *S. capillifolium*	97
tenellum (Brid.) Pers. ex Brid.			223
tenerum Sull. & Lesq. in Sull.		= *S. capillifolium*	92
tenerum Warnst.		= *S. capillifolium*	92
tenue (Nees & Hornsch.) Dozy		= *S. fimbriatum*	111
teres (Schimp.) Ångstr. in C. Hartm.			117
teres	var. *reticulata* C. Jens.		121
	var. *squarrulosum* (Schimp.) Warnst.		121
torreyanum Sull.			170
Truncata Russow		= Rigida	228
Truncata Husn.		= Acutifolia	65
viridum Flatberg		= *S. cuspidatum*	170
warnstorfianum Du Rietz in Sjörs		= *S. warnstorfii*	98
warnstorfii Russow			98
wulfianum Girg.			131

BIBLIOGRAPHY

Åberg, G. 1937. Untersuchungen über die Sphagnum-Arten der Gruppe Subsecunda in Europa. *Ark. Bot. A*, **29**, 1–77.

Allorge, P. 1934. Muscinées du nord et du centre de l'Espagne. *Revue bryol. lichénol.*, **7**, 267.

Allorge, V. 1955. Catalogue preliminaire des muscinées du pays basque français et espagnol. *Revue bryol. lichénol.*, **24**, 96–131, 248–333.

Allorge, V. 1970. Une localité nouvelle pour le *Sphagnum pylaesii* Brid. en Espagne. *Revue bryol. lichénol.*, **37**, 685–688.

Allorge, V. & Casas de Puig, C. 1962. Au sujet des bryophytes récoltés au cours de l'excursion de l'Association Internationale de Phytosociologie dans les Pyrénées franco-espagnoles (22–29 mai, 1960). *Revue bryol. lichénol.*, **31**, 213–238.

Almquist, E. 1962. *Th. O. B. N. Krok och S. Almquist: Svensk Flora för skolor. II Kryptogamer utom ormbunkväxter*. 7th ed. Stockholm.

Amann, J. 1928. *Materiaux pour la Flore Cryptogamique Suisse. Vol. 6, part 2: Bryogeographie de la Suisse.* Zurich: Fretz Frères.

Amann, J., Meylan, C. & Culmann, P. 1912. *Flore des mousses de la Suisse.* Lausanne.

Anderson, L. E. 1963. Modern species concepts: mosses. *Bryologist*, **66**, 107–119.

Andrews, A. L. 1910. Dr. Röll's proposals for the nomenclature of *Sphagnum*. *Bryologist*, **13**, 4–6.

Andrews, A. L. 1911–1961. Notes on North American *Sphagnum*. 1911 I. *Bryologist*, **14**, 72–75; 1912a II. **15**, 1–9; 1912b III. **15**, 63–66, 70–74; 1913a IV. **16**, 20–24; 1913c V. **16**, 59–62, 74–76; 1915 VI. **18**, 1–6; 1917 VII. **20**, 84–89; 1919 VIII. **22**, 45–49; 1921 IX. **24**, 81–86; 1958 X. Review **61**, 269–276; 1959 XI. *Sphagnum subsecundum*, **62**, 87–96; 1960 XII. *Sphagnum cyclophyllum*, **63**, 229–234; 1961 XIII. *Sphagnum pylaesii*, **64**, 208–214.

Andrews, A. L. 1913b. Order Sphagnales. Family 1. Sphagnaceae. *N. Am. Flora*, **15**, 1–31.

Andrews, A. L. 1933. What is *Sphagnum americanum? Annls bryol.*, **6**, 1–6.

Andrews, A. L. 1937–1951. Notes on the Warnstorf *Sphagnum* Herbarium (IV–VI. Studies in the Warnstorf *Sphagnum* Herbarium); 1937 I. *Annls bryol.*, **9**, 3–12; 1941a II. The section Malacosphagnum. *Bryologist*, **44**, 97–102; 1941b III. The subgenus Inophloea in South America, **44**, 155–159; 1947 IV. The group Acutifolia in South America, **50**, 181–186; 1949 V. The group Cuspidata in South America, **52**, 124–130; 1951 VI. The subgenus Inophloea in the eastern hemisphere, **54**, 83–91.

Andrews, A. L. 1938. The North American Altantic species of *Sphagnum*. *Annls bryol.*, **11**, 15–20.

Andrus, R. E. 1980a. *Sphagnum subtile* (Russow) Warnst. and allied species in North America. *Syst. Bot.*, **4**, 351–362.

Andrus, R. E. 1980b. Sphagnaceae (peat moss family) of New York State. In: *Contributions to a flora of New York State III*, edited by R. S. Mitchell. Albany, New York

Ångström, J. 1842. *Dispositio muscorum in Scandinavia hucusque cognitorum.* Upsaliae Holmiae.

Ångström, J. 1864. Om några mindre kanda eller omtvistade Sphagna. *Öfvers. K. VetenskAkad. Förh.*, **21**, 197–203.

Anon. 1964. Resolutions. International Botanical Congress, 10th, Edinburgh, 1964. *Taxon*, **13**, 282–292.

Arne, H. W. & Jensen, C. 1896. Ein bryologischer Ausflug nach Tasjö. *Bih. K. svenska VetenskAkad. Handl.*, ser. 3, **21** (10), 1–64.

Augier, J. 1966. *Flore des bryophytes.* Paris: Paul Lechevalier.

Austin, C. F. 1870. Musci appalachiani: *tickets of specimens of mosses collected mostly in the eastern part of North America.* New Jersey: Closter.

Bakalar, S. & Balogh, M. 1979. *Sphagnum girgensohnii,* ein neues boreales Florenelement in Ungarn. *Bot. Közl.,* **66,** 11–14.
Barkman, J. J. & Glas, P. 1960. *Sphagnum dusenii,* een nieun veenmos voor Nederland. *Belmontia,* ser. 2, **4–5,** 230–238.
Bauer, E. 1903a. Musci europaei exsiccati. Schedae nebst kritischen Bemerkungen zur ersten Serie. *Sber. dt. naturw.-med. Ver. Böhm. "Lotos",* **51,** 117–142.
Bauer, E. 1903b. *Musci europaei exsiccati. Die Laubmoose Europas unter Mitwirkung namhafter Bryologen und Floristen,* ser. 1, nos 1–50. Prague.
Beauvois, See **Palisot (de) Beauvois.**
Belkevich, P. I., Gayduk, K. A. & Christova, L. R. 1976. Possibility of using peat for natural environment protection. *Proc. int. Peat Congr. 5th,* **1,** 328–348.
Bertsch, K. 1959. *Moosflora von Sudwestdeutschland.* Stuttgart: Eugen Ulmer.
Boesen, D. F., Lewinsky, J. & Rasmussen, L. 1975. A check list of the bryophytes of the Faeroes. *Lindbergia,* **3,** 69–78.
Bomansson, J. O. & Brotherus, V. F. 1894. *Herbarium Musci Fennici. Enumeratio plantarum Musci Fennici qvam edidit Societas pro Fauna et Flora Fennica. II. Musci.* 2nd ed. Helsinki.
Boros, A. 1953. *Magyaroszág Mohá.* Budapest: Akademiai Kiado.
Boros, A. 1968. *Bryogeographie und bryoflora Ungarns.* Budapest: Akademiai Kiado.
Bottini, A. 1913. Sfagni d'Italia. *Webbia,* **4,** 107–141.
Bottini, A. 1920. Sfagnologia Italiana. *Memorie R. Accad. naz. Lincei. Cl. Sci. fis. nat.,* ser. 5, **13,** 1–87.
Braithwaite, R. 1871–1875. On bog mosses. 1871. *Mon. microsc. J. Trans.,* **6,** 1–5, 268–273; 1872, **7,** 55–58, 256–258; **8,** 3–4, 157–158; 1873a, **9,** 12–15, 214–216; **10,** 14–15, 218–221; 1874, **11,** 155–158, 255–257; **12,** 11–13, 168–170; 1875, **13,** 61–64, 229–232; **14,** 47–49.
Braithwaite, R. 1877. *Sphagnaceae britannicae exiccatae.* Handwritten labels, nos 1–53. London: David Bogue.
Braithwaite, R. 1880. *The Sphagnaceae or peat-mosses of Europe and North America.* London.
Braithwaite, R. 1881 *Sphagnum subbicolor,* Hampe. *J. Bot., Lond.,* **19,** (n.s. **10**), 116.
Breutel, J. C. 1824. Beitrag zu der Moosgattung *Sphagnum. Flora, Jena,* **7,** 433–441.
Bridel (Brideri), S. E. 1798. *Muscologiae recentorium seu analysis, historia, et descriptio methodica omnium muscorum frondosorum, hucusque cognitorum, ad normam Hedwigii.* II, I–XI. Paris, Gothae.
Bridel (Brideri), S. E. 1806. *Muscologiae recentorium supplementum seu species muscorum. I-VIII.* Paris, Gothae.
Bridel (Brideri), S. E. 1819. *Muscologiae recentorium supplementum. IV. Seu Mantissa generum specierumque muscorum frondosorum universa. Methodus nova muscorum ad naturae normam melius instituta et Muscologiae recentiorum accomodata. XVIII.* Gothae.
Bridel (Brideri), S. E. 1826–1827. *Bryologia universa seu systematica ad novam methodium dispositio, historia et descriptio omnium muscorum frondosorum hucusque cognitorum cum synonymia et auctoribus probatissimus. I-XLVI.* Lipsiae.
Bruch, P. 1825. Ueber Sphagna, nebst Bemerkungen zu den in Nr 88 de Botan. Zeitung fur 1824 durch Herrn Inspecktor Breutel mitgetheilten Beobachtungen. *Flora, Jena,* **8,** 625–635.
Bryan, V. 1955. Chromosome studies in the genus *Sphagnum. Bryologist,* **58,** 15–39.
Buen, H. 1961 *Sphagnum riparium* in Norway. *Nytt Mag. Bot.,* **9,** 25–31.
Bullock, A. A. 1966. Limitation of valid publication by reference to earlier descriptions. *Taxon,* **15,** 74–77.
Cardot, J. 1884. Notes sphagnologiques. Descriptions de quelques variétiés nouvelles. *Revue bryol.,* **11,** 54–56.

Cardot, J. 1886. Les sphaignes d'Europe, revision critique des espèces et étude de leurs variations. *Bull. Soc. r. Bot. Belg.*, **25,** 19–136.

Cardot, J. 1887. Revision des sphaignes de l'Amérique du Nord. *Bull. Soc. r. Bot. Belg.*, **26,** 39–61.

Cardot, J. 1897. Répertoire sphagnologique. Catalogue alphabétique de toutes les espèces et variétiés du genre *Sphagnum* avec la synonymie, la bibliographie et la distribution géographique, d'après les travaux les plus recents. *Bull. Soc. Hist. nat. Autun.*, **10,** 235–432.

Cardot, J. 1907. Mousses nouvelles du Japon et de Corée. *Bull. Herb. Boissier,* 2nd ser., **7,** 709–717.

Casares-Gil, A. 1925. Los esfagnales de la Peninsula Ibérica. *Mems R. Soc. esp. Hist. nat.,* **13,** 5–81.

Castelli, L. 1955. Contribution à la flora bryologique du massif de la Vanoise. *Revue bryol. lichénol.,* **24,** 227–238.

Chater, A. O. & Brummitt, R. K. 1966. Subspecies in the works of Friedrich Ehrhart. *Taxon,* **15,** 95–106.

Clymo, R. S. 1963. Ion exchange in *Sphagnum* and its relation to bog ecology. *Ann. Bot.,* **27,** 309–324.

Clymo, R. S. 1973. The growth of *Sphagnum:* some effects of environment, *J. Ecol.,* **61,** 849–869.

Clymo, R. S. & Hayward, P. M. 1982. The ecology of *Sphagnum.* In: *Bryophyte ecology,* edited by A. J. E. Smith, 30–78. London: Chapman & Hall.

Corley, M. F. V., Crundwell, A. C., Düll, R., Hill, M. O. & Smith, A. J. E. 1981. Mosses of Europe and the Azores: annotated list of species, with synonyms from the recent literature. *J Bryol.,* **11,** 609–689.

Coster, I. & Pankow, H. 1968. Illustrierter Schlüssel zur Bestimmung einiger mitteleuropäischer Sphagnum-Arten. *Wiss. Z. Univ. Rostock,* **17,** 285–323.

Couderc, J. M. & Goff, P. le. 1969. Etude geographique et floristique des Sphaignes de Touraine. Les Sphaignes de la Gatine Tourangelle. *Revue bryol. lichénol.,* **36,** 509–530.

Courtejaire, J. 1962. Quelques remarques phytogeographiques sur les Sphaignes des etangs du massif du Madrès (Pyrenées-Orientales). *Revue bryol. lichénol.,* **31,** 157–167.

Courtejaire, J. 1962. La microaire francaise de *Sphagnum pylaiei.* Bryologist, **65,** 38–47.

Crome, G. E. W. 1803. *Sammlung deutscher Laub-Moose.* Schwerin.

Crum, H. A. 1975. Comments on *Sphagnum capillaceum. Contr. Univ. Mich. Herb.,* **11,** 89–93.

Crum, H. A. 1976, *Mosses of the Great Lakes Forest.* Ann Arbor: University Herbarium.

Crum, H. A. 1984. *Sphagnopsida, Sphagnaceae. (North American Flora, Series II, Part 11).* Bronx, New York: New York Botanical Garden.

Crum, H. A. & Anderson, L. E. 1981. *Mosses of eastern North America,* Sphagnum. Vol. 1, 21–66. New York: Columbia University Press.

Crum, H. A. & Snider, J. A. 1977. *Sphagnum quinquefarium* in the American West. *Bryologist,* **80,** 156–158.

Crum, H. A., Steere, W. C. & Anderson, L. E. 1965. A list of the mosses of North America. *Bryologist,* **68,** 377–432.

Dalby, D. H. 1969. *A key to the genus* Sphagnum *in Europe.* London: Imperial College.

Dalla Torre, K. W. von & Sarnthein, L. G. von. 1904. *Die Moose (Bryophyta) von Tirol, Vorarlberg und Liechtenstein.* Innsbruck.

Deighton, F. C., Stevenson, J. A. & Cummins, G. B. 1962. Formae speciales and the code. *Taxon,* **11,** 70–71.

Demaret, F. 1941. Revision espèces belges de *Sphagnum* Dill, sous-section Cymbifolia Lindb. *Bull. Jard. bot. État Brux.,* **16,** 267–286.

De Notaris, J. 1836. Mantissa muscorum ad Floram pedemontanum. *Mem. Reg. Scient. Acad. Taurin.*, **39,** 211–261.

De Zuttere, P. 1966. Les Sphaignes de la section Cuspidata Schlieph. en Belgique. *Buxbaumia*, **20,** 15–26.

De Zuttere, P. 1974. Les Sphaignes de Belgique. *Naturalistes belg.*, **55,** 258–282.

Dickson, J. H. 1973. *Bryophytes of the Pleistocene. The British record and its chronological and ecological implications.* Cambridge: Cambridge University Press.

Dierssen, K. 1973. Erganzungen zur Moosflora Islands. *Herzogia*, **3,** 83–109.

Dierssen, K. 1982. *Die wichtigsten Pflanzengesellschaften der Moore NW-Europas.* Génève: Conservatoire et Jardin botaniques.

Dillenius, J. J. 1718. *Catalogus plantarum sponte circa Gissam nescentium; cum observationibus botanicis, synonymiis necessariis, tempore et locis, in quibus plantae reperiuntur. Praemittitur praefatio et dissertatio brevis de variis plantarum methodis, ad calcem vero adjicitur fungorum et muscorum methodica recensio hactenus desiderata.* XVI. Frankfurt.

Dillenius, J. J. 1741. *Historia muscorum in qua circiter sexcentae species veteres et novae ad sua genera relatae describuntur et iconibus genuinis illistrantur: cum appendice et indice synonymorum.* XVI. Oxford.

Dismier, G. 1928. Flore des Sphaignes de France. *Archs Bot. Bull. mens.*, **1,** 1–63.

Dixon, H. N. 1924. *The student's handbook of British mosses.* 3rd ed. Eastbourne: Sumfield & Day.

Dokturovsky, V. 1912. K flore mkhov Amurskoy oblasti. (Ref: Zur Moosflora des Amurgebietes). *Izvest. imp. S.-Peterb. bot. Sada*, **12,** 105–120.

Dolezal, R. 1976. Pripevek k rozsireni raselinku na Morave. *Preslia*, **48,** 369–373.

Dombrovskaya, A. V., Koreneva, M. M. & Tyuremnov, S. N. 1959. *Atlas rastitel' nykh ostatkov, vstrechaemykh v torfe.* Moscow; Leningrad.

Dozy, F. 1854. Bijdrage tot de anatomie en phytographie der Sphagna. *Verh. K. ned. Akad. Wet.*, **2,** 3–11.

Dozy, F. & Molkenboer, J. H. 1851. *Prodromus florae Batavae. II: I. (Plantae cellulares. Musci frondosi et Hepaticae).* Leyden.

Dull, R. 1969. Ubersicht zur Bryogeographie Sudwestdeutschlands unter besonderer Berücksichtigung der Arealtypen. *Herzogia*, **1,** 215–320.

Duncan, U. K. 1962. Illustrated key to *Sphagnum* mosses. *Trans. Proc. bot. Soc. Edinb.*, **39,** 290–301.

Du Rietz, G. E. 1945. Nàgra namnfràgor inom släktet *Sphagnum. Svensk bot. Tidskr.*, **39,** 151–152.

Du Reitz, G. E. 1964. *Sphagnum nemoreum* Scop. or *Sph. capillaceum* (Weis) Schrank. *Svensk bot. Tidskr.*, **58,** 167–171.

Dusen, K. F. 1887. *Om sphagnaceernas utbrening i Skandinavien. En vaxtgeorgrafisk studie.* Thesis, University of Uppsala.

Eddy, A. 1977. *Sphagnum subsecundum* agg. in Britain. *J. Bryol.*, **9,** 309–319.

Ehrhart, J. 1780. Versuch eines Verzeichnisses der um Hannover wild wachsenden Pflanzen. *Hann. Mag.*, 209–240.

Ehrhart, J. 1788a. *Plantae cryptogamae Linn. quas in locis carum natabilis collegit et exsiccavit.* Dec. VIII, nos 71–80. Hannover.

Ehrhart, J. 1788b. *Beiträge zur Naturkunde, und den damit verwandten Wissenschaften besonders der Botanik, Chemie, Haus- und Landwirtschaft, Arzneigelahrtheit und Apothekerkunst. III.* Hannover.

Ehrhart, J. 1793. *Plantae cryptogamae Linn. quas in locis carum natabilis collegit et exsiccavit.* Dec. XXV & XXVI, nos 241–260. Hannover.

Ekman, E. 1969. On the use of peat in oil pollution control. *Suo*, **20,** 61–65.

Eriksson, P. A. 1976. *Sphagnum strictum* funnen i Dalarna. *Svensk bot. Tidskr.*, **70,** 56.
Eriksson, P. A. 1979. Nya lokaler för *Sphagnum angermanicum* i Västerdalarna. *Svensk bot. Tidskr.*, **73,** 202.
Fearnsides, M. 1938. Graphic keys for the identification of *Sphagna*. *New Phytol.*, **37,** 409–424.
Ferguson, P., Lee, J. A. & Bell, J. N. B. 1978. Effects of sulphur pollutants on the growth of *Sphagnum* species. *Environ. Pollut.*, **16,** 151–162.
Flatberg, K. I. 1983. Typification of *Sphagnum capillifolium* (Ehrh.) Hedw. *J. Bryol.*, **12,** 503–508.
Flatberg, K. I. 1984. A taxonomic revision of the *Sphagnum imbricatum* complex. *K. norske Vidensk. Skr.*, **3,** 1–80.
Flatberg, K. I. 1986. Taxonomy, morphovariation, distribution and ecology of the *Sphagnum imbricatum* complex with main reference to Norway. *Gunneria*, **54,** 1–118.
Flatberg. K. I. 1988a. *Sphagnum skyense* sp. nov. *J. Bryol.*, **15,** 101–107.
Flatberg, K. I. 1988b. *Sphagnum viridum* sp. nov., and its relation to *S. cuspidatum*. *K. Norske Vidensk. Selsk, Skr.*, **1,** 1–64.
Flatberg, K. I. 1988c. Taxonomy of *Sphagnum annulatum* and related species. *Ann. bot. fenn.*, **25,** 303–350.
Flatberg, K. I. & Frisvoll, A. A. 1984. *Sphagnum arcticum* sp. nov. *Bryologist*, **87,** 143–148.
Fleischer, M. 1904. *Die Musci der Flora von Buitenzorg (zugleich Laubmoosflora von Java). I. Sphagnales; Bryales (Arthrodontei (Haplolepideae)). Flore de Buitenzorg.* Leiden: Jardin Botanique de l'Etat.
Fuchs, H. P. 1958. Historische Bemerkungen zum Begriff der Subspecies. *Taxon,* **7,** 44–52.
Gams, H. 1957. *Kleine Kryptogamenflora. IV. Die Moos und Farnplfanzen (Archegoniaten).* 4th ed. Stuttgart: Gustav Fischer.
Gams, H. 1962. Remarques ultérieures sur la phylogénie des sphaignes. *Revue bryol. lichénol* , n.s. **31,** 1–4.
Gauthier, R. 1980. *La végétation des tourbières et les sphaignes du parc des Laurentides, Quebec.* (Études écologiques). Quebec: Université Laval.
Gaume, R. 1955, 1956. Catalogue des Muscinées de Bretagne d'après les documents inédite du Dr. F. Camus. I. *Revue bryol. lichénol.*, **24,** 1–28, 183–192; II **25,** 1–115.
Girgensohn, G. K. 1860. Naturgeschichte der Laub- under Lebermoose Liv-, Est.- u. Kurlands nebst kurzer Charakteristik derjenigen Gattungen und Arten welche in den genannten Provinzen noch gefunden werden könnten, so wie derjenigen, welche in den übrigen Theilen Russlands bisher aufgefunden sind. *Arch. Naturk. Liv.-, Est.-u. Kurlands,* **2,** ser. 2, 1–488.
Glowacki, J. 1905. Beitrag zur Laubmoosflora von Gmünd in Karnten. *Jb. naturh. Landesmus.*, **27,** 93–128.
Goossens, M. & de Sloover, J. 1981. Etude taxonomique et synecologique des éspèces du genre *Sphagnum* section Subsecunda dans une tourbière de Haute Ardenne. *Bull. Soc. r. Bot. Belg.*, **114,** 89–105.
Gorham, E. & Pearsall, W. H. 1956. Acidity, specific conductivity and calcium content of some bog and fen waters in northern Britain. *J. Ecol.*, **44,** 129–141.
Grabarz, M, 1969. Rozmieszczenie gutunkow rodzaju *Sphagnum* L. na Lubelszczyznie. (Distribution of species of *Sphagnum* in the Lublin District). *Annls Univ. Mariae Curie-Sklodowska, Ser, C,* **24,** 21–39.
Green, B. H. 1968. Factors influencing the spatial and temporal distribution of *Sphagnum imbricatum* Hornsch ex. Russ. in the British Isles. *J. Ecol.*, **56,** 47–58.
Hagen, I. 1899. Musci Norvegiae borealis. *Tromsø Mus. Arsh.*, **21.**
Hampe, E. 1847. Eine Referat über die columbischen Moose, welche Herr Moritz gesammelt, und dem Königl. Herbarium in Schöneberg bei Berlin überliefert hat. *Linnaea,* **20,** 65–98.
Hampe, E. 1880. Ein neues *Sphagnum* Deutschlands. *Flora, Jena,* **63,** 440–441.

Hansen, B. 1961. Sphagnaceae (Studies in the flora of Thailand 4). *Dansk bot. Ark.*, **20,** 89–108.
Hartman, C. J. 1838. *Handbok i Skandinaviens flora, innefattande Sveriges och Norriges vexter, till och med mossorna,* 3rd ed. *II. Floran.* Stockholm, (1843. 4th ed. Stockholm).
Hartman, C. 1849, 1858, 1861. *Handbok i Skandinaviens flora, innefattande Sveriges och Norriges vexter till och med mossorna; ordnande efter prof Fries' system, af C. J. Hartman.* 5th ed. Stockholm; 7th ed. Stockholm; 8th ed. Stockholm.
Hartman, C. 1861. *Handbok i Skandinaviens flora, innefattande Sveriges och Norges växter, till och med mossorna, af C. J. Hartman.* 9th ed. Stockholm.
Hebrard, J P. 1980. Contribution a l'étude des muscinées du parc national des ecrins. Observations floristiques et ecologiques. *Cryptogam., Bryol., Lichenol.,* **1,** 339–397.
Hedwig, J. 1782, *Fundamentum historiae naturalis muscorum frondosorum concernens eorum flores, fructus, seminalem propagationem adjecta generum dispositione methodica, iconibus illustratis.* Parts I and II. Leipzig: Crusius.
Hedwig, J. 1801. *Species muscorum frondosum (ed. F. Schwaegrichen).* Leipzig.
Hennezel, F. D' & Coupal, B. 1972. Peatmoss—a natural absorbent for oil spills. *Bull. Can. Inst. Min. Metall.,* **65,** 51–53.
Herzog, T. 1921. Die Bryophyten meiner zweiten Reise durch Bolivia. *Biblthca bot.,* **88,** 1–31.
Herzog, T. 1926. *Geographie der Moose.* Jena.
Hesselbo, A. 1918. The bryophyta of Iceland. In: *Botany of Iceland,* I. 395–677.
Hill, M. O. 1975. *Sphagnum subsecundum* Nees. and *S. auriculatum* Schimp. in Britain. *J. Bryol.,* **8,** 435–441.
Hill, M. O. 1976. A critical assessment of the distinction between *Sphagnum capillaceum* (Weiss) Schrank and *S. rubellum* Wils. in Britain. *J. Bryol.,* **9,** 185–191.
Hill, M. O. 1977, *Sphagnum flexuosum* and its varieties in Britain. *Bull. Br. bryol. Soc.,* **29,** 19.
Hill, M. O. 1978, Sphagnopsida. In: *The moss flora of Britain and Ireland,* edited by A. J. E. Smith, 30–78. Cambridge, Cambridge University Press.
Hill, M. O. 1988. *Sphagnum imbricatum* ssp. *austinii* (Sull.) Flatberg and ssp. *affine* (Ren. & Card.) Flatberg in Britain and Ireland. *J. Bryol.,* **15,** 109–115.
Hoffmann, G. F. 1795–1796. *Deutschlands Flora oder botanisches Taschenbuch für das Jahr 1795. II. Cryptogamie.* Erlangen.
Holmen, K. 1955. Chromosome numbers of some species of *Sphagnum. Bot. Tidsskr.,* **52,** 37–42.
Holmen, K. 1964. Additions to the *Sphagnum* flora of Greenland. *Bryologist,* **67,** 458–460.
Holmen, K. & Lange, B. 1958. *Sphagnum wulfianum* and *Sphagnum centrale:* their morphology and occurrence in Greenland. *Bot. Tidsskr.,* **54,** 379–386.
Holmen, K. & Scotter, G. W. 1971. Mosses of the Reindeer Preserve, Northwest Territories, Canada. *Lindbergia,* **1,** 34–56.
Hornschuch, F. 1820. Uber die von Chamisso und Bergius gesammelten Moose. *Flora, Jena,* **3,** 511–522.
Horrell, E. C. 1900. The European Sphagnaceae (after Warnstorf). *J. Bot., Lond.,* **38,** 110–122, 161–167, 215–224, 252–258, 303–315, 338–353, 383–392, 422–426, 469–480.
Horton, D. G., Vitt, D. H. & Slack, N. G. 1979. Habitats of circumboreal-subarctic *Sphagna:* 1. A quantitative analysis and review of species in the Caribou Mountains, northern Alberta. *Can. J. Bot.,* **57,** 2283–2317.
Hübener, J. W. P. 1833. *Muscologia germanica oder Beschreibung der deutschen Laubmoose. Im erweiterten Umfange nach dem jetzigen Stande der Wissenschaft, nebst Erörterung der Standorter und ihrer Entdecker, der Synonyme seit Hoffmann und Roth, mit erlauternden Anmerkungen.* Leipzig.
Hübener, J. W. P. & Genth, C. F. F. 1837. *Deutschlands Lebermoose in getrockneten Exemplaren.* III. Leif., nos 51–75. Mainz.

Hult, R. 1881. Försök till analytisk behandling af växtformationena. *Meddn. Soc. Fauna Flora fenn.*, **8**, 1–155.
Hunt, G. E. 1867. On mosses new to Britain. *Mem. Proc. Manchr lit. phil. Soc.*, **23** (3rd ser. **3**), 231–244.
Husnot, T. 1882. *Sphagnicola europaea. Descriptions et figures des sphaignes de l'Europe.* Caen.
Ireland, R. R., Bird, C. D., Brassard, G. R., Schofield, W. B. & Vitt, D. H. 1980. Checklist of tne mosses of Canada. *Publ. Bot. Nat. Mus. natur. Sci. (Can.)*, **8**, 1–75.
Isoviita, P. 1966. Studies on *Sphagnum* L. I. Nomenclatural revision of the European taxa. *Ann. bot. fenn.*, **3**, 199–264.
Isoviita, P. 1970. Studies on *Sphagnum* L. II. Synopsis of the distribution in Finland and the adjacent parts of Norway and the USSR. *Ann. bot. fenn.*, **7**, 157–162.
Jalas, J. 1965. Die zonale und regionale Gliederung der fennoskandischen Vegetation. *Rev. roum. Biol. ser. Bot.*, **10**, 109–113.
Jelenc, F. 1977. Herborisation du 5 Septembre 1977; quelques biotypes à Sphaignes de la region de Châlus (Haute-Vienne). *Bull. Soc. bot. Centre-Ouest*, n.s. **8**, 148–151.
Jensen, C. 1890. *De danske Sphagnum-arter. Festskrif, udgivet af den Botaniske Forening i Kjøbenhavn i anledning af dens halvhundredaarsfest, den 12 April 1890.* København.
Jensen, C. 1915. *Danmarks mosser eller beskrivelse af de i Danmark med Faeroerne fundne bryofyter. I. Hepaticales, Anthocerotales og Sphagnales.* København.
Karczmarz, K. & Sokolowski, A. W. 1977. New data to bryophytes flora of north-eastern Poland *Annls Univ. Mariae Curie-Sklodowska, C*, **32**, 45–52.
Kaule, G. 1973. Zum vorkommen von *Sphagnum centrale* Jensen und *Sphagnum subnitens* Russow et Warnst. in Sudbayern. *Herzogia*, **2**, 423–435.
Klinggraff, H. von. 1872. Beschreibung der in Preussen gefundenen Arten und Varietäten der Gattung *Sphagnum. Schr, phys.-ökon. Ges. Königsb.*, **13**, 1–10.
Klinggraff, H. von. 1880. *Versuch einer topographischen Flora der Provinz Westpreussen.* Danzig.
Klinggraff, H. von. 1893. *Leber- und Laubmoose West- und Ostpreussens.* Danzig.
Koppe, F. 1964. Die Moose des Neidersachsischen Tieflandes. *Abh. naturw. Ver. Bremen*, **36**, 237–424.
Krause, H. L. 1921. *Rostocker Moosflora. Verzeichnis der bis 1920 aus der Nordostecke Mecklenburgs bis Bukspitze, Warnow, Güstrow, Sülze bekannt gewordenen Moosarten.* Rostock.
Krisai, R. 1977. Sphagnologische notizen aus Osterreich. *Herzogia*, **4**, 235–247.
Lag, J. 1958. *Sphagnum wulfianum* found in Sor-Varanger, northern Norway. *Blyttia*, **16**, 176.
Lamarck, J. B. A. P. M. De & Candolle, A. P. De. 1805. *Flore française, ou descriptions succinctes de toutes les plantes qui croissent naturellement en France, disposées selon une nouvelle méthode d'analyse, et précedées par un exposé des principes élémentaires de la botanique.* 3rd ed. Paris.
Lange, B. 1952a. A revision of the *Sphagnum* flora of Iceland. *Bot. Tidsskr.*, **49**, 192-195.
Lange, B. 1952b. The genus *Sphagnum* in Greenland. *Bryologist*, **55**, 117–126.
Lange, B. 1955a. *Sphagnum tenerum* Sull. & Lesq. and *Sphagnum tenerum* (Aust.). Warnst. *Bot. Tidsskr.*, **52**, 43–47.
Lange, B. 1955b. The genus *Sphagnum* in alpine zones at Abisko, North Sweden and some other arcto-alpine areas. *Mitt. thüring. bot. Ges.*, **1**, 145–150.
Lange, B. 1963. Studies in the *Sphagnum* flora of Iceland and the Faeroes. *Bot. Tidsskr.*, **59**, 220–213.
Lange, B. 1969. The distribution of *Sphagnum* in northernmost Scandinavia. *Bot. Tidsskr.*, **65**, 1–43.
Lange, B. 1973. The *Sphagnum* flora of hot springs in Iceland. *Lindbergia*, **2**, 81–93.

Lange, B. 1977. Additional notes on the distribution of *Sphagnum pylaesii. Bryologist,* **80,** 527–529.
Lange, B. 1982. Key to northern boreal and arctic species of *Sphagnum,* based on characteristics of the stem leaves. *Lindbergia,* **8,** 1–29.
Lepage, E. 1945, 1946. Les lichens, les mousses et les hépatiques du Québec et leur rôle dans la formation du sol arable dans la région du bas de Québec, de Lévis à Gaspé. IV. Inventaire des espèces du Québec: 2. Les mousses. *Naturaliste can.,* **72,** 241–265, 315–338; **73,** 33–56, 101–134, 207–232, 395–411.
Lesquereux, L. 1868. A catalogue of the species of mosses found up to the present time on the north-west coast of the United States, and especially in California. *Mem. Calif. Acad. Sci.,* **1,** 1–38.
Lesquereux, L. & James, T. P. 1879. Descriptions of some new species of North American mosses. (With a supplement by W. P. Schimper). *Proc. Am. Acad. Arts Sci.,* **14,** 133–141.
Lesquereux, L. & James, T. P. 1884. *Manual of the mosses of North America.* Boston
Lid, J. 1925. An account of the Cymbifolia group of the *Sphagna* of Norway. *Nyt. Mag Naturvid.,* **63,** 224–259.
Lid, J. 1929. *Sphagnum strictum* Sulliv. and *Sphagnum americanum* Warnst. in Scotland. *J. Bot., Lond.,* **67,** 170–175.
Lid, J. 1932. A list of some Norwegian *Sphagna. K. norske Vidensk. Selsk. Forh.,* **4,** 172–174.
Limpricht, K. G. 1881, 1882. Zur Systematik der Torfmoose. *Bot. Zbl.,* **7,** 311–319; **10,** 214–222.
Limpricht, K. G. 1885–1890. Die Laubmoose Deutschlands, Oesterreichs und der Schweiz, I. Sphagnaceae, Andreaeceae, Archidiaceae, Bryineae (Cleistocarpae, Stegocarpae (Acrocarpae)). In: *Kryptogamen-Flora von Deutschland, Oesterreich und der Schweiz,* edited by L. Rabenhorst. 2nd ed. Leipzig.
Limpricht, K. G. & Limpricht, W. 1895–1904. Die laubmoose Deutschlands, Oesterreichs und der Schweiz. Unter Berücksichtigung der übrigen Länder Europas u. Sibiriens. III. Hypnaceae u. Nachträge, Synonymen-Register u. Litteratur-Verzeichniss. In: *Kryptogamen-Flora von Deutschland, Osterreich und der Schweiz,* edited by L. Rabenhorst. Vol. 4. 2nd ed. Leipzig.
Lindberg, H. 1899. Bidrag till kannedomen om de till *Sphagnum cuspidatum-gruppen* hörande arternas utbredning i Skandinavien och Finland. *Acta Soc. Fauna Flora fenn.,* **18,** 1–26.
Lindberg, H. 1900. Om *Sphagnum annulatum* Lindb. fil. *Meddn Soc. Fauna Flora fenn.,* **24,** 66–67.
Lindberg, H. 1903. Kritische Bestimmungstabellen der europäische *Sphagna cuspidata. Lotos,* 1903, 124.
Lindberg, S. O. 1862. Torfmossornas byggnad, utbredning och systematiska uppstallning. *Öfvers. K. VetenskAkad. Förh.,* **19,** 113–156.
Lindberg, S. O. 1871. Revisio critica iconum in opere Flora danica muscos illustrantium. *Acta Soc. Sci. fenn.,* **10,** 1–118.
Lindberg, S. O. 1872. Contributio ad floram cryptogamam Asiae boreali-orientalis. *Acta Soc. Sci. fenn.,* **10,** 222–280.
Lindberg, S. O. 1874. *Manipulus muscorum* secundus. *Not. sallsk. Fauna Flora fenn. Forh.,* **13,** 351–417.
Lindberg, S. O. 1879. *Musci scandinavici in systemate novo naturali dispositi.* Uppsala.
Lindberg, S. O. 1882. *Europas och Nord Amerikas hvitmossor (Sphagna) jämte en inledning om utvecklingen och organbildningen inom mossornas alla tre hugvudgrupper.* Thesis, Alexanders-Universitet, Helsinki.
Lindberg, S. O. 1883. *Kritsk granskning af mossorna uti Dillenii Historia muscorum.* Helsinki.
Linnaeus (von Linné), C. 1753. *Species plantarum 2.* Holmiae.
Lobley, E. M. & Fitzgerald, J. W. 1970. A revision of the genus *Sphagnum* L. in a flora of the north-east of Ireland. *Ir. Nat. J.,* **16,** 357–365.

Maas, W. S. G. 1965a. *Sphagnum dusenii* and *Sphagnum balticum* in Britain. *Bryologist*, **68**, 211–217.
Maas, W. S. G. 1965b. Zur Kenntnis des *Sphagnum angermanicum* in Europa. *Svensk bot. Tidskr.*, **59**, 332–344.
Maas, W. S. G. 1966. Studies on the taxonomy and distribution of *Sphagnum*. I. *Sphagnum pylaesii* and *Sphagnum angermanicum* in Quebec and some phytogeographic considerations. *Bryologist*, **69**, 95–100.
Maas, W. S. G. 1967. Studies on the taxonomy and distribution of *Sphagnum*. IV. *Sphagnum majus, Sphagnum annulatum, Sphagnum mendocinum* and *Sphagnum obtusum* in North America. *Nova Hedwigia*, **14**, 187–213.
Maas, W. S. G. & Harvey, M. J. 1973. Studies on the taxonomy and distribution of *Sphagnum*. VII. Chromosome numbers in *Sphagnum*. *Nova Hedwigia*, **24**, 193–205.
Malmer, N. 1966. *De svenska* Sphagnum *artenas systematik och ekologi*. 3rd ed. University of Lund.
Martensson, O. & Nilsson, E. 1974. On the morphological colour of bryophytes. *Lindbergia*, **2**, 145–159.
Matsuda, Y. 1981. Comparison of morphological characters among *Sphagnum apiculatum, S. amblyphyllum* and *S. angustifolium*. *Hikobia*, Suppl. 1, 403–412.
Melin, E. 1913. Sphagnologische Studien in Tiveden. *Ark. Bot.*, **13**, 9, 1–59.
Melin, E. 1919. *Sphagnum angermanicum* n. sp. *Svensk bot. Tidskr.*, **13**, 21–25.
Milde, J. 1869. *Bryologia silesiaca. Laubmoos-Flora von Nord- und Mittel-Deutschland, unter besonderer Berücksichtigung Schlesiens und mit Hinzunahme der Floren von Jütland, Holland, der Rheinpfalz, von Baden, Franken, Böhmen. Mähren und der Umgegend von München*. Leipzig.
Mitten, W. 1869. Musci austro-americani. Enumeratio muscorum omnium austro-americanorum auctori huqusque cognitorum. *J. Linn. Soc. Bot.*, **12**, 1–659.
Moen, A. & Synnott, D. 1983. *Sphagnum subfulvum* in Ireland compared with the occurrence in Norway. *J. Bryol.*, **12**, 331–336.
Moldenhawer, J. J. P. 1812, *Beyträge zur Anatomie der Pflanzen*. Kiel.
Mougeot, J. B., Nestler, C. & Schimper, W. P. 1854. *Stirpes cryptogamae vogeso-rhenanae; quas in Rheni superioris inferiorisque, nec non Vogesorum praefecturis collegerunt*. Fasc. XIV, nos 1301–1400. Bruyères.
Müller, C. 1848–1849. *Synopsis muscorum frondosorum omnium huqusque cognitorum. I. Musci vegetationis acrocarpieae*. Berlin.
Müller, C. 1853. *Deutschlands Moose oder Anleitung zur Kenntnis der Laubmoose Deutschlands, der Schweiz, der Niederlande und Däemarks für Anfänger sowohl wie für Forscher bearbeitet*. Halle.
Müller, C. 1874. Novitates Bryotheca Mullerianae. I. Musci philippinensis praesertim Wallisiani adjectis nonnullis muscis aliis indicis. *Linnaea*, **38**, 545–572.
Müller, C. 1887. Sphagnorum novorum descriptio. *Flora, Jena*, **70**, 403–422.
Müller, C. 1901. *Genera muscorum frondosorum. Classes Schistocarporum, Cleistocarporum, Stegocarporum complectentia, exceptis Orthotrichaceis et Pleurocarpis. Gattungen und Gruppen der Laubmoose in historischer und systematischer Beziehung, sowie nach ihrere geographischen Verbreitung unter Berücksichtigung der Arten*. Leipzig.
Necker, N. J. de. 1771. *Methodus muscorum per classes, ordines, genera ac species cum synonymis, nominibus trivialibus, locis natalibus, observationibus digestorum, aeneisque figuris illustratorum*. Mannheim.
Nees von Esenbeck, C. G., Hornschuch, F. & Sturm, J. 1823. *Bryologia germanica, oder Beschreibung der in Deutschland und in der Schweiz wachsenden Laubmoose. I.* Nurnberg.

Nichols, G. E. 1920. *Sphagnum* moss: war substitute for cotton in absorbent surgical dressings. *Rep. Smithson Instn 1918,* 221–234.

Norrlin, J. P. 1874. Öfversigt af Torneå (Muonio) och angränsande delar af Kemi Lappmarkers mossor och lafvar. *Not. sallsk. Fauna Flora fenn. Forh.,* **13,** 271–348.

Nyholm, E. 1969. *Illustrated moss flora of Fennoscandia.* Fasc. 6. Stockholm.

Osvald, H. 1940. *Sphagnum flavicomans* (Card.) Warnst. Taxonomy, distribution and ecology. *Acta phytogeogr. suec.,* **13,** 39–49.

Pakarinen, P. 1978. Distribution of heavy metals in the *Sphagnum* layer of bog hummocks and hollows. *Ann. Bot. fenn.,* **15,** 287–292.

Pakarinen, P. & Tolonen, K. 1976. Studies on the heavy metal content of ombrotrophic *Sphagnum* species. *Proc. int. Peat Congress, 5th, Poznan,* **2,** 264–275.

Palisot (de) Beauvois, A. M. F. J. 1805. *Prodrome des cinquième et sixième familles de l'aethéogamie. Les mousses. Les lycopides.* Paris.

Palmstruch, J. W. 1803. *Svensk botanik. II.* Stockholm.

Papp, C. 1967. *Briofitele din Republic Socialista Romania.* Iasi.

Paul, H. 1923. Sphagnaceae (Torfmoose). In: *Die naturlichen Pflanzenfamilien nebst ihren Gattungen und wichtigeren Arten insbesondere den Nutapflanzen unter Mitwirkung zahlreicher hervorrangender Fachgelehrten begründet von A. Engler und K. Prantl,* edited by A. Engler, **10,** 1, 105–125. 2nd ed.

Paul, H. 1931. Sphagnales (Torfmoose). In: *Die Süsswasser-Flora mitteleuropas. 14. Bryophyta (Sphagnales – Bryales – Hepaticae),* edited by A. Pascher, 1–46. 2nd ed. Jena.

Pavletic, Z. 1955. *Prodromus Flore Briofita Jugoslavije.* Zagreb.

Persson, H. 1949. Studies in the broyphyte flora of Alaska – Yukon. *Svensk bot. Tidskr.,* **43,** 491–533.

Petrov, S. 1975. *Bryophyta Bulgarica Clavis Diagnostica.* Sofia: Academia Scientarum Bulgarica.

Pierrot, R. B. 1973. Clé des *Sphagnum* de la région Poitou-Charentes-Vendée. *Bull. Soc. bot. Centre-Ouest,* **4,** 20–22.

Pilous, Z. 1971. Bryophyta, Mechorosty, Sphagnidae – Mechy raselinikove. In: *Flora CSSR, Ser. C, Bryophyta,* edited by A. Pilat, I. Praha.

Poetsch, J. S. & Schiedermayr, K. B. 1872. *Systematische Aufzählung der im Erzherzogthume Oesterreich ob der Enns bisher beobachteten samenlosen Pflanzen (Kryptogmen).* Wien: K. K. Zool.-Bot. Gesellschaft.

Pohle, R. R. 1915. Materialy diya poznaniya rastitel'nosti severnoy Rossii. 1. K. flora mkhov severnoy Rossii. *Trudy imp. S.-peterb. bot. Sada (Acta Horti Petropolit.),* **33** (1), 1–148.

Proctor, M. C. F. 1955. A key to the British species of *Sphagnum. Trans. Br. bryol. Soc.,* **2,** 552–560.

Prokhanov, Y. 1964. Proposals (on botanical nomenclature). *Taxon,* **13,** 25–28.

Puustjarvi, V. 1968. Cation exchange capacity in *Sphagnum* mosses and its effect on nutrient and water absorption. *Peat Plant News,* (4) **1,** 54–58.

Rabenhorst, L. 1848. *Deutschlands Kryptogamen-Flora oder Handbuch zur Bestimmung der kryptogamischen Gewächse Deutschlands, der Schweiz, des lombardisch-venetianischen Königreichs und Istriens. II: 3. Leber-,Laubmoose und Farn.* Leipzig.

Raffaelli, M. 1976. Gli Sfagni Tosco-Emilani. *Webbia,* **30,** 159–175.

Rahman, S. M. A. 1972. Taxonomic investigations on some British *Sphagna.* I. *Sphagnum subsecundum* sensu lato. *J. Bryol.,* **7,** 169–179.

Ratcliffe, D. A. 1964. Mires and bogs. In: *The vegetation of Scotland,* edited by J. H. Burnett, 426–478. Edinburgh: Oliver & Boyd.

Renauld, F. & Cardot, J. 1885. Notice sur quelques mousses de l'Amérique du Nord. *Revue Bryol.,* **12,** 44–47.

Roivainen, H. 1937. Bryological investigations in Tierra del Fuego. With diagnoses of many new species by Edwin B. Bartram. *Ann. bot. Soc. zool. bot. fenn. 'Vanamo'*, **9** (2).
Röll, J. 1885, 1886. Zur Systematik dur Torfmoose. *Flora, Jena,* **68,** 569–580, 585–598; **69,** 33–44, 73–80, 89–94, 105–111, 129–137, 179–187, 227–242, 328–337, 353–370, 419–427, 467–476.
Röll, J. 1888. "Artentypen" und "Formereihen" bei den Torfmoosen. *Bot. Zbl.,* **34,** 310–314, 338–342, 374–377, 385–389.
Röll, J. 1889. Die Torfmoos-Systematik und die Descendenz-Theorie. *Bot. Zbl.,* **39,** 305–311, 337–344.
Röll, J. 1907. Beitrag zur Moosflora des Erzgebirges. *Hedwigia,* **46,** 185–245.
Röll, J. 1910. Die Benennung der Sphagna-Arten nach den Regeln des internat. botan. Kongresses von Wien 1905. *Allg. bot. Z.,* **16,** 70–71.
Rønning, O. I. 1958. Studies in *Sphagnum molle* Sull. and related forms. *Acta boreal. A. Scientia,* **14,** 1–24.
Rønning, O. I. 1965. Sphagnum-artenes utbredelse i Nord-Norge. University of Trondheim.
Roth, A. W. 1800. *Tentamen florae germanicae. III. Continens synonyma et adversaria ad illustrationem florae germanicae.* Leipzig.
Roth, G. 1906. *Die europäischen Torfmoose. Nachtragsheft zu den Europäischen Laubmooser.* Leipzig.
Roth, G. 1908. Neuere Torfmoosformen. *Hedwigia,* **47,** 321–329.
Roth, G. & Leipzig, W. V. von. 1906. Die europäischen Torfmoose. *Bryologist,* **9,** 102-103.
Rüling, J. P. 1786. Verzeichnis der an und auf dem Harz wildwachsenden Bäume, Gesträuche, und Kräuter; nach dem Sexual-System des Hrn Ritters von Linnegeordnet. In: *Anleitung den Hartz und andere Bergwerke mit Nuzen zu bereisen,* edited by C. W. J. Gattere, **2,** 186–247, Gottingen.
Russow, E. 1865. *Beitrage zur Kenntniss der Torfmoose.* Dissertation, Dorpat. (Published in *Arch. Naturk. Liv.-, Est.- u. Kurlands,* **2,** ser. 7, (1877), 83–162).
Russow, E. 1887a. Über den gegenwärtigen Stand meiner seit dem Frühling 1886 wieder auf genommenen Studien an den einheimischen Torfmoosen. *Sber. naturfGes. Univ. Dorpat,* **8,** 305–325.
Russow, E. 1887b. Zur Anatomie resp. physiologischen und vergleichenden Anatomie der Torfmoose. *Schr. NaturfGes. Univ. Dorpat,* **3,** 1–35.
Russow, E. 1888. Über den Begriff "Art" bei den Torfmoosen. *Sber. naturfGes. Univ. Dorpat,* **8,** 413–426.
Russow, E. 1889. Über die Ergebnisse meiner fortgesetzten sphagnologischen Studien. *Sber. naturfGes. Univ. Dorpat,* **9,** 94–113.
Russow, E. 1894. Zur Kenntnis der Subsecundum- und Cymbifolium gruppe europäischer Torfmoose, nebst, einem Anhang, enthaltend eine Aufzählung der bisher im Ostbalticum beobachteten Sphagnum-Arten und einen Schlüssel zur Bestimmung dieser Arten. *Arch. Naturk. Liv.-, Est.- u. Kurlands,* **10,** ser. 2, 361–527.
Savicz(-Ljubitzkaja), L. 1936. *Sphagnales partis europeae URSS.* (In Russian). Moscow.
Savicz(-Ljubitzkaja), L. 1952. Sphagnales. (In Russian). *Flora plantarum cryptogamarum URSS. I. Musci frondosi.* Moscow.
Savicz-Ljubitzkaja, L. I. & Abramov, I. I. 1959. The geological annals of Bryophyta. *Revue bryol. lichénol.,* n.s., **28,** 330–342.
Savicz-Ljubitzkaja, L. I. & Smirnova, Z. N. 1968. *The handbook of Sphagnaceae of the USSR.* (In Russian). Leningrad: Akad. Nauk SSSR, Komarov Botanical Institute.
Sayre, G., Bonner, C. E. B. & Culberson, W. L. 1964. The authorities for the epithets of mosses, hepatics and lichens. *Bryologist,* **67,** 113–135.
Scagel, R, F, R., Bandoni, R. J., Rouse, G. E., Schofield, W. B., Stein, J. R. & Taylor, T. M. C. 1965. *An evolutionary survey of the plant kingdom.* Belmont.

Schimper, W. P. 1857. Memoire pour servir à l'histoire naturelle des sphaignes (*Sphagnum* L.). *Mém. prés. div. Sav. Acad. Sci. Inst. Fr., Sci. Math. Phys.*, **15,** 1–97.

Schimper, W. P. 1858. *Versuch einer Entwicklungs-Geschichte der Torfmoose (Sphagnum) und einer Monographie der in Europa vorkommenden Arten dieser Gattung.* Stuttgart.

Schimper, W. P. 1880. *Synopsis muscorum europaeorum praemissa introductione de elementis bryologicis tractante.* Stuttgart. 1876, 2nd ed. Stuttgart.

Schkuhr, C. 1810. *Deutschlands kryptogamische Gewächse. II. Vier und zwanzigste Pflanzen-Klasse, die enthält deutschen Moose.* Wittenberg.

Schleicher, J. C. 1804. *Plantae cryptogramicae Helvetiae quas in locis earum natalibus collegit et exsiccavit.* Cent. II. Bex.

Schliephacke, K. 1865. Beiträge zur Kenntnis der *Sphagna. Verh. zool.-bot.* Ges. Wien, **15,** 383–414.

Schmidel, C. C. (ed. Bischoff). 1797. *Icones plantarum et analyses partium aeri incisae atque vivis coloribus insignitae adjectis indicubus nominum necessariis figurarum explicationibus et brevibus animadversionibus.* 2nd ed. Erlangen.

Schrader, H. A. 1794. *Spicilegium florae germanicae. I.* Hannover.

Schrader, H. A. 1796. *Systematische Sammlung kryptogamischer Gewachse. I.* Göttingen.

Schrader, H. A. 1801. Correspondenz-Nachrichten. I. Auszug aus einem Schreiben vom Herrn Prof. Swartz. *J. Bot., Gött.,* **1,** (1800), 397–400.

Schrank, F, von P. 1789. *Baierische Flora. II.* Munchen.

Schultz, K. F. 1819. *Prodromi florae stargardiensis supplementum primum.* Neubrandenburg.

Schultze-Motel, W. 1962. Das moosherbar von Carl Warnstof. *Willdenowia,* **3,** 289–313.

Schumacher, A. 1939. *Sphagnum strictum* in Europa. *Annls bryol.,* **12,** 143–153.

Scopoli, J. A. 1760. *Flora carniolica exhibens plantas Carniolae indigenas et distributas in classes naturales cum differentis specificis, synonymis recentiorum, locis natalibus, nominibus incolarum, observationibus selectis, viribus medicis.* Wien.

Scopoli, J. A. 1772. *Flora carniolica exhibens plantas Carnioliae indigenas et distributas in classes, genera, species, varietates, ordine linnaeano. II.* 2nd ed. Wien.

Sherrin, W. R. 1927. *An illustrated handbook of the British* Sphagna *(after Warnstorf).* London: Taylor and Francis.

Simola, L. K. 1977. The tolerance of *Sphagnum fimbriatum* towards lead and cadmium. *Ann. bot. fenn.,* **14,** 1–5.

Sjörs, H. 1943. Nàgra myrtyper vid Mjölkvattnet. *Sver. Nat.,* 1943, 81–88.

Sjörs, H. 1944. *Sphagnum subfulvum* n.sp. and its relations to *S. flavicomans* (Card.) Warnst and *S. plumolosum* Röll. *Svensk bot. Tidskr.,* **38,** 403–426.

Sjörs, H. 1949. Om *Sphagnum lindbergii* i sodra delen av Sverige. *Svensk bot. Tidskr.,* **43,** 568–585.

Sjörs, H. 1966. *Sphagnum angermanicum* found in northern Dalarna, Sweden. *Bot. Notiser,* **119,** 361–364.

Skogen, A. 1970. A new locality for *Sphagnum angermanicum,* and its distribution in Norway. *Nytt Mag. Bot.,* **17,** 7–10.

Slater, F. M. & Slater, E. J. 1978. The changing status of *Sphagnum imbricatum* Hornsch. ex Russ. on Borth Bog, Wales. *J. Bryol.,* **10,** 155–161.

Smarda, J. 1970. Complements à la flore muscinale de la Bulgarie. *Revue bryol. lichénol.,* **37,** 33–46.

Smirnova, Z. N. 1929. The distribution of *Sphagnum contortum* Schultz and *Sphagnum quinquefarium* (Lindb.) Warnst. in USSR. *Annls bryol.,* **2,** 107–116.

Smith, A. J. E. 1978. *The moss flora of Britain and Ireland.* Cambridge: Cambridge University Press.

Smith, A. J. E. & Newton, M E. 1966. Chromosome studies on some British and Irish mosses. I. *Trans. Br. bryol. Soc.,* **5,** 117–130.
Sobotka, D. 1975, Rozmieszczemie *Sphagnum wulfianum* Girgens. w Polsce. (Distribution of *Sphagnum wulfianum* Girgens. in Poland). *Fragm. flor. goebot.,* **21,** 143–145.
Sorsa, V. 1955. Outlines of meiosis in the moss genus *Sphagnum. Hereditas,* **41,** 250–258.
Sorsa, V. 1956. The quadripolar spindle and the change of orientation of the chomosomes in meiosis of *Sphagnum. Ann. Acad. Sci. fenn.,* ser. A, 4, **33,** 1–64.
Stearn, W. T. 1945. Please leave article 32 alone! *Taxon,* **3,** 141–143.
Stefureac, I. 1962. Relictes subarctiques dans la bryoflore du marais eutrophe de Dragoisa Carpathes orientales. *Revue bryol. lichénol.,* **31,** 68–73.
Stefureac, T. I. 1976, 1977. Nouvelles contributions a l'écologie et à la corologie des Sphaignacées en Roumanie. *Studii Comun. Muz. Stiint. nat. Bacau, Biol. Veg.,* **9–10,** 97–112.
Sturm, J. 1819. *Deutschlands Flora in Abbildungen nach der Natur mit Beschreibungen.* II. Nürnberg.
Sullivant, W. S. 1845. *Musci alleghaniensis, sive spicilegia muscorum atque hepaticorum quos in itinere a Marylandia usque ad Georgiam per tractus montium a.d. MDCCCLII,* decerpserunt Asa Gray et W. A. Sullivant (interjectis nonnullis aliunde collectis). Fasc. 2, nos 135–292. Columbus, Ohio.
Sullivant, W. S. 1849. Contributions to the bryology and hepaticology of North America, *II. Mem. Am. Acad. Arts Sci.,* n.s. **4,** 169–176.
Sullivant, W. S. 1856. The musci and hepaticae of the United States east of the Mississippi River. In: *Gray's manual of botany,* edited by S. F. Gray. 2nd ed. New York.
Suzuki, H. 1955. A list of *Sphagnum* species from Hokkaido with descriptions of the new additions to the Japanese flora. *J. Sci. Hiroshima Univ., ser. B, div. II,* **7,** 63–89.
Susuki, H. 1958. Taxonomical studies on the Subsecunda group of the genus *Sphagnum* in Japan, with special reference to variation and geographical distribution. *Jap. J. Bot.,* **16,** 227–268.
Suzuki, H. 1956. Studies on the Palustria group of the *Sphagna* of Japan. *J. Sci. Hiroshima Univ. Ser. B, div. II,* **7,** 153–172.
Szafran, B. 1946. Próba wyjaśnienia zwiazku filogenetycznego miedzy sekcjami torfowców. (Résumé: Un essai d'explication des relations phylogénétiques entre les sections des sphaignes). *Acta Soc. Bot. Pol.,* **17,** 219–237.
Szafran, B. 1949. Pochodzenie torfowców. The origin of the *Sphagna. Acta Soc. Bot. Pol.,* **20,** 35–44.
Szafran, B. 1957. Mchy (Musci). I. Flora polska. Rośliny zarodnikowe Polski i ziem ościennych. Warzawa.
Szafran, B. 1963. Bryophyta. I. Musci-Mchy. (Flora Slodkowodna Polski no. 16). Warzawa.
Tallis, J. H. 1962. The identification of *Sphagnum* spores. *Trans. Br. bryol. Soc.,* **4,** 209–213.
Taylor, J. 1953. *Sphagnum capillaceum* (Weiss) Schrank, *Kew Bull.,* **2,** 277–278.
Terasmae, J. 1955. On the spore morphology of some *Sphagnum* species. *Bryologist,* **58,** 306–311.
Tjuremnov, S. N. 1963. On the distribution of *Sphagnum imbricatum.* (In Russian). *Byull. Mosk. Obshch. Ispyt., Prir. otd. Biol.,* **68,** 98–109.
Tolf, R. 1891. Öfversigt af Smålands mossflora. *Bih. K. svenska VetenskAkad. Handl.,* ser. 3, **16** (9), 1–98.
Touffet, J. 1966. La flore sphagnologique des montagnes noires de Bretagne. *Bot. Rhedonica,* ser. A, no. 2, 87–98.
Touffet, J. 1968. Répartition et écologie du *Sphagnum pylaiei* en Bretagne. *Revue bryol. lichénol.,* **36,** 203–212.
Tryon, R. 1962. A commentary on superfluous names. *Taxon,* **11,** 116-120.

Tuomikoski, R. 1946. Suomen rahkasammalista ja niiden tuntemisesta ilman mikroskooppia. (Finnish peat mosses and their identification without a microscope). I. Luonnon Ystävä, **50,** 113–117; II. **50,** 150–159.
Tuomikoski, R. 1958. Uber den heutigen Stand der Laubmoosesystematik. *Uppsala Univ. Årsskr.,* no. 6, 65–69.
Vitt, D. H. & Andrus, R. E. 1975. *Sphagnum aongstroemii* in North America. *Bryologist,* **78,** 463–467.
Vitt, D. H. & Andrus, R. E. 1977. The genus *Sphagnum* in Alberta. *Can. J. Bot.,* **55,** 331–357.
Vitt, D. H., Crum, H. & Snider, J. 1975. The vertical zonation of *Sphagnum* species in hummock-hollow complexes in northern Michigan. *Mich. Bot.,* **14,** 190–200.
Voss, E. G. 1965. On citing the names of publishing authors. *Taxon,* **14,** 154–160.
Waldheim, S. 1944. Die Torfmoosvegetation der Provinz Närke. *Acta Univ. Lund.,* n.s., ser. 2, **40** (6), 1–91.
Warburg, E. F. 1963. *Census catalogue of British mosses.* 3rd ed. Ipswich.
Waren, H. 1926. Untersuchungen uber sphagnumreiche Pflanzengesellschaften der Moore Finnlands. *Acta Soc. Fauna Flora fenn.,* **55,** 1–133.
Warncke, E. 1979. Danske torvemosser. *Natur Mus. Arhus,* **19,** 1–18.
Warnstorf, C. 1877. Zwei neue europäische moosformen. *Bot. Ztg,* **35,** 478–479.
Warnstorf, C. 1881. *Die europäischen Torfmoose. Eine Kritik und Beschreibung derselben.* Berlin.
Warnstorf, C. 1882a. Bryologische Notizen aus Westpreussen. *Hedwigia,* **21,** 1–2.
Warnstorf, C. 1882b. Die Torfmoose im königlichen botanischen Museum zu Berlin, Eine bryologische Studie. *Bot. Zbl.,* **9,** 96–102, 131–136, 166–173.
Warnstorf, C. 1882c. Neue deutsche Sphagnumformen. *Flora, Jena,* **65,** 205–208.
Warnstorf, C. 1882d. Einige neue Sphagnumformen. *Flora, Jena,* **65,** 464–466.
Warnstorf, C. 1882e. Die Sphagnumformen der Umgegend von Bassum in Hannover. *Flora, Jena,* **65,** 547–553.
Warnstorf, C. 1883. Die Torfmoose des v. Flotow'schen Herbarium in königl. bot. Museum in Berlin. *Flora, Jena,* **66,** 371–380.
Warnstorf, C. 1884a. *Sphagnotheka europaea.* (Mimeographed labels). Neuruppin.
Warnstorf, C. 1884b. Sphagnologische Rückblicke. *Flora, Jena,* **67,** 469–483, 485–516, 597–611.
Warnstorf, C. 1886. Zwei Artentypen der Sphagnum aus der Acutifolium-gruppe. *Hedwigia,* **25,** 221–231.
Warnstorf. C. 1888a, 1890a, 1892a. *Europäische Torfmoose.* Ser. I, Mimeographed labels, nos 1–100. Neuruppin; ser. II, nos 101–200. Neuruppin; ser. III, nos 201–300. Neuruppin.
Warnstorf, C. 1888b. Die Acutifoliumgruppe der europäischen Torfmoose. Ein Beitrag zur Kenntnis der *Sphagna. Verh. bot. Ver. Prov. Brandenb.,* **30,** 79–127.
Warnstorf, C. 1888c. Revision der *Sphagna* in der Bryotheca europaea von Rabenhorst und in einigen älteren Sammlungen. *Hedwigia,* **27,** 265–276.
Warnstorf, C. 1889. *Sphagnum crassicladum* Warnst., ein neues Torfmoos fur Europa aus der Subsecundagruppe. *Bot. Zbl.,* **40,** 165–167.
Warnstorf, C. 1890b. *Sphagnum degenerans* var. *immersum,* ein neues europäisches Torfmoos. *Bot. Zbl.,* **42,** 102–105.
Warnstorf, C. 1890c. Beiträge zur Kenntnis exotischer *Sphagna. Hedwigia,* **29,** 179–210, 213–258.
Warnstorf, C. 1890d. Contributions to the knowledge of North American *Sphagna.* I. *Bot. Gaz.,* **15,** 127–140; II. **15,** 189–198; III. **15,** 217–227; **15,** 242–255.
Warnstorf, C. 1890e. Die Cuspidatum-Gruppe der europäischen *Sphagna.* Ein Beitrag zur Kenntnis der Torfmoose. *Abh. Bot. Ver. Prov. Brandenb.,* **32,** 173–231.
Warnstorf, C. 1891. Beiträge zur Kenntnis exotischer *Sphagna. Hedwigia,* **30,** 12–46, 127–178.

Warnstorf, C. 1892b. Einige neue exotischer *Sphagna. Hedwigia,* **31,** 174–182.
Warnstorf, C. 1893a. Beitrage zur Kenntnis exotischer *Sphagna. Hedwigia,* **32,** 1–17.
Warnstorf, C. 1893b. Charakteristik und Uebersicht der europäischen Torfmoose nach dem heutigen Standpunkte der Sphagnologie (1893). *Schr. nat. wiss. Ver. Harzes,* **8.**
Warnstorf, C. 1894. Characteristik und Uebersicht der nord-, mittel- und südamerikanischen Torfmoose nach dem heutigen Standpunkte der Sphagnologie (1893). *Hedwigia,* **33,** 307–337.
Warnstorf, C. 1895. Beiträge zur Kenntnis exotischer *Sphagna. Allg. bot. Z.,* **1,** 92–95, 115–117, 134–136, 172–174, 187–189, 203–206, 227–230.
Warnstorf, C. 1897. Die moor-Vegetation der Tucheler Heide, mit besonderer Berücksichtigung der Moose. Bericht über die im Auftrage des Westpr. Bot.-Zool. Vereins in der Zeit vom 4 bis 29 Juli 1896 ausgeführte bryologische Forschungsreise. *Schr. naturf. Ges. Danzig,* n.s. **9,** 111–179.
Warnstorf, C. 1898. Beiträge zur Kenntnis exotischer und europäischer Torfmoose. *Bot. Zbl.,* **76,** 385–390, 417–423.
Warnstorf, C. 1899. Neue Beiträge zur Kryptogamenflora der Mark Brandenburg. Verzeichnis der in der Niederlausitz beobachteten Moose nebst kritschen Bemerkungen zu verschiedenen Arten, sowie Mitteilungen über neue Beobachtungen aus anderen Teilen der Mark. II Specieller Teil. *Abh. Bot. Ver. Prov. Brandenb.,* **41,** 19–80.
Warnstorf, C. 1900a. Weitere Beiträge zur Kenntnis der Torfmoose. *Bot. Zbl.,* **82,** 7–14, 39–45, 65–76.
Warnstorf, C. 1900b. Sphagnaceae (Torfmoose). In: *Die natürlichen Pflanzenfamilien nebst ihren Gattungen und wichtigeren Arten insbesonderer den Nutzpflanzen unter Mitwirkung zahlreicher hervorrangender Fachgelehrten begrundet von A. Engler und K. Prantl, I,* edited by A. Engler, **3,** 1, 248–262. Danzig.
Warnstorf, C. 1902, 1903. *Leber-und Torfmoose. Kryptogamenflora der Mark Brandenburg und angrenzenden Gebiete herausgegeben von dem Botanischen Verein der Provinz Brandenburg, I.* Leipzig.
Warnstorf, C. 1907, 1908. Neue europäische und aussereuropäische. Torfmoose. *Hedwigia,* **47,** 76–124.
Warnstorf, C. 1911. *Sphagnales – Sphagnaceae (Sphagnologia universalis).* (Das Pflanzenreich. Regni vegetabilis conspectus 51). Leipzig.
Watson, W. 1918. *Sphagna,* their habitats, adaptations and associates. *Ann. Bot.,* **32,** 535–551.
Weber, F. & Mohr, D. M. H. 1804. *Naturhistorische Reise durch einen Theil Schwedens.* Göttingen.
Weber, E. & Mohr, D. M. H. 1807. *Botanisches Taschenbuch auf das Jahr 1807.* (Deutschlands kryptogamische Gewachse). Kiel.
Wiemarck, H. 1937. *Förteckning över Skandinaviens växter utgiven av Lunds Botaniska Förening. 2: Mossor.* 2nd ed. Lund: Lund Botanical Research.
Weis(s), F. W. 1770. *Plantae cryptogamicae florae göttingensis.* Göttingen.
Wijk, R. van der. 1949. Het geslacht *Sphagnum* in Nederland. *Ned. kruidk. Archf.,* **56,** 83–159.
Wikj, R. van der & Margadant, W. D. 1965. New combinations in mosses. VIII. *Taxon,* **14,** 196–198.
Wijk, R. van der, Margadant, W. D. & Florschutz, P. A. 1959. Index muscorum. I (A–C). *Regnum veg.,* **17,** i–xxviii, 1–548.
Willis, J. H. 1953. Systematic notes on Victorian mosses. 2. Victor. *Naturalist, Hull,* **70,** 55–57
Wilson, W. 1855. *Bryologia britannica; containing the mosses of Great Britain and Ireland, systematically arranged and described according to the method of Bruch and Schimper, with illustrative plates: being a new (third) edition, with many additions and alterations, of the Muscologia britannica of Messrs Hooker and Taylor.* London.

Wilson, W. & Hooker, J. D. 1845, 1847. Musci. In: *The botany of the Antarctic voyage of H.M. Discovery Ships Erebus and Terror in the years 1839–1843, under the command of Captain Sir James Clark Ross*, edited by J. D. Hooker. I: **1,** 117–143. II: **1,** 395–423. London.
Wylie, A. P. 1957. The chromosome numbers of mosses. *Trans. Br. bryol. Soc.*, **3,** 260–284.
Zenker, J. C. & Dietrich, F. D. 1821. *Musci thuringici, vivis exemplaribus exhibuerunt et illustraverunt.* Fasc. 1. Jena.
Zepf, S. 1952. Uber sie Differenzierung des *Sphagnum* Blattes. *Z. Bot.*, **40,** 87–118.
Zickendrath, E. 1900. Beiträge zur Kenntnis der Moosflora Russlands. II. *Byull. mosk. Obshch. Ispyt, Prir, Otd, Biol.*, n.s. **14,** 241–366.